Signaling Mechanisms of Oxygen and Nitrogen Free Radicals

Signaling Mechanisms of Oxygen and Nitrogen Free Radicals

Igor B. Afanas'ev

CRC Press
Taylor & Francis Group
Boca Raton London New York

CRC Press is an imprint of the
Taylor & Francis Group, an **informa** business

CRC Press
Taylor & Francis Group
6000 Broken Sound Parkway NW, Suite 300
Boca Raton, FL 33487-2742

First issued in paperback 2019

ISBN-13: 978-1-4200-7374-4 (hbk)
ISBN-13: 978-0-367-38556-9 (pbk)

Library of Congress Cataloging-in-Publication Data

Afanas'ev, Igor B., 1935-
    Signaling mechanisms of oxygen and nitrogen free radicals / Igor B. Afanas'ev.
        p. cm.
    "A CRC title."
    Includes bibliographical references and index.
    ISBN 978-1-4200-7374-4 (hard back : alk. paper)
    1. Active oxygen in the body. 2. Nitrogen in the body. 3. Cellular signal transduction.
I. Title.

QP535.O1A416 2009
616.07'1--dc22
                                                                      2009010931

Contents

Preface

The major aim of *Signaling Mechanisms of Oxygen and Nitrogen Free Radicals* is to present an overview of the regulatory functions of reactive oxygen and nitrogen radicals in physiological and pathophysiological states. The second part of the twentieth century in this field was characterized by extensive studies of free radical-mediated biological processes under in vitro and in vivo conditions and in whole organisms including animals and humans. However, most published works investigated the damaging effects of reactive oxygen and nitrogen species (ROS and RNS) and their roles in various pathologies (heart disease, cardiovascular diseases, cancer, inflammation, diabetes, arthritis, and many others). These studies stimulated interest in the antioxidant treatment of patients suffering from such pathologies. Important theories of the role of free radicals in the development of aging and senescence were developed during this time.

Interest in the search for the most dangerous ROS and RNS was well justified (see, for example, "Who Is the Villain?" by I.B. Afanas'ev)[1] for the development of proper antioxidant and chelating treatments involving the so-called free radical pathologies. However, it soon became very clear that ROS and RNS also possess important regulatory functions in living organisms. Of course, their bactericidal phagocyte functions were known from the initial years of free radicals studies from 1970 to 1980, but real interest in the regulatory role of ROS and RNS came much later. Puzzling findings of participation of physiological free radical superoxide in heterolytic enzymatic reactions led to an understanding of its signaling functions (and those of other ROS and RNS) in these processes.

At present, the study of ROS and RNS signaling is an important and major focus of biological free radical investigations. In this book, I note some of the main areas of inquiry today regarding signaling mechanisms of ROS and RNS in enzymatic processes. As a biochemist, I am also trying to draw scientists' attention to the chemical mechanisms of these reactions. I should like to point out that although the mechanisms of many biological processes, as a rule, are too complicated to allow us to adequately describe all the chemical reactions taking place, there are reliable specific methods (ESR, ESR spin trapping, chemiluminescence, and some others) that permit the study of free radicals in many enzymatic processes. (I'd like to mention the well-known fact that the presence of an unpaired electron responsible for the enhanced reactivity and the short lifetime of free radicals make them paramagnetic and detectable by those selective methods, whereas, unfortunately, there are no similar methods for detection of ionic intermediates).

This book contains seven chapters. In Chapter 1, "Introduction: Free Radical Damage versus Free Radical Signaling," the difference between signaling functions and damaging effects of ROS and RNS in biological systems is discussed. Chapter 2, "Free Radical Signaling in Reactions Catalyzed by Enzymes—Producers

---

[1] In *Superoxide Ion*, vol. 2, Chapter 1 (Boca Raton, FL: CRC Press, 1989).

of Reactive Oxygen and Nitrogen Species," is dedicated to free radical signaling in processes catalyzed by enzymes, producers of superoxide and nitric oxide. These enzymes (xanthine oxidase, NADPH oxidase, and NO synthase) are able to use produced ROS and RNS as signaling species in their own catalytic processes. Similarly, Chapter 3, "Reactive Oxygen and Nitrogen Species Signaling in Mitochondria," considers ROS and RNS signaling produced by mitochondrial enzymes. Reactive oxygen and nitrogen species produced by these enzymes are, of course, able to start many damaging processes in cells and tissues; these processes were reviewed earlier by Denisov and Afanas'ev (2005).[2] In Chapter 4, "ROS and RNS Signaling in Catalysis of Heterolytic Reactions by Kinases, Phosphatases, and Other Enzymes," signaling functions of ROS and RNS are considered in enzymatic heterolytic reactions, first of all, hydrolysis, and etherification. This chapter supplies possibly the most abundant and important data on ROS and RNS signaling in catalysis by enzymes that do not produce free radicals by themselves.

In Chapters 5, "ROS and RNS Signaling in Apoptosis," and Chapter 6, "ROS and RNS Signaling in Senescence and Aging," signaling by ROS and RNS in two important biological processes, apoptosis and aging/senescence, are examined. Finally, Chapter 7, "Mechanisms of ROS and RNS Signaling in Enzymatic Catalysis," covers mechanisms of free radical signaling in enzymatic processes.

All comments, criticisms, and suggestions are welcomed by the author.

Igor B. Afanas'ev
Vitamin Research Institute
Moscow

---

[2] E.T. Denisov and I.B. Afanas'ev, *Oxidation and Antioxidants in Organic Chemistry and Biology* (Boca Raton, FL: Taylor & Francis, 2005).

Author

Igor B. Afanas'ev was born in Moscow and graduated from Moscow Chemico-Technological Institute in 1958. He received his Ph.D. and D.Sci. degrees (chemistry) in 1963 and 1972, respectively. He became a professor of chemical kinetics and catalysis in 1982.

Professor Afanas'ev has been chairman of the Russian Committee of Society for Free Radical Research (SFRR) (1989–1998), a member of the International Editorial Board, *Free Radicals in Biology & Medicine* (1994–1998), and the organizer of International Meetings on Free Radicals in Chemistry and Biology in Leningrad (1991), Moscow (1994), Kiev (1995), and Moscow–Yaroslavl (together with Professors A. Azzi and H. Sies; 1998). He was also chairman of 14 regional meetings in the former USSR on free radicals in chemistry and biology (1984–1998). At present, he is the editor of the special issue *Free Radicals in Biology and Medicine: Current Theories and Experimental Findings* in *Frontiers in Bioscience* (2008).

Professor Afanas'ev has been an invited lecturer in many international conferences on free radicals from 1984 to 2009. He is the author of several books and chapters of books published by CRC Press including "Radical Reactions of Aromatic Compounds" (in *Chemical Kinetics of Small Organic Radicals*, vol. 4, ed. Z.B. Alfassi, 1998), *Superoxide Ion: Chemistry and Biological Implications* (vols. 1 and 2, 1989–1990), and *Oxidation and Antioxidants in Organic Chemistry and Biology* (together with Professor E.T. Denisov, 2005). He is author and coauthor of works published in Russian and English (approximately 120 publications).

His major scientific interests are chemistry (the experimental studies and correlation analysis of reactivity of free radicals), biology (the study of mechanisms of lipid peroxidation and the investigation of mechanisms of antioxidant activity of flavonoids and vitamins under in vitro and in vivo conditions), and medicine (the application of nontoxic natural antioxidants for the treatment of patients suffering from free radical pathologies). At present, his major interest is the investigation of signaling functions of free radicals in heterolytic enzymatic processes.

List of Abbreviations

AA	arachidonic acid
Abl	tyrosine kinase
AC5	5-adenylyl-cyclase
AGEs	advanced glycation endproducts
AICD	activation-induced T-cell death
Akt/PKB	serine/threonine protein kinase B
ALS	amyotrophic lateral sclerosis
AMPK	5′-AMP-activated protein kinase
Ang-1	angiopoietin-1
AngII	angiotensin II
ANT	adenine nucleotide translocase
AOPPs	advanced oxidation protein products
AP-1	transcription factor activator protein
APC	anoxia preconditioning
AR	aldose reductase
ASK1	apoptosis signal-regulating kinase 1
ATF-1	activating transcription factor 1
BAEC	bovine aortic endothelial cells
Bap	benzo(a)pyrene
BCA	2'-benzoyl-oxcinnamaldehyde
CAD	coronary artery disease
CaM	calmodulin
CCR	cytochrome *c* release
cGMP	cyclic guanosine monophosphate
CHF	chronic heart failure
c-Jun	N1/2 amino-terminal kinase JNK
CL	chemiluminescence
CLA	conjugated linoleic acid
CMEC	coronary microvascular endothelial cell
COX-2	cyclooxygenase-2
cPLA2	cytosolic phospholipase A2
CREB	cAMP response element-binding proteins
CRP	C-reactive protein
DAG	diacylglycerol
1-EBIO	1-ethyl-2-benzimidazolinone
ECE	endothelin converting enzyme
ECM	extracellular matrix
EDRF	endothelium-derived relaxing factor
EFS	electrical field stimulation
EGF	epidermal growth factor
EGFR	epidermal growth factor receptor

eNOS	endothelial NOS III
eNOSox	ferrous oxygenase domain
EPC	endothelial progenitor cell
ER	endoplasmic reticulum
ERK1/2	extracellular signal-regulated kinase 1/2
eSOD	extracellular superoxide dismutase
ESW	extracorporeal shock wave
ET-1	endothelin-1
FasL	Fas ligand
FMLP	formyl methionyl leucyl phenylalanine
FPI	fluid percussion brain injury
FSS	fluid shear stress
G6PD	glucose-6-phosphate dehydrogenase
GD	glucose depletion
GM-CSF	granulocyte-macrophage colony-stimulating factor
GST	glutathione transferase
GTP	guanosine monophosphate
GzmM	Granzyme M
HAEC	human aortic endothelial cell
HaRas	small GTP-binding-protein Ras
HASMC	human aortic smooth muscle cells
H$_4$B	tetrahydrobiopterin
Hcy	homocysteine
12-HETE	12(S)-hydroxyeicosatetraenoic acid
HG	high glucose
HHcy	hyperhomocysteinemia
HMEC-1	human microvascular endothelial cell
HO-1	heme oxygenase-1
HPV	hypoxic pulmonary vasoconstriction
H/R	hypoxia-reoxygenation
HSC	hepatic stellate cell
Hsp 27	heat shock protein 27
IC	intermediate conductance
IDPm	mitochondrial NADP+-dependent isocitrate dehydrogenase
IFM	interfibrillar mitochondria
ILK	integrin-linked kinase
IMM	inner mitochondrial membrane
iNOS	inducible enzyme NOS II
InsP3	inositol 1,4,5-trisphosphate
IPC	preconditioning
I/R	ischemia/reperfusion
IRS-1	insulin receptor substrate-1
JAK	Janus kinase
JNK	c-Jun amino-terminal kinase
LA	α-lipoic acid
5-LO	5-lipoxygenase

LPC	lysophosphatidylcholine
LV	left ventricle
MAPK	mitogen-activated protein kinase
2ME	2-methoxyestradiol
MGO	methylglyoxal
MI	myocardial infarction
MKK6	mitogen-activated protein kinase kinase 6
MLK3	lineage kinase 3
MMP-9	metalloproteinase-9
MnTBAP	Mn(III)-tetrakis(4-benzoic acid) porphyrin
MPT	mitochondrial permeability transition
MR	mineralocorticoid receptor
mtNOS	mitochondrial NO synthase
NFAT	nuclear factor of activated T cells
NF-κB	nuclear factor-kappa B
NGF	nerve growth factor
NHE-1	Na+/H+ exchanger isoform 1
NHEK	human epidermal keratinocytes
NMDA	N-methyl-D-aspartate
nNOS	neuronal NOS I
NOHLA	N-hydroxy-L-arginine
NOX	NADPH oxidase enzymes
3-NPA	3-nitropropionic acid
Nrf2	nuclear erythroid factor 2
NSAID	nonsteroidal anti-inflammatory drugs
ODC	ornithine decarboxylase
6-OHDA	6-hydroxydopamine
OZ	opsonized zymosan
p38	MAPK kinase
PAD	pial artery dilation
PAEC	porcine aortic endothelial cells
PAF	platelet-activating factor
PAH	pulmonary arterial hypertension
PAK1	member of family of serine/threonine protein kinase
PCAEC	porcine coronary artery endothelial cell
PDE	phosphodiesterase
PDGF	platelet-derived growth factor
PDTC	pyrrolidine dithiocarbamate
PECAM-1	platelet endothelial cell adhesion molecule-1
PIA2	phospholipase A2
PI3-K	phosphatidylinositol 3-kinase
PI(3)P	phosphatidylinositol-3-phosphate
PKAI and PKAII	cAMP-dependent protein kinases I and II
PKC	protein kinase C
PKD	protein kinase D

PKG-1	cGMP-dependent protein kinase I
PL	phospholipase
PMA	phorbol 12-myristate 13-acetate
POX	proline oxidase
PP	serine/threonine protein phosphatase
PPAR-α	peroxisome proliferator-activated receptor-α
Prx III	peroxiredoxin
PTK	protein tyrosine kinase
PTP	protein-tyrosine phosphatase
PVAT	perivascular adipose tissue
PYK2	Ca-dependent tyrosine kinase
RhoA	small GTPase protein
RMC	rat mesangial cell
ROCK	RhoA-activated kinase
ROS	reactive oxygen species
RV	right ventricle
SC	small conductance
SCD	sickle cell disease
SDH	succinate dehydrogenase
SFR	slow force response
sGC	soluble isoform of guanylate cyclase
SHR	spontaneously hypertensive rats
SOD	superoxide dismutase
SOD1	superoxide dismutase-1
Src	the member of Src family of proto-oncogenic tyrosine kinases
SSM	subsarcolemmal mitochondria
STAT	signal transducer of transcription
SW	shock waves
TGF-β	transforming growth factor-β
Tiam1	activator of the small GTPase Rac
TCR	T-cell receptor
TH	tryptophan hydroxylase
TNFα	tumor necrosis factor alpha
TPr	thromboxane A2 receptor
TRAF4	TNF receptor-associated factor 4
UCP1	uncoupled protein 1
VASP	vasodilator-stimulated phosphoprotein
Vav2	guanine nucleotide exchange factor
VDAC	voltage-dependent anion channel
VEGF	vascular endothelial growth factor
VSMC	vascular smooth muscle cell
XDH	xanthine dehydrogenase
XIAP	X-linked inhibitor of apoptosis protein
XO	xanthine oxidase

1 Introduction

Free Radical Damage versus Free Radical Signaling

In chemistry (not to speak of biology), free radicals (i.e., species with an unpaired electron or, in other words, compounds containing a three-valent carbon atom) were considered to be freaks, if they existed at all, because the four-valent carbon atom is an obligatory element of all organic compounds. However, the existence of free radicals became a proven fact when in 1900 Moses Gomberg reported his results on the reaction of triphenylmethyl halides with metals. The yellow reaction product was identified as free triphenylmethyl radical Ph_3C. This radical was a stable compound due to the delocalization of an unpaired electron on three phenyl groups. However, already in 1929, F.A. Patheth and W. Hofeditz demonstrated in their classic experiment that highly reactive methyl radicals were formed when the vapors of tetramethyllead, $Pb(CH3)4$, were mixed with gaseous hydrogen during its passage through a silica tube.

It was not surprising that, due to their reactivity, free radicals were usually considered by biologists as life-damaging factors because the first free radicals identified in chemical works were reactive short-living alkyl radicals. Furthermore, in the early years, biologists could not imagine free radicals to be the metabolites of normal physiological processes, and believed that they were only formed by various environmental factors (contamination by toxic chemical compounds, irradiation, etc.). From 1905 to 1910, two prominent biologists, Leonor Michaelis and Maud Menten, suggested that semiquinones were the intermediates of the mitochondrial respiratory chain. However, they are reactive free radicals and able to participate only in electron transfer reactions, and cannot be engaged in damaging attacks by free radicals.

A beginning was made in free radical studies in biology by the work of McCord and Fridovich, who suggested that the free radical anion of dioxygen $O_2^{\cdot-}$ (superoxide) was formed in the reactions catalyzed by xanthine oxidase (XO). This discovery (see Chapter 2), which was confirmed by excellent experimental results, completely changed our views on the role and functions of free radicals in biology. Subsequent discovery of the second physiological free radical nitric oxide (NO) in 1986 by Furchgott, Ignarro, Murad, and others led to much clearer understanding of free radical-mediated processes in biology.

It is not surprising that, from the start, everyone's attention was drawn to the study of the damaging effects of free radicals in cells, tissues, and whole organisms. First, it was thought that $O_2^{\cdot-}$ was the "super-oxidant" able to oxidize biomolecules. However, the thermodynamics of its reactions with organic compounds indicates that $O_2^{\cdot-}$ is not an oxidant but a moderate reductant. Therefore, it soon became clear that $O_2{\cdot}^-$ and NO are really precursors of reactive oxygen (ROS) and nitrogen (RNS) species.

This is now ancient history when it was believed that $O_2{\cdot}^-$ was able to react with hydrogen peroxide (H_2O_2) to form hydroxyl radicals (the Haber–Weiss cycle):

$$O_2{\cdot}^- + H_2O_2 \Rightarrow O_2 + HO{\cdot} + HO{\cdot}^- \tag{1.1}$$

Subsequent experiments showed that $O_2{\cdot}^-$ can be a source of hydroxyl radicals, but only in the presence of iron ions or complexes (i.e., through the Fenton reaction):

$$O_2{\cdot}^- + Fe^{3+} \Rightarrow O_2 + Fe^{2+} \tag{1.2}$$

$$Fe^{2+} + H_2O_2 \Rightarrow Fe^{3+} + HO{\cdot} + HO^- \tag{1.3}$$

After the identification of NO as a physiological product of NO synthases, another way to reactive hydroxyl radicals through the reaction of $O_2{\cdot}^-$ with NO was demonstrated:

$$O_2{\cdot}^- + ({\cdot})NO \Rightarrow -OONO \tag{1.4}$$

$$-OONO + H^+ \Rightarrow HOONO \tag{1.5}$$

$$HOONO \Rightarrow HO{\cdot} + {\cdot}NO_2 \tag{1.6}$$

Now it has been demonstrated that the generation of ubiquitous and relatively harmless $O_2{\cdot}^-$ and NO by enzymatic processes in normal physiological states may initiate the side formation of numerous harmful ROS and RNS such as hydroxyl radicals, H_2O_2, hypochlorous acid (HOCl), peroxynitrite ($ONOO^-$), and various other reactive nitrogen free radicals.

At present, numerous free radical-mediated damaging processes are identified under in vitro and in vivo conditions: lipid, peroxidation oxidation of proteins, DNA oxidative damage, and so on. It was also shown that the development of many pathologies (cardiovascular diseases, hypertension, diabetes mellitus, cancer and carcinogenesis, inflammation, and others) and aging is associated with overproduction of free radicals.

Unfortunately, until now there have been some unanswered questions concerning the structure of free radicals or other reactive species responsible for the initiation of damaging processes. Hydroxyl radicals are, of course, the most probable suspects, but due to their exclusive reactivity, it is difficult to understand how these radicals can achieve target biomolecules without annihilation at their meeting with neighboring molecules. Two explanations used to be offered: (1) actual reagents in the oxidation are not free hydroxyl radicals but ferryl or perferryl complexes formed in the reaction of ferrous and ferric ions with H_2O_2, and (2) hydroxyl radicals are formed by a "site-specific" mechanism through the reaction of H_2O_2 with "iron fingers," that is, with iron ions adsorbed on lipid, protein, or DNA molecules. There was, and is, a heated discussion about possible reactivity of such hydroxyl-like reactive species,

TABLE 1.1

Publications on Free Radical Signaling from 1980 to 2007

Works	1980	1990	1998	2000	2002	2007
Free radical signaling (number)	6	86	639	1062	1573	7262
Free radical signaling (percentage of content)	0.4	2.1	6.8	7.7	11.6	12.8
Free radical (total number)	1425	4144	9413	13812	13499	56600

but until now it has been impossible to get experimental evidence supporting one or another mechanism.

Another mechanism damaging processes is the free-radical-mediated nitration of various substrates. It has been proposed that nitration reagents might be RNS such as NO, ONOO$^-$, nitrate radical, and probably the other RNS. The formation of nitrated products (for example, nitrotyrosine) was demonstrated under various conditions, but the mechanisms of these free-radical reactions are also uncertain. It is known that the classic mechanism of the nitration of aromatic compounds is the electrophilic addition of the nitronium ion (NO$_2$$^+$) to aromatic nuclei. The formation of NO$_2$$^+$ in biological systems is unknown and very unlikely, and another possible nitration species is the nitrate radical formed by the decomposition of ONOO$^-$ via Reaction 6. However, at present, we have no experimental data supporting any mechanism of free radical-mediated nitration under in vitro or in vivo conditions. (Damaging effects of free radicals in biological systems have been reviewed in detail in the recent book *Oxidation and Antioxidants in Organic Chemistry and Biology* [1]).

Approximately 10 years later after the beginning of extensive studies on free-radical processes in biology from 1968 to 1970, "free radical scientists" became interested in the signaling functions of free radicals. The number of works on free-radical signaling compared to the total number of free-radical publications from 1980 to 2007 is presented in Table 1.1 (from Medline). It is seen that the study of free-radical signaling began approximately in 1980 and from 2001 to 2002 up to the present time became about 10% of the content of all publications dedicated to free radicals in biology and medicine.

Up-to-date works on free radical signaling are considered in Chapter 7. In this chapter, it would be useful just to comment on some peculiarity of this problem. Strictly speaking, the completely correct determination of free radical signaling is impossible. From my point of view, a main feature of the development of this concept must be the investigation of regulatory functions of physiological free radicals O$_2$$^-$ and NO, and their reactive diamagnetic derivatives H$_2$O$_2$ and ONOO$^-$, in enzymatic processes. Thus, the signaling activity of ROS and RNS is opposite to their damaging activity through the direct H abstraction or addition reactions of free radicals with biomolecules. However, the difference between damaging and regulatory signaling reactions of free radicals is not always easy to determine because free radicals also accomplish their signaling functions in enzymatic cascades, leading to the initiation of pathological disorders. For example, ROS and RNS play an important role as signaling molecules in apoptosis (Chapter 5) and aging (Chapter 6). Therefore, it might be useful to consider ROS and RNS as signaling species in any physiological or pathological enzymatic process, and as damaging species when free radicals directly destroy biomolecules.

Another question: Could reactive short-living free radicals such as hydroxyl and peroxyl possess signaling activity? However, as will be seen from Chapter 7, the main function of signaling species is to participate in enzymatic catalysis, accelerating or slowing it without its disturbance. It seems to me that this condition could not be applied to such reactive species, and therefore, they cannot fulfill signaling functions.

REFERENCE

1. ET Denisov and IB Afanas'ev. 2005. *Oxidation and antioxidants in organic chemistry and biology.* Boca Raton, FL: Taylor & Francis.

2 Free Radical Signaling in Reactions Catalyzed by Enzymes—Producers of Reactive Oxygen and Nitrogen Species

2.1 REACTIVE OXYGEN SPECIES (ROS) SIGNALING IN REACTIONS CATALYZED BY XANTHINE OXIDASE

2.1.1 MECHANISM OF CATALYTIC ACTIVITY OF XANTHINE OXIDASE

There is some justice in the general opinion that contemporary free radical studies in biology and medicine began after the discovery by McCord and Fridovich [1–3] of superoxide production by xanthine oxidase (XO). These authors showed that XO catalyzed cytochrome c reduction, which was inhibited by the enzyme superoxide dismutase (SOD). (Actually, McCord and Fridovich applied another enzyme, bovine erythrocyte carbonic anhydrase containing SOD as an admixture). In 1968–1970, superoxide ($O_2 \cdot^-$) was positively identified in biological systems by biological, physico-chemical, and spectral analytical methods [4–6].

XO and xanthine dehydrogenase (XDH) are two forms of the enzyme xanthine oxi-doreductase. These forms are interconvertible, and they depend on the oxidation state of protein thiols. Xanthine oxidoreductase catalyzes the oxidation of xanthine (X) to urate with concomitant reduction of dioxygen or NAD. The enzyme exists as a homodimer with subunits containing one FAD, one molibdopterin, and two 2F/2S clusters [7].

Production of ROS, $O_2 \cdot^-$, and hydrogen peroxide (H_2O_2) by XO has been thoroughly studied. In 1974, Olson et al. [8] showed that, during the oxidation of xanthine to urate, six electrons are transferred from fully reduced enzyme through four redox centers. Catalysis by XO is accompanied by the reduction of dioxygen into H_2O_2 by two-electron reduction mechanism and into $O_2 \cdot^-$ by one-electron reduction mecha-nism (Figure 2.1). Although it was believed earlier that only XO, and not XDH, is able to generate $O_2 \cdot^-$, it was now shown that XDH is also a producer of $O_2 \cdot^-$, although not as effective as XO [9,10]. Uric acid, the final reaction product of XO reaction, is capable of inhibiting the catalytic activity of xanthine oxidase [11]. Inhibition of XO by uric acid apparently occurs under both in vitro and in vivo conditions. Therefore, physiological levels of uric acid might inhibit XO activity in human plasma [12].

$$XO(6) \implies XO(4) \implies XO(2) \implies XO(1) \implies XO(0)$$

$$\Downarrow \qquad\qquad \Downarrow \qquad\qquad \Downarrow \qquad\qquad \Downarrow$$

$$H_2O_2 \qquad H_2O_2 \qquad O_2{}^{\cdot-} \qquad O_2{}^{\cdot-}$$

FIGURE 2.1   Mechanism of ROS generation by xanthine oxidase.

It was widely accepted that $O_2{}^{\cdot-}$ is the only oxygen radical produced by xanthine oxidase. However, as early as in 1970, Beauchamp and Fridovich [13] proposed that XO produced hydroxyl radicals in addition to $O_2{}^{\cdot-}$ through the reaction of $O_2{}^{\cdot-}$ with H_2O_2 (the Haber–Weiss reaction). It is now known that this reaction proceeds only in the presence of iron ions (the Fenton reaction); therefore, XO can produce hydroxyl radicals only indirectly by the catalysis of adventitious iron ions. Nonetheless, Kuppusamy and Zweier [14] also suggested that XO is able to generate hydroxyl radicals directly by the reduction of H_2O_2. However, the subsequent works by Lloyd and Mason [15] and Britigan et al. [16] have shown that the source of hydroxyl radicals in Kuppusamy and Zweier's experiments was also adventitious iron, which presented in commercial XO samples and buffers.

Tissue distribution of XO is an important factor of $O_2{}^{\cdot-}$-induced injury. XO levels are practically undetectable in normal human plasma, but the levels of plasma-circulating XO and the ability of circulating XO to bind to vascular cells of various organs sharply increase during some pathological states such as reperfusion injury, hepatitis, adult respiratory distress syndrome, and atherosclerosis. In 1992, Partridge et al. [17] demonstrated that pulmonary microvascular endothelial cells constitutively released XO and XDH under hypoxic conditions. Houston et al. [18] have shown that endothelial-bound XO is able to generate $O_2{}^{\cdot-}$, which rapidly reacts with nitric oxide, producing peroxynitrite (ONOO⁻).

2.1.2   INHIBITORS AND ACTIVATORS OF XANTHINE OXIDASE

Xanthine oxidase activity depends on many factors. Hassoun et al. [19] has found that XO was inhibited by nitric oxide. Ichimori et al. [20] suggested that nitric oxide (NO) inhibited xanthine oxidoreductase under anaerobic conditions through the interaction with the sulfhydryl groups of the reduced molybdenum center of XO and XDH. In contrast, Houston et al. [21] showed that NO did not inhibit XO, and that the apparent decrease in XO activity observed in earlier works is explained by the suppression of uric acid formation by ONOO⁻. These findings were supported by Lee et al. [22] who also found that NO is unable to inhibit xanthine oxidase, while ONOO⁻ irreversibly inhibited XO activity through the disruption of its molybdenum center.

Dioxygen is not the only compound that is reduced by xanthine oxidase. Millar [23] has shown that XO catalyzed the conversion of dioxygen and nitrite to ONOO⁻ when both reagents were simultaneously presented in the medium. Saleem and Ohshima [24] proposed that, under anaerobic conditions, XO reduced NO into NO− (HNO), which inhibited the XO activity.

H_2O_2 is another inhibitor of xanthine oxidase. It decreased XDH protein expression in bovine aortic endothelial cells (BAEC) and stimulated conversion of XDH to XO.

Endothelial XO and XDH expression also strongly depend on H_2O_2 and calcium [25]. Among various inhibitors of XO, allopurinol draws enhanced attention, being a medicine for the treatment of cardiovascular diseases. It is known that XO converts the prodrug oxypurinol into allopurinol. However, Galbusera et al. [26] recently demonstrated that this conversion is accompanied by additional $O_2\cdot^-$ production.

2.1.3 SIGNALING PROCESSES CATALYZED BY XANTHINE OXIDASE

$O_2\cdot^-$ and H_2O_2 are major ROS performing signaling functions in many enzymatic processes. In many works, XO was applied as the ROS producer for studying the signaling functions of ROS during various processes catalyzed by heterolytic enzymes (see Chapter 4). However, it is noteworthy that, under in vivo conditions, XO is not only an $O_2\cdot^-$ producer due to the presence in the cells of other enzymes such as NADPH oxidase (Nox) and NO synthase (NOS).

In 1992, Wu et al. [27] showed that $O_2\cdot^-$ produced by XO activated phospholipase A2 (PIA$_2$), a lipolytic enzyme that caused phospholipid hydrolysis and arachidonic acid release in the rat corpus luteum. In addition to $O_2\cdot^-$, H_2O_2 probably also participates in the PIA$_2$ activation because catalase was an even more efficient inhibitor of this process than $O_2\cdot^-$. Accordingly, Gustafson et al. [28] has shown that H_2O_2 stimulated the release of radiolabeled arachidonic acid (14C-AA) in cultured intestinal epithelial cells. These findings indicated that H_2O_2 may stimulate phospholipase A2-mediated AA release from human intestinal epithelial cells.

Zang et al. [29] found that tumor necrosis factor alpha (TNF-α) induced ceramide-dependent activation of c-Jun N-terminal kinase mitogen-activated protein kinase (JNK MAPK) and subsequent production of $O_2\cdot^-$, which inhibited endothelium-dependent NO-mediated dilation of coronary arterioles. It was also shown that XO was only an $O_2\cdot^-$ producer in these experiments because its formation was inhibited by $O_2\cdot^-$ scavenger or inhibitors of ceramide-activated protein kinase JNK and was insensitive to the inhibitors of p38 and extracellular signal-regulated (ERK) protein kinases, Nox, or mitochondrial respiratory chain.

Ross and Armstead [30] have shown that $O_2\cdot^-$ produced by XO activated protein tyrosine kinase (PTK) and protein kinase ERK contributing to impairment of ATP-sensitive K+ (KATP) and calcium-activated potassium (KCa) channel pial artery dilation (PAD) in pigs. Wang and Doerschuk [31] reported that ROS generated by XO caused the activation of p38 MAPK in pulmonary microvascular endothelial cells. Brzezinska et al. [32] has shown that $O_2\cdot^-$ produced by XO directly affected the function of ion channels in vascular endothelium. Gaitanaki et al. [33] investigated the effects of L-ascorbic acid, catalase, and SOD on the p38-MAPK activation induced by XO or H_2O_2 in the isolated perfused amphibian heart. H_2O_2-induced activation of p38-MAPK was totally inhibited by L-ascorbic acid or catalase. p38-MAPK was also activated by perfusing amphibian hearts with the ROS-generating system of X/XO. Finally, these findings showed that X/XO induced phosphorylation of the potent p38-MAPK substrates MAPKAPK2 and HSP27, and this activation was reduced by the simultaneous use of catalase and SOD.

Matesanz et al. [34] investigated stimulation by an X/XO system of proliferative and hypertrophic response of human aortic smooth muscle cells (HASMC). It was

$$XO \longrightarrow O_2^{\cdot-} + H_2O_2 \longrightarrow \text{phospholipase A2 (PIA}_2) \longrightarrow AA \qquad [27]$$

$$TNF\alpha \longrightarrow JNK \longrightarrow XO \text{ } activation \longrightarrow O_2^{\cdot-} \longrightarrow \text{ } NO\text{-}mediated \text{ } dilation$$
$$\text{of coronary arterioles } [29]$$

$$XO \longrightarrow ROS \longrightarrow p38 \text{ } activation \text{ } (microvascular \text{ } pulmonary \text{ } ECs) \qquad [31]$$

$$XO \longrightarrow O_2^{\cdot-} \longrightarrow PKC \text{ } activation \qquad [35]$$

$$XO \longrightarrow O_2^{\cdot-} \longrightarrow PKC \text{ } activation \text{ } (via \text{ } thiol \text{ } oxidation) \qquad [36]$$

$$XO \longrightarrow O_2^{\cdot-} \longrightarrow PTK \text{ } and \text{ } ERK \text{ } activation \qquad [30]$$

$$XO \longrightarrow O_2^{\cdot-} \longrightarrow Activation \text{ } of \text{ } c\text{-}Jun \text{ } N \text{ } and \text{ } p38 \text{ } MAPK \text{ } kinases \qquad [34]$$

Legend: PTK: protein tyrosine kinase, ERK1/2: extracellular signal-regulated kinase 1/2, p38: MAPK kinase, c-Jun N: MAPK kinase, PKC: protein kinase C.

FIGURE 2.2 Signaling functions of xanthine oxidase.

found that the X/XO system exhibited hypertrophic properties for human vascular smooth muscle, which were mediated by redox-sensitive pathways involving JNK and p38-MAPK activation but not extracellular signal-regulated kinase 1/2 (ERK1/2) or serine/threonine protein kinase B (Akt/PBK) activity. Finally, the effects of X/XO on MAPK activation, AP-1 activity, and cell size were dependent on the extracellular release of $O_2^{\cdot-}$ by XO, as they were prevented by both SOD and allopurinol.

$O_2^{\cdot-}$ generated by XO is able to activate protein kinase C. Larsson and Cerutti [35] demonstrated that the treatment of mouse epidermal JB6 cells with low concentrations of $O_2^{\cdot-}$ but not of H_2O_2 resulted in the activation of the in vitro phosphorylating capacity of cytoplasmic extracts containing PKC. Knapp and Klann [36] have shown that $O_2^{\cdot-}$ stimulated autonomous PKC activity via thiol oxidation and the release of zinc from the cysteine-rich region of PKC. Silva et al. [37] has shown that XO-generated $O_2^{\cdot-}$ increased Cl$^-$ absorption in isolated perfused medullary thick ascending limbs through the activation of protein kinase C.

Zhen et al. [38] has recently shown that XO-produced $O_2^{\cdot-}$ upregulated the expression of eNOS at both transcriptional and translational levels in human coronary artery endothelial cells. It was suggested that this effect could be mediated by limiting the availability of NO, thereby exerting a negative feedback influence on NOS expression through activation of NFκB. Examples of ROS signaling in the processes catalyzed by XO are presented in Figure 2.2.

2.1.4 ENHANCEMENT OF ROS PRODUCTION BY XO IN PATHOLOGICAL STATES

ROS signaling increases sharply in many pathological disorders, and XO frequently becomes one of the major toxic factors. There is practically no plasma XO activity under normal conditions, but oxidative stress may increase it by several times. Thus, De Long et al. [39] demonstrated that the activity of XO increased in hypertrophied and failing heart. They found that the expression of functional XO was elevated in failing but not in hypertrophic ventricles, suggesting its potential role in the transition from cardiac hypertrophy into heart failure.

Saavedra et al. [40] pointed out the importance of the imbalance between XO- and NOS-signaling pathways, which might be the cause of uncoupling in the failing heart. Landmesser et al. [41] found that the activity of XO bound to the endothelium increased by about 200% in patients with chronic heart failure (CHF), together with sharp decrease in eSOD. Both factors were contributing to endothelial dysfunction in patients with CHF. Duncan et al. [42] demonstrated a direct link between chronic inhibition of XO activity and preservation of cardiac function. Their findings show that XO activity is elevated in the early asymptomatic stages of cardiac dysfunction and that chronic inhibition of XO activity is possible with oral administration of allopurinol, which results in improved contractile function. The relationship between XO and neuronal nitric oxide synthase (nNOS) was shown in the sarcoplasmic reticulum of cardiac myocytes [43]. Deficiency of nNOS (but not eNOS) led to a sharp increase in XO-mediated $O_2 \cdot^-$ production.

XO-dependent $O_2 \cdot^-$ overproduction can contribute to the development of atherosclerosis. Landmesser et al. [44] has found that angiotensin II (Ang II) substantially increased endothelial XO protein levels and XO-dependent $O_2 \cdot^-$ production in cultured endothelial cells. These findings are relevant to endothelial dysfunction in patients with coronary disease. There are findings suggesting an increase in XO-dependent $O_2 \cdot^-$ production in aging. Thus, Newaz et al. [45] reported that XO activity and XO expression increased in the aorta of aging rats. Similarly, Aranda et al. [46] found positive correlation between XO activity and the age in human and rat plasma.

2.2   ROS SIGNALING BY PHAGOCYTE AND NONPHAGOCYTE NADPH OXIDASES

The family of Nox enzymes is one of the most important $O_2 \cdot^-$ producers in phagocytic and nonphagocytic cells. The discovery of phagocyte Nox (originally named gp91^{phox}, and now Nox2) was made as early as 1977 [47]. Now six homologs of phagocyte Nox have been identified in nonphagocytic cells: Nox1, Nox3, Nox4, Nox5, Duox1, and Duox2. These enzymes transfer electrons across the plasma membrane and generate $O_2 \cdot^-$ and other ROS. (Both dual oxidases Duoxes 1 and 2 have a domain homologous to the Nox of the phagocyte Nox gp91^{phox}/Nox2 and a domain homologous to thyroid peroxidase.) We have no need here to consider in detail the structure, properties, and mechanism of $O_2 \cdot^-$ generation by Nox enzymes, which were presented and discussed in numerous reviews (see, for example, References [48,49,50,51]). For example, DeLeo et al. [50] discussed Nox activation and assembly during phagocytosis. In an excellent comprehensive review, Quinn and Gauss [51] considered the structures and regulation of phagocyte and nonphagocyte oxidases.

2.2.1   PHAGOCYTE NADPH OXIDASE

Phagocyte NADPH oxidase Nox2 does not exist as a whole in "dormant" state, and is synthesized only after stimulation with various organic and inorganic compounds

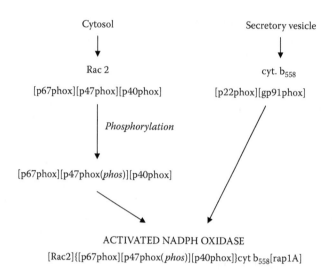

FIGURE 2.3   Synthesis of phagocyte NADPH oxidase upon activation.

and particles, while nonphagocyte NADPH oxidases Nox1, Nox3, Nox4, Nox5, Duox1, and Duox2 exist in cells as complete enzymes. Correspondingly, the modes of oxygen radical production in these cells differ significantly. The stimulation of Nox in phagocytes results in "oxidative burst," that is, release of a relatively big amount of oxygen radicals, which is extinguished during a short time (5–30 min), while non-phagocytic cells produce oxygen radicals continuously but in smaller quantities.

All Noxs catalyze the overall Reaction 2.1:

$$2\ O_2 + NADPH \Leftrightarrow 2\ O_2{}^{\cdot-} + NADP^+ + H^+ \qquad (2.1)$$

The mechanism of activation of phagocyte Nox is well established [48]. The core enzyme consists of five components: $p40^{phox}$, $p47^{phox}$, $p67^{phox}$, $p22^{phox}$, and $gp91^{phox}$. In dormant cells, three of these five components—$p40^{phox}$, $p47^{phox}$, and $p67^{phox}$—exist in the cytosol as a complex. The other two, $p22^{phox}$ and $gp91^{phox}$, are located in the membranes of secretory vesicles and specific granules as a heterodimeric flavo-hemoprotein cytochrome b558. When dormant cells are exposed to the stimuli, the cytosolic component $p47^{phox}$ becomes phosphorylated and the entire cytosolic complex migrates to the membrane where it associates with cytochrome b558 to form an active oxidase capable of transferring electrons from the substrate to dioxygen (Figure 2.3).

2.2.2 Signaling Processes in Activation of Phagocyte NADPH Oxidase

One of the major activation steps that start the activation process of phagocyte Nox is the phosphorylation of its components catalyzed by various enzymes. In 1991, Nauseef et al. [52] showed that protein kinase C (PKC)-dependent phosphorylation of $p47^{phox}$ component, followed by the association of $p47^{phox}$ with the cytoskeleton and with translocation of $p47^{phox}$ and $p67^{phox}$ to the plasma membrane, resulted in the assembly of Nox. El Benna et al. [53] demonstrated that PKC catalyzed phosphorylation of the p47^{phox} serine residues. Beck-Speier et al. [54] has shown that sulfite stimulated Nox of human neutrophils via protein kinase C and calcium/calmodulin pathways. Dang et al. [55] found that PKCζ, an atypical PKC isoform, participated in Nox activation by formyl-methionyl-leucyl-phenylalanine (FMLP) receptors, but not by phorbol-12 myristate 13-acetate (PMA), and probably regulated the bactericidal function of leukocytes.

Watson et al. [56] has found that Nox of human neutrophils can be activated by both protein-kinase-C-dependent and protein-kinase-C-independent ways. The initial burst of $O_2\cdot^-$ production activated by FMLP, opsonized zymosan (OZ) and latex beads, was independent of the activity of protein-kinase-C-dependent intracellular activation processes, but the activity of this kinase was required to extend or sustain the duration of $O_2\cdot^-$ production. Curnutte et al. [57] demonstrated that protein kinase C and type 1 and/or 2A protein phosphatases are involved in a continuous phosphorylation of Nox p47^{phox} of neutrophils that maintains the oxidase in the assembled/active state. Recently, Miller et al. [58] examined the source of ROS in paraquat-stimulated microglial cells. Paraquat-induced $O_2\cdot^-$ formation followed by translocation of the p67^{phox} cytosolic subunit of Nox to the membrane, was mediated by PKCδ and ERK protein kinases.

It has been found [59] that homocysteine (Hcy) induced $O_2\cdot^-$ production in monocytes through PKC-dependent phosphorylation of p47^{phox} and p67^{phox} subunits of Nox that supposedly played an important role in Hcy-induced inflammatory response during atherogenesis. It was established that several protein kinase C isoforms phosphorylated p47^{phox}, resulting in its membrane translocation and the activation of Nox. However, Cheng et al. [60] reported that FMLP-induced PKCδ phosphorylation is essential in the catalysis of p47^{phox} phosphorylation. Thus, functionally active PKCδ is required for p47^{phox} phosphorylation and reconstitution of Nox.

Several other enzymes participate in Nox activation. Cytosolic phospholipase A2 (cPLA2) catalyzes the formation of arachidonic acid (AA), which activates phagocyte Nox [61]. Henderson et al. [62] has shown that AA, released by phospholipase A2, is necessary for both the activation and maintenance of $O_2\cdot^-$ generation by PMA-stimulated Nox. Cherny et al. [63] demonstrated that AA was an activator of both Nox and proton channels in human eosinophils.

Hazan et al. [64] examined the role of cPLA2 in the activation of Nox by OZ in human neutrophils. They suggested that the activation of cPLA2 by PMA or OZ was mediated by ERK MAPK kinases. Zhao et al. [65] concluded that the cPLA2 activity or AA is required for p67^{phox} and p47^{phox} translocation in human monocytes but cPLA2 did not influence phosphorylation of these components. These results suggest that cPLA2 regulates Nox activity by control of the arachidonate-sensitive assembly of the complete oxidase complex through the translocation of p67^{phox} and p47^{phox}.

Dana et al. [66] showed that cPLA2 is located downstream of protein kinase C in the process of Nox activation. O'Dowd et al. [67] demonstrated that the β-adrenergic agonist adrenaline inhibited FMLP-stimulated but not PMA-stimulated $O_2\cdot^-$ production in PMN upstream of the Nox through PKA protein kinase and cPLA2. Zhang et al. [68] examined the role of AA in platelet-derived growth factor (PDGF)-stimulated ROS generation in human lens epithelial cells. They found that

PDGF-stimulated AA release depended on both active cPLA2 and ERK1/2 protein kinase. Chenevier-Gobeaux et al. [69] has shown that cPLA2 participated in the regulation of human NADPH oxidase Nox2 of synovial cells from patients with rheumatoid arthritis and osteoarthritis. They found that cPLA2-stimulated AA production is an important cofactor of synovial Nox activity.

Phosphatidylinositol 3-kinase (PI3-K) is another activator of phagocyte Nox [70–76]. Ding et al. [70] and Vlahos et al. [71] have shown that PI3-K kinase catalyzed phosphorylation of p47^{phox} component of neutrophils. Chen et al. [72] demonstrated the existence of Akt-mediated phosphatidylinositol 3-kinase-dependent p47^{phox} phosphorylation, which contributed to respiratory burst activity in human neutrophils. Yamamori et al. [73] also showed that PI3-K/p38 MAPK/Rac pathway took place in the activation of Nox in bovine neutrophils. Suh et al. [74] showed that p40^{phox} and phosphatidylinositol-3-phosphate (PI3P) activated Nox during FcγR-mediated phagocytosis. Poolman et al. [75] reported that the antiinflammatory compound resveratrol inhibited PI3-K-dependent activation of NADPH oxidase in human monocytes. Gao et al. [76] showed that the class IA PI3-K regulated CD11b/CD18-integrin-dependent PMN adhesion and activation of Nox that led to ROS production at sites of PMN adhesion.

It has already been shown that the Akt-mediated PI3-K-dependent pathway is responsible for p47^{phox} phosphorylation. Hoyal et al. [77] also demonstrated that Akt protein kinase activated Nox by phosphorylation of serines Ser304 and Ser328 of p47^{phox}. These findings support the importance of the Akt/B protein kinase cascade in the activation of leukocyte Nox.

Numerous studies suggest that MAPK also play an important role in the activation and priming of polymorphonuclear leukocytes. It has been suggested that the signal transduction pathways activated by the cytokines TNF-α and granulocyte-macrophage colony-stimulating factor (GM-CSF) led to priming of PMNs through the stimulation of multiple MAPK cascades, including ERKs, JNKs, and p38-MAPK. McLeish et al. [78] suggested that TNF-α and GM-CSF activated ERK and p38-MAPK by different signal transduction pathways. Both ERK and p38 MAPK cascades contributed to the ability of TNF-α and GM-CSF to prime Nox in human PMNs.

Zu et al. [79] has shown that exposure of human neutrophils to FMLP, PMA, or ionomycin rapidly induced the activation of p38 and p44/42 MAPK kinases, but stimulation with inflammatory cytokine TNF-α triggered only the activation of p38 kinase. Brown et al. [80] considered the relative role of p38-MAPK in the priming and activation processes. They found that it plays a critical role in TNF-α priming of human and porcine Nox for $O_2^{.-}$ production in response to complement-opsonized zymosan (OPZ), but little, if any, role in neutrophil priming by platelet-activating factor (PAF). The OPZ-mediated activation process was independent of p38-MAPK activity, in contrast to oxidase activation by FMLP. Furthermore, the incubation of neutrophils with TNF-α resulted in the p38-MAPK-dependent phosphorylation of p47^{phox} and p67^{phox} components.

Dewas et al. [81] demonstrated that ERK1/2 was involved in p47^{phox} phosphorylation in FMLP-stimulated human neutrophils. Moreover, FMLP triggered both the ERK1/2-MAPK and PKC pathways to phosphorylate p47^{phox}. Later on, these authors also showed that proinflammatory cytokines GM-CSF and TNF-α stimulated MAPK pathways of phosphorylation of Ser345 on p47^{phox}, a cytosolic component of Nox, in human neutrophils [82]. Seru et al. [83] has shown that the small GTP-binding-protein Ras (HaRas) activated Nox in human neuroblastoma cells via the ERK1/2 pathway.

It was shown that the small GTPase protein RhoA stimulates $O_2^{\cdot-}$ generation during phagocytosis of serum-(SOZ) and IgG-opsonized zymosan particles (IOZ). Kim et al. [84] suggested that there is a signal pathway regulated by Rho GTPase where RhoA, ROCK (RhoA-activated kinase), p38-MAPK, ERK1/2, and p47^{phox} are subsequently activated, leading to the activation of Nox. Alba et al. [85] has shown that 5′-AMP-activated protein kinase (AMPK) modulated Nox activity in human neutrophils. Activated AMPK significantly diminished both PMA- and FMLP-stimulated $O_2^{\cdot-}$ release by human neutrophils and reduced translocation to the cell membrane and phosphorylation of p47^{phox}, a cytosolic component of Nox.

The small GTPase Rac1, a required catalytic subunit of Nox, translocates to the membrane p47^{phox} and p67^{phox}. An important effector molecule for Rac1 is the serine/threonine kinase p21-activated kinase (PAK1), which is able to stimulate Nox-dependent $O_2^{\cdot-}$ production. Martyn et al. [86] demonstrated that PAK1 activated the Nox complex in human neutrophils, binding directly to p22^{phox}. It suggests that flavocytochrome b is the oxidase-associated membrane target of this kinase. Similarly, Roepstorff et al. [87] showed that PAK1 function was required for FMLP stimulation of Nox-mediated $O_2^{\cdot-}$ production. El Bekay et al. [88] recently demonstrated that Ang-II-stimulated $O_2^{\cdot-}$ production by neutrophil Nox involved the G-protein Rac2 activation through multiple signaling pathways such as Ca(2)(+)/calcineurin and phosphorylation of the p38 MAPKs, ERK1/2, and JNK1/2.

Lysophosphatidylcholine (LPC), an oxidized phospholipid presented in micromolar concentrations in blood and inflamed tissues, inhibits $O_2^{\cdot-}$ generation by FMLP- and PMA-stimulated neutrophils. Lin et al. [89] suggested that LPC-induced elevation of intracellular cAMP is partially responsible for the inhibition of FMLP-stimulated phosphorylation of ERK and Akt and membrane translocation of p67^{phox} and p47^{phox}. Fay et al. [90] has shown that the activation of Ca(2+)-activated potassium channels of small and intermediate conductance caused the production of $O_2^{\cdot-}$ and H_2O_2 by neutrophils and granulocyte-differentiated PLB-985 cells. Shima et al. [91] suggested that some calcium channel blockers can suppress cytokine-induced neutrophil activation, preventing possible progression of atherosclerosis through the activation of ERK and PI3K/Akt pathways, induced by TNF-α but not by GM-CSF.

Hendersen et al. [92] has shown that the activation of the $O_2^{\cdot-}$-generating Nox depolarized the membrane potential of cytoplasts prepared from human neutrophils; that is, that phagocyte Nox must be an electrogenic enzyme. Schrenzel et al. [93] confirmed that phagocyte Nox is a generator of electron currents across the plasma membrane of eukaryotic cells. The effects of PMA- and AA-activated Nox on proton and electron currents in human eosinophils have been shown by DeCoursey et al. [63,94,95]. Gauss et al. [96] demonstrated that TNF-α treatment of human monocytic cells and isolated human monocytes resulted in upregulation of the Nox gene. These findings suggest that TNF-α produced by activated macrophages could serve as an autocrine/paracrine regulator of Nox, increasing or prolonging $O_2^{\cdot-}$ production.

Many pathological states are characterized by the activation of phagocyte Nox and the enhancement of ROS production. Thus, Alvarez-Maqueda et al. [97] investigated the role of hyperhomocysteinemia (Hcy) in essential functions of human neutrophils. They found that Hcy increased $O_2\cdot^-$ release by neutrophils to the extracellular medium, and that this effect was inhibited by SOD and diphenyleneiodonium (DPI), an inhibitor of Nox activity. Enzyme from rat peritoneal macrophages displayed a similar response. These effects were accompanied by the enhanced translocation of p47^{phox} and p67^{phox} subunits of Nox to the plasma membrane. Hcy also increased the activation and phosphorylation of MAPKs, p38-MAPK, and ERK1/2. Siow et al. [98] showed that homocysteine treatment caused a significant elevation of intracellular $O_2\cdot^-$ in monocytes and induced the activation of Nox, which was regulated through PKC-dependent phosphorylation of p47^{phox} and p67^{phox}.

Sickle cell disease (SCD) is a chronic inflammatory condition characterized by high leukocyte counts, altered cytokine levels, and endothelial cell injury. It has been shown that mononuclear leukocytes from SCD patients release higher amounts of $O_2\cdot^-$ compared to normal controls. Marcal et al. [99] suggested that this is explained by the enhanced levels of gp91^{phox} gene expression and p47^{phox} phosphorylation in monocytes from SCD patients. Signaling processes in phagocyte Nox activation are presented in Figure 2.4.

2.2.3 Signaling Processes Catalyzed by Phagocyte NADPH Oxidase

The major function of phagocyte Nox is the production of $O_2\cdot^-$ upon activation during phagocytosis and microbial killing. However, this enzyme also participates in cell-signaling processes. It is of importance that signaling by phagocyte Nox could result in the cycling overproduction of ROS. After synthesis at the cell membrane, phagocyte Nox releases $O_2\cdot^-$ and H_2O_2 into extracellular space. However, the ability of phagocyte Nox to initiate signaling processes indicates that a certain part of ROS is produced by enzyme inside the cell as has already been suggested.

Anrather et al. [100] demonstrated that the ubiquitous transcription factor NFκB was a downstream target of $O_2\cdot^-$ produced by Nox. It was also found that gp91^{phox} expression was dependent of NFκB; therefore, there is the possibility of a positive feedback loop in which NFκB activation by oxidative stress leads to further radical production via Nox (Figure 2.5). Coffey et al. [101] has shown that ROS produced by Nox regulated the activity of 5-lipoxygenase (5-LO) by expression of the key leukotriene in alveolar macrophages. Zalba et al. [102] has found that phagocytic Nox-dependent $O_2\cdot^-$ production stimulated matrix metalloproteinase-9 (MMP-9) in human monocytes and that this relationship may be relevant in the atherosclerotic process. Vega et al. [103] demonstrated that ROS produced by Nox are involved in the upregulation of cyclooxygenase-2 (COX-2), the enzyme responsible for prostaglandin (PG) synthesis. This process was mediated by MAPKs, p38, and ERK 1/2, and the transcription factor NFκB (Figure 2.5).

2.2.4 Nonphagocyte NADPH Oxidase

Nonphagocyte Nox contains the same components as phagocyte Nox [104], but their distribution depends on cell types (endothelial cells, mesangial cells, fibroblasts, and others). Thus, it is known that the core component gp91^{phox}-p22^{phox} heterodimer

PKC $\Rightarrow$ *phosphorylation of* (Ser) p47^{phox} $\Rightarrow$ *assembly of* NADPH oxidase   [52]

PKC + protein phosphatases 1/2 $\Rightarrow$ *phosphorylation of* p47^{phox} [57]

PKCδ + ERK $\Rightarrow$ p67^{phox} *translocation*   [58]

FMLP $\Rightarrow$ *phosphorylation of* PKCδ $\Rightarrow$ *phosphorylation of* p47^{phox} [60]

cPLA2 $\Rightarrow$ AA $\Rightarrow$ NADPH *activation*   [61–63]

cPLA2 $\Rightarrow$ AA $\Rightarrow$ p67^{phox} and p47^{phox} *translocation but not phosphorylation*   [65]

PI3-K $\Rightarrow$ Akt $\Rightarrow$ *phosphorylation of* p47^{phox} [72–78]

PI3-K $\Rightarrow$ p38 MAPK $\Rightarrow$ Rac $\Rightarrow$ NADPH oxidase *activation*   [75]

Akt $\Rightarrow$ *Ser304, 328* p47^{phox} *phosphorylation* $\Rightarrow$ NADPH oxidase *activation*   [77]

TNFα *or* GM-CSF $\Rightarrow$ ERK, p38 *activation* $\Rightarrow$ *priming* NADPH oxidase   [78]

FMLP, PMA, *or* ionomycin $\Rightarrow$ p38 *and* p44/42 *activation*   [79]

TNFα *or* GM-CSF $\Rightarrow$ *Ser345 phosphorylation on* p47^{phox} [82]

HaRas $\Rightarrow$ ERK(1/2) $\Rightarrow$ NADPH oxidase *activation*   [83]

RhoGTPase $\Rightarrow$ RhoA $\Rightarrow$ ROCK $\Rightarrow$ p38 $\Rightarrow$ ERK(1/2) $\Rightarrow$ p47^{phos}

$\Rightarrow$ NADPH oxidase *activation*   [84]

p21-*activated* kinase PAK1 $\Rightarrow$ NADPH oxidase *activation*   [86]

FMLP $\Rightarrow$ PAK1 $\Rightarrow$ NADPH oxidase *activation*   [87]

AngII $\Rightarrow$ Ca(2)/calcineurin $\Rightarrow$ p38, ERK1/2, c-Jun *phosphorylation*

$\Rightarrow$ Rac2 activation $\Rightarrow$ NADPH oxidase *activation*   [88]

Legend: AA: arachidonic acid, InsP3: 1,4,5-trisphosphate, ERK1/2: extracellular signal-regulated kinase ½, PI3K: phosphatidylinositol 3-kinase, ATF-1: activating transcription factor 1, NOX1: NADPH oxidase enzyme, VEGF: vascular endothelial growth factor, Akt/PKB: serine/threonine protein kinase B, GTP: guanosine monophosphate, JAK: Janus kinase, STAT: signal transducer of transcription, CREB: cAMP response element-binding protein, EGFR: epidermal growth factor receptor, PKG-1: cGMP-dependent protein kinase I, TRAF4: TNF receptor-associated factor 4, H/R: hypoxia-reoxygenation.

FIGURE 2.4 Signaling processes in activation of phagocyte NADPH oxidase.

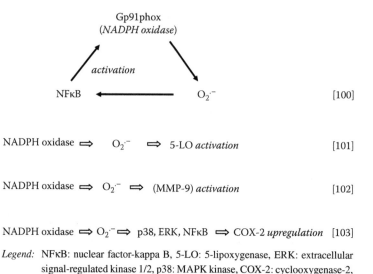

Gp91phox
(*NADPH oxidase*)

activation

NFκB ⟵———————— $O_2{}^{\cdot-}$ [100]

NADPH oxidase ⟹ $O_2{}^{\cdot-}$ ⟹ 5-LO *activation*   [101]

NADPH oxidase ⟹ $O_2{}^{\cdot-}$ ⟹ (MMP-9) *activation*   [102]

NADPH oxidase ⟹ $O_2{}^{\cdot-}$ ⟹ p38, ERK, NFκB ⟹ COX-2 *upregulation* [103]

Legend: NFκB: nuclear factor-kappa B, 5-LO: 5-lipoxygenase, ERK: extracellular
signal-regulated kinase 1/2, p38: MAPK kinase, COX-2: cyclooxygenase-2,
MMP-9: metalloproteinase-9.

FIGURE 2.5 Phagocyte NADPH oxidase signaling.

(cytochrome b558) responsible for the enzyme activity of phagocyte Nox is located at
the plasma membrane in neutrophils. However, in endothelial cells, Nox is localized
in intracellular space rather than plasma membrane bound, and a significant amount
of the Nox subunits in unstimulated cells are present as fully assembled functional
$O_2{}^{\cdot-}$-generating complexes [105]. Endothelial Nox is not only "constitutively" active
but also responds to stimulation by various agonists (such as Ang II, phorbol ester,
and cytokines). Potential mechanisms of such stimulation could be an increase in the
total number of fully assembled complexes, the translocation of p47^{phox} or Racl to
partially assembled complexes, changes in phosphorylation of subunits, or changes
in the amount of NADPH available to the enzyme [105].

Nonphagocyte Nox is expressed differently in different cells. Thus, Gorlach et al.
[106] demonstrated that endothelial cells but not smooth-muscle cells are able to
express Nox. Nonetheless, it was found that vascular smooth-muscle cells (VSMCs)
can show prominent Nox activity, although they lack Nox1 gp91^{phox} oxidase. Ellmark
et al. [107] suggested that, under resting conditions, Nox activity in mouse-cultured
VSMCs was largely dependent upon Nox4 expression. On the other hand, San Martin
et al. [108] have shown that Nox1 NADPH oxidase produced $O_2{}^{\cdot-}$ upon the activation
of VSMCs by advanced glycation end products.

2.2.5 LOCALIZATION AND MECHANISMS OF ACTIVATION OF NONPHAGOCYTE NADPH OXIDASE

It has been shown that vascular (nonphagocyte) Noxs produce $O_2{}^{\cdot-}$ inside the cells.
However, $O_2{}^{\cdot-}$ is not released in the cytosol but rather remains inside vesicles, which
may serve to confine its toxic action. Therefore, under physiological conditions, ROS

production is not accompanied by oxidative stress but performs certain signaling functions. Probably, it could be a cause of the specific subcellular localization of the enzyme within a particular signaling domain. For example, Nox1 is located at the cell periphery in VSMCs, whereas Nox4 is associated with focal adhesions [109].

Yang and Rizzo [110] showed that subunits of endothelial Nox were preassembled in resting cells and that the enzyme operated in membrane rafts, specifically in caveolae. Stimulation with TNF-α induced additional recruitment of the p47^{phox} regulatory subunit to the raft and enhanced O_2·$^-$ production. TNF-α also activated eNOS presented in the same membrane compartment. Dual activation of O_2·$^-$ and NO-generating enzymes resulted in the formation of ONOO$^-$ and nitration of tyrosine containing proteins localized to rafts.

Recently, Miller et al. [111] introduced a new concept of Nox regulation. They demonstrated that Nox1 is localized in endosomes of VSMCs and, when stimulated by cytokines, the oxidase generates O_2·$^-$ inside these vesicles, leading to activation of NFκB. The most important finding of these authors is that Nox1 activity and signaling require ClC-3, a chloride/proton exchanger that is also localized in the endosomes [112]. ClC-3 is apparently required for charge neutralization of the electron flow generated by Nox1 across the membrane of signaling endosomes.

2.2.6 ACTIVATORS AND INHIBITORS OF ENDOTHELIAL NADPH OXIDASE

Numerous compounds are able to activate nonphagocyte Nox, and its activation is frequently a starting point of various pathological disorders. It has been shown that Ang II, an important toxic initiator of cardiovascular pathologies, is an effective activator of Nox [113,114]. Activation of nonphagocyte Nox by Ang II was reported in VSMCs, endothelial cells, cardiomyocytes, tissue, and animals [115–118]. Li and Shah [119] showed that Ang II-induced O_2·$^-$ generation by endothelial cells depended on the Nox subunit p47^{phox}. They found that the serine phosphorylation of p47^{phox} followed by p47^{phox} translocation and binding of phospho-p47^{phox} to p22^{phox} are initiated steps of Ang II-induced Nox activation. Vasoconstrictor peptide endothelin-1 (ET-1), which is able to induce cardiac hypertrophy, is also an activator of Nox. Li et al. [120] showed that ET-1 stimulated O_2·$^-$ production in vitro in carotid arteries of normotensive rats, which increased in DOCA-salt hypertensive rats supposedly through the endothelin$_A$-Nox pathway. Similarly, Ang II-stimulated expression of ET-1 in adventitial fibroblasts was mediated, at least in part, by Nox [121]. Loomis et al. [122] showed that ET-1 induced O_2·$^-$ production and vasoconstriction through activation of endothelial Nox and uncoupled NOS in the rat aorta.

PMA and TNF-α stimuli, widely used for activation of phagocyte Nox, can also be applied for stimulation of nonphagocyte Nox. Li et al. [123] reported that the stimulation with PMA of coronary microvascular endothelial cells (CMECs) significantly increased NADPH-dependent ROS production, and that this response was absolutely dependent of the presence of p47^{phox} subunit. Similar results were obtained with TNF-α. It is interesting that, without PMA and TNF-α, endothelial Nox was able to produce significant amounts of O_2·$^-$ in the absence of the p47^{phox} subunit, whereas this subunit was absolutely necessary in the presence of these agonists.

TNF-α also activated the Nox1 NADPH oxidase in mouse fibroblasts when cells underwent necrosis [124].

Both NADPH and NADH activate endothelial Nox. However, it has been shown [125] that only NADPH stimulated the same strong lucigenin-amplified chemi-luminescence (CL) when both small (5 μM/L) and high (400 μM/L) lucigenin concentrations were applied, whereas NADH-dependent CL was detectable only at high lucigenin concentration and was not inhibited by $O_2\cdot^-$ inhibitors. As lucigenin-amplified CL is an extremely sensitive assay of $O_2\cdot^-$ [126], these findings demonstrate that NADH-dependent $O_2\cdot^-$ production appears to be an artifact. Manea et al. [127] showed that the transcription factor NFκB is a regulator of p22^{phox} subunit and Nox activity in human aortic smooth-muscle cells.

As in the case of phagocyte Nox (see earlier text in this chapter), homocysteine sharply enhanced the activity of endothelial Nox and the levels of $O_2\cdot^-$ and ONOO$^-$ in aortas isolated from hyperhomocysteinemic rats [128]. Yi et al. [129] demonstrated that homocysteine stimulated de novo ceramide synthesis and thereby induced Nox activation by increase of Rac GTPase activity in rat mesangial cells (RMCs). It was also shown that guanine nucleotide exchange factor (GEF) Vav2 importantly contrib-uted to Hcys-induced increase in Rac1 activity and consequent activation of Nox in RMCs via ceramide-associated tyrosine phosphorylation. Becker et al. [130] showed that hyperhomocysteinemia (HHcy) caused a 100% increase in the p22phox subunit of Nox in rats and mice. It was suggested that $O_2\cdot^-$ produced by Nox reduced the ability of NO to regulate mitochondrial function in the myocardium.

H_2O_2 is a well-known oxidant that is supposedly able to directly oxidize many biomolecules. It induces oxidative stress that contributes to endothelial dysfunction. However, it was also found that, in smooth-muscle cells, H_2O_2 stimulated the produc-tion of $O_2\cdot^-$ by the activation of Nox. Coyle et al. [131] showed that H_2O_2 stimulated $O_2\cdot^-$ formation in porcine aortic endothelial cells (PAEC) by the activation of NOS and Nox.

NO modulated $O_2\cdot^-$ production by Nox in human vascular endothelial cells by inducing heme oxygenase-1 [132]. Selemidis et al. [133] also showed that NO caused suppression of Nox-dependent $O_2\cdot^-$ production in human endothelial cells by S-nitrosylation of p47^{phox}. In mesangial cells, NO-generated products markedly decreased $O_2\cdot^-$ production as well as Nox1 mRNA and protein levels of Nox [134]. Heumuller et al. [135] recently showed that apocynin, a widely used inhibitor of Nox, is not really an inhibitor but an antioxidant.

2.2.7   SIGNALING MECHANISMS OF ACTIVATION AND CATALYTIC ACTIVITY OF ENDOTHELIAL NADPH OXIDASE

It has already been shown that phagocyte Nox is able to participate in signaling processes through ROS production inside the cells. Signaling is a major function of endothelial Nox in many catalytic processes. Similar to its phagocyte analog, the activation of endothelial Nox might lead to the cyclical generation of $O_2\cdot^-$, but in this case, the significance of $O_2\cdot^-$ overproduction is possibly of much more importance.

Recurrent ROS generation is an important factor of signaling processes catalyzed by endothelial Nox, and ROS signaling occurs in many enzymatic reactions. Hu et al. [136] showed that H_2O_2 produced by Nox in human aortic endothelial cells decreased the threshold concentration of inositol 1,4,5-trisphosphate (InsP3) required for release of intracellularly stored calcium but increased the sensitivity of intracellular calcium stores to InsP3. Fan et al. [137] demonstrated that the transaction of epidermal growth factor (EGF) receptor and the activation of ERK1/2, PI3 kinase, and ATF-1 are included in signaling pathways involved in the upregulation of Nox1. Ruhul Abid et al. [138] also showed that Nox activity was required for vascular endothelial growth factor (VEGF) activation of the PI3K-Akt-forkhead cascade and p38 MAPK, but not ERK1/2 or JNK in human umbilical vein endothelial cells and human coronary artery endothelial cells. PI3K-Akt-forkhead signaling was mediated at post-VEGF receptor level, and involved the nonreceptor tyrosine kinase Src.

Sadok et al. [139] demonstrated that Nox1-dependent $O_2 \cdot^-$ production stimulated colon adenocarcinoma cell migration. It was found that Nox1 activation by AA controlled cell migration through 12-lipoxygenase and protein kinase Cδ. Harraz et al. [140] pointed out that, whereas the mechanisms of Nox activation are well studied, the mechanisms of its downregulation remained unclear. These authors proposed that ROS produced by Nox may autoregulate the enzyme by inhibiting Rac activity. They found that mitogen-activated protein kinase kinase 6 (MKK6) downregulated Nox through the enhancement of Rac-GTPase activity.

Inoue et al. [141] studied the mechanism of calcium regulation in mast cells by ROS produced by Nox. They found that antigen and thapsigargin (a receptor-independent agonist) caused the release of ROS, leukotriene C(4), TNF-α, and interleukin (IL)-13. These findings suggest that ROS, which are produced upstream of calcium influx by Nox, and downstream of calcium influx by the mitochondria, regulated the proinflammatory response of mast cells.

Bendall et al. [142] found that $O_2 \cdot^-$ production was significantly elevated in the aortas of Nox2 transgenic (Nox2-Tg) mice. Increased $O_2 \cdot^-$ production from endothelial Nox2 overexpression led to increased endothelial NOS protein and ERK1/2 phosphorylation in transgenic aortas. Li et al. [143] investigated the role of Nox-derived ROS in cardiac differentiation in mouse embryonic stem cells. They demonstrated that Nox participated in the regulation of developmental processes such as cardiac differentiation through an MAPK kinase–dependent pathway.

Activation of Nox by Ang II was considered earlier. In 1998, mechanisms of ROS-mediated signaling cascades initiated by Ang II were studied by Ushio-Fukai et al. [144]. They showed that the treatment of rat VSMCs with Ang II induced a rapid increase in intracellular H_2O_2 and the phosphorylation of both p42/44 MAPK and p38 MAPK kinases. However, exogenous H_2O_2 activated p38-MAPK only but not p42/44 MAPK. The Ang II/p38 pathway was mediated by $O_2 \cdot^-$ because diphenylene iodonium, a Nox inhibitor, inhibited it. The authors concluded that p38-MAPK is a critical component of ROS-mediated signaling pathways activated by angiotensin II in VSMCs, which play a crucial role in vascular hypertrophy.

Schieffer et al. [145] investigated the role of $O_2 \cdot^-$ generated by Nox in the Ang II activation of the JAK/STAT cascade. Ang II stimulation of rat aortic smooth muscle cells induced a rapid increase in $O_2 \cdot^-$ formation, phosphorylation of JAK2, STAT1a/b, and STAT3. These results suggest that stimulation of the JAK/STAT cascade by Ang II requires $O_2 \cdot^-$ generated by the Nox. Chan et al. [146] showed that Ang II-induced phosphorylation of both p38 MAPK and ERK1/2 through the activation of Nox in the rostral ventrolateral medulla (RVLM). They concluded that the PKCβ/Nox/ERK1/2/ CREB/c-fos cascade represents a novel signaling cascade by Ang II in the RVLM.

Ding et al. [147] has shown that Ang II induced mesangial cell proliferation via the JNK-AP-1 pathway. These authors identified ROS and the epidermal growth factor receptor (EGFR) as the upstream mediators of JNK activation. In cultured human mesangial cells (HMCs), Ang II increased intracellular $O_2 \cdot^-$ production, which led to translocation of cytosolic p47^{phox} and p67^{phox} subunits of Nox to the membrane. Taken together, these data suggest that the ROS/EGFR/JNK cascade is a transducing pathway for the Ang II proliferative effect in cultured HMCs. Hingtgen et al. [148] demonstrated that $O_2 \cdot^-$ generated by Nox2-containing Nox in primary neonatal rat cardiomyocytes mediated Ang II-induced Akt activation and cardiomyocyte hypertrophy. Feliers et al. [149] proposed that ROS produced by Nox mediated the Ang II stimulation of VEGF mRNA translation in proximal tubular epithelial cells.

Nox-produced $O_2 \cdot^-$ plays an important role in cardiac disorders. Thus, Bendall et al. [118] showed that Nox2 (gp91^{phox}-containing) NADPH oxidase was an important factor in the development of Ang II-induced cardiac hypertrophy in mice. Similarly, Nox-derived $O_2 \cdot^-$ mediated Ang II-induced cardiac hypertrophy in neonatal rat cardiac myocytes [150].

It has been hypothesized that high glucose (HG) is a factor contributing to long-term complications of diabetes mellitus through the generation of ROS. Hua et al. [151] demonstrated that HG-stimulated ROS formation through Nox and diacylglycerol-sensitive PKC suppressed Ca^{2+} signaling in response to ET-1 in rat mesangial cells, leading to diabetic complications. High glucose stimulated expression and secretion of VEGF by mesangial cells [152]. This effect of high glucose depended on $O_2 \cdot^-$ generated by Nox and mediated by PKCβ1 and PKCz isoforms of protein kinase C.

It was shown that reduced levels of cGMP-dependent protein kinase I (PKG-I) in vasculature contribute to diabetic vascular dysfunctions. Liu et al. [153] demonstrated that glucose-mediated downregulation of PKG-I expression in rat aorta VSMCs occurred through PKC-dependent activation of Nox-derived $O_2 \cdot^-$ production. Iwashima et al. [154] showed that aldosterone, a potential cardiovascular risk hormone, induced $O_2 \cdot^-$ generation via mineralocorticoid receptor (MR)-mediated activation of Nox and Rac1 in endothelial cells, contributing to the development of aldosterone-induced vascular injury.

Harfouche et al. [155] studied the role of ROS in signaling of the angiopoietin-1 (Ang-1)/tie-2 receptor pathway. Exposure of human umbilical vein endothelial cells to Ang-1 induced rapid production of ROS, particularly $O_2 \cdot^-$ produced by Rac-1-dependent Nox. $O_2 \cdot^-$-mediated tie-2 receptor signaling was involved in the regulation of ERK1/2 and p38-MAPK phosphorylation but not Akt activation. It was also found that Ang-1 stimulated significant PAK-1 phosphorylation at Ser199/204 and Thr423, which was inhibited by antioxidants. Chen et al. [156] also investigated the role of endothelial Nox-derived ROS in Ang-1-induced angiogenesis. Ang-1 increased ROS production in porcine coronary artery endothelial cells (PCAEC), which was inhibited by Nox inhibitors diphenylene iodinium and apocynin. These inhibitors also significantly attenuated Ang-1-induced Akt and p44/42 MAPK phosphorylation. These authors suggested that endothelial Nox-derived ROS played a critical role in Ang-1-induced angiogenesis. Datla et al. [157] proposed that Nox4 NADPH oxidase is an important factor of angiogenic responses in human microvascular endothelial cells. Overexpression of Nox4 enhanced receptor tyrosine kinase phosphorylation and activation of ERK1/2.

TNF-α receptor-associated factors (TRAFs) play important roles in TNF-α signaling by interaction with MAPKs, probably through ROS-dependent pathways. Li et al. [158] investigated TNF-α-induced p47^{phox} phosphorylation of human microvascular endothelial cells (HMEC-1), which increased p47^{phox}-TRAF4 association, membrane translocation of p47^{phox}-TRAF4, and NADPH-dependent $O_2\cdot^-$ production. The formation of TRAF4-p47^{phox} complex was accompanied by the enhancement of the activities of ERK1/2 and p38-MAPK, which was inhibited by the $O_2\cdot^-$ scavenger tiron.

It has been recently shown that ROS mediate myocardial ischemia-reperfusion (I/R) and angiogenesis through the MAPKs and the serine–threonine kinase Akt/protein kinase B pathways. Chen et al. [159] showed that hypoxia-reoxygenation (H/R) stimulated Nox-derived ROS, resulting in Akt and ERK1/2 activation and angiogenesis in PCAECs. Schemes of ROS signaling by endothelial Nox are presented in Figure 2.6.

2.2.8 CYCLICAL GENERATION OF SUPEROXIDE IN SIGNALING PROCESSES CATALYZED BY ENDOTHELIAL NOX

It was already pointed out that, in some cases, ROS signaling by phagocyte Nox resulted in the cycling overproduction of $O_2\cdot^-$. Such a signaling function could be of even more importance for endothelial Nox, which produces $O_2\cdot^-$ inside the cells. Cui et al. [160] showed that AA induced JNK activation in rabbit proximal tubular epithelial cells through a redox cascade, in which Nox was responsible, at least in part, for AA-induced $O_2\cdot^-$ generation and JNK activation. Mitsushita et al. [161] studied the signaling role of $O_2\cdot^-$ produced by Nox in the activation of the Ras oncogene, which transforms various mammalian cells and is responsible for the development of a high population of malignant human tumors. They demonstrated that the Ras oncogene upregulated the expression of Nox1 NADPH oxidase through the MAPK pathway. Thus, Nox1 may function as a critical mediator downstream of Ras, presenting a new molecular mechanism for Ras transformation.

Pancreatic adenocarcinoma is an aggressive human malignancy that is characterized by resistance to apoptosis. Mochizuki et al. [162] showed that NADPH oxidase (Nox4)-mediated $O_2\cdot^-$ generation led to antiapoptotic activity in pancreatic cancer cells. These authors suggested that $O_2\cdot^-$ produced by Nox4 may enhance cell survival signals and diminish apoptosis through the Akt–ASK1 pathway in pancreatic cancer cells. The signaling function of $O_2\cdot^-$ depended on the activation of cell survival kinase Akt and the Ser83 phosphorylation of apoptosis signal-regulating kinase 1 (ASK1), resulting in the suppression of apoptosis.

El Jamali et al. [163] investigated the mechanism of the activation of calcium-regulated NADPH oxidase Nox5 by H_2O_2 in K562 cells expressing Nox5. They found that

NADPH oxidase $\Rightarrow$ H_2O_2 $\Rightarrow$ InsP3 $\downarrow$ [136]

EGF *transaction* $\Rightarrow$ [ERK1/2, PI3 kinase, ATF-1] *activation* $\Rightarrow$ NOX1 $\uparrow$ [137]

                  activation   *activation*

NADPH oxidase $\Rightarrow$ VEGF $\Rightarrow$ PI3K-Akt *forkhead cascade* $\Rightarrow$ p38 MAPK   [138]

MKK6 kinase $\Rightarrow$ Rac-GTPase *activation* $\Rightarrow$ NADPH oxidase $\downarrow$ [140]

Nox2 $\Rightarrow$ $O_2^{\cdot-}$ $\Rightarrow$ NO synthase $\uparrow$ and ERK1/2 *phosphorylation*   [142]

AngII $\Rightarrow$ NADPH oxidase $\Rightarrow$ $O_2^{\cdot-}$ $\Rightarrow$ p38 MAPK   [144]

AngII $\Rightarrow$ NADPH oxidase $\Rightarrow$ $O_2^{\cdot-}$ $\Rightarrow$ JAK2 + STATS *phosphorylation*   [145]

AngII $\Rightarrow$ PKCβ $\Rightarrow$ NADPH oxidase $\Rightarrow$ ERK1/2 $\Rightarrow$ CREB $\Rightarrow$ c-fos [146]

AngII $\Rightarrow$ NADPH oxidase $\Rightarrow$ $O_2^{\cdot-}$ $\Rightarrow$ EGFR $\Rightarrow$ [147]

Ang II $\Rightarrow$ Nox2 $\Rightarrow$ $O_2^{\cdot-}$ $\Rightarrow$ Akt *activation*   [148]

Legend: AA: arachidonic acid, InsP3: 1,4,5-trisphosphate, ERK1/2: extracellular signal-regulated kinase 1/2, PI3K: phosphatidylinositol 3-kinase, ATF-1: activating transcription factor 1, NOX1: NADPH oxidase enzyme, VEGF: vascular endothelial growth factor, Akt/PKB: serine/threonine protein kinase B, GTP: guanosine monophosphate, JAK: Janus kinase, STAT: signal transducer of transcription, CREB: cAMP response element-binding protein, EGFR: epidermal growth factor receptor, PKG-1: cGMP-dependent protein kinase I, TRAF4: TNF receptor-associated factor 4, H/R: hypoxia-reoxygenation, HG: high glucose.

FIGURE 2.6 Endothelial NADPH oxidase activation.

the initial calcium influx generated by H_2O_2 was amplified by a feedback mechanism that involved Nox5-dependent $O_2^{\cdot-}$ production; that is, cyclical generation of $O_2^{\cdot-}$. Taken together, these findings demonstrate Nox5 activation by protein kinase-dead c-Abl through a calcium-mediated, redox-dependent signaling pathway (Figure 2.6).

2.2.9 NONPHAGOCYTE NOX IN PATHOLOGICAL STATES

At present, the role of nonphagocyte Nox in the initiation or mediation of various pathologies is widely discussed (see, for example, Reference [164]). Although ROS signaling in pathological states is usually considered a damaging factor, the opposite ROS effects had to be taken into consideration as well. For example, ROS can facilitate or mimic insulin action; therefore, ROS forming in response to insulin stimulation of target cells may be second messengers in the insulin action cascade leading to the enhancement of insulin signal transduction. Such a proposal seems to disagree with earlier data that chronic exposure to relatively high levels of ROS cause β-cell impairment and the chronic complications of diabetes. However, it is now known that ROS are able to suppress the protein-tyrosine phosphatases (PTPs),

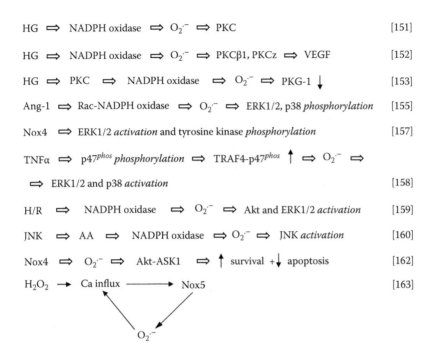

FIGURE 2.6 (continued)

negative regulators of insulin action (see Chapter 4). Thus, Goldstein et al. [165] showed that NADPH oxidase Nox4 is able to mediate the insulin-stimulated generation of cellular ROS and inhibit protein phosphatase PTP1B.

Wei et al. [166] studied Ang II-induced skeletal muscle insulin resistance in rats, which depended on ROS generation. They suggested that Ang II induced NFκB activation and TNF-α-ROS production, and decreased insulin-stimulated Akt Ser473 phosphorylation. However, despite possible positive signaling effects of ROS in diabetes, their overproduction remains an important damaging factor. For example, Thallas-Bonke et al. [167] recently showed that the advanced glycation endproducts (AGEs) stimulated diabetic nephropathy in rats by $O_2^{\cdot-}$ produced through a protein kinase Cα activation of Nox. Shi et al. [168] demonstrated that chronic accumulation of advanced oxidation protein products (AOPPs) occurring in diabetes may promote renal inflammation, possibly through the activation of renal Nox.

Numerous studies demonstrate an important role of endothelial Nox in cardiovascular and heart diseases. Qin et al. [169] showed that, in rabbits with heart failure, Nox inhibition reduced myocyte apoptosis and improved cardiac function in heart failure after myocardial infarction. Hong et al. [170] demonstrated that atorvastatin protected against cerebral infarction in halothane-anesthetized adult male Sprague–Dawley rats via the inhibition of Nox-derived $O_2^{\cdot-}$.

Guzik et al. [171] showed that Nox-mediated $O_2^{\cdot-}$ production was greater in coronary arteries from patients with coronary artery disease (CAD) even in vessels without overt atherosclerotic plaques. It was also found that Nox (60%) and XO

(25%) are primary sources of $O_2{\cdot}^-$ in CAD. Activation of Nox and XO was supposedly mediated through Ang II and PKC-dependent pathways. Nediani et al. [172] studied events triggered by increased Nox activity in the left (LV) and right (RV) ventricles of end-stage failing human hearts. They found that diseased RV and LV showed a significant increase in $O_2{\cdot}^-$ produced by Nox that positively correlated with p47^{phox} membrane translocation. MDA content and ERK and p38 protein kinases, but not JNK, were also activated. Oudot et al. [173] suggested that endothelial Nox is a major contributor to age-related cardiovascular deterioration in mice, the renin–angiotensin system being involved in the enhancement of vascular $O_2{\cdot}^-$ production during aging. Gupte et al. [174] demonstrated that the high level of Nox, stimulated by glucose-6-phosphate dehydrogenase (G6PD)-derived NADPH, generated enhanced $O_2{\cdot}^-$ concentration in failing hearts of patients with ischemic cardiomyopathy. The PKC-Src kinase signaling pathway was apparently responsible for the coordination of activation of G6PD and Nox in human cardiac muscle.

The role of ROS in cancer is an ambiguous one; however, ROS signaling probably plays an important role in both cancer development and suppression. It has been proposed that ROS mediate mitogenic signal and cancer induction in some nonphagocytic cells. In 1999, Suh et al. [175] showed that mox1 (nonphagocyte-specific gp91^{phox} homolog), which encoded a homolog of the catalytic subunit of Nox, is expressed in the colon, prostate, uterus, and vascular smooth muscle, increasing $O_2{\cdot}^-$ generation and cell growth.

Edderkaoui et al. [176] showed that ROS mediated the pro-survival effect of the extracellular matrix (ECM) in human pancreatic cancer cells. It was also found that Nox4 presented in human pancreatic adenocarcinoma tissues and that ROS were produced through 5-LO (5-lipoxygenase)/Nox cascade. Laurent et al. [177] reported that ROS produced by cytosolic Nox were responsible for an increased mitotic rate and cell transformation in nontransformed cells. In this case, the signaling cascade possibly included MAPK kinases, Akt/protein kinase B, and phospholipase D. On the other hand, in tumor cells, the increased ROS generation originated from elevated mitochondrial production and decrease in antioxidant activity. In contrast to nontransformed cells, exogenous H_2O_2 decreased proliferation and induced death in cancer cells.

Gupta et al. [178] demonstrated the important role of elevated ROS levels in tumor progression through ERK1/2 and p38-MAPK activation in the malignant progression of mouse keratinocytes. Cho et al. [179] provided evidence that oncogenic H-Ras activated DNA repair capacity by ROS-mediated signaling pathway through Ras/PI3K/Rac1/Nox cascade. Arnold et al. [180] found that the Nox1 protein overexpression could be a reversible signal for cellular proliferation with relevance for a common human tumor.

Chen et al. [181] showed that, in the kidney, $O_2{\cdot}^-$ production by Nox depended on the availability of molecular oxygen (dioxygen). Therefore, $O_2{\cdot}^-$ generation may be limited in the kidney, both in the normal renal medulla and in the cortex of hypertensive and diabetic kidneys. It has been shown that liver I/R injury is mediated by the overproduction of ROS. Furthermore, it is widely recognized that the AP-1 transcription complex is an important redox-activated factor involved in I/R liver injury.

Marden et al. [182] demonstrated that JunD, an AP-1 component, reduced I/R injury to the liver by the inhibition of JNK1 activation probably through the regulation of Noxs. Liu et al. [183] found that chronic hypoxia (10% O_2 for 3 weeks) caused a sharp increase in gp91^{phox}-dependent $O_2\cdot^-$ production in intrapulmonary arteries in mice. Chen et al. [184] has studied the role of Nox-generated $O_2\cdot^-$ in PCAECs subjected H/R and in the mouse myocardial I/R model. It was found that Nox and its subunit p47^{phox} activated Akt and ERK1/2 protein kinases, enhanced angiogenic growth factor expression, and angiogenesis in myocardium undergoing I/R.

As might be seen from previous studies, Nox is an important factor of hypertension in animals and humans. For example, Dai et al. [185] demonstrated that $O_2\cdot^-$ production was elevated in sympathetic neurons in DOCA-salt hypertension via Nox activation. They found that mRNAs for Nox subunits p47^{phox}, p22^{phox}, gp91^{phox}, Nox1, and Nox4 were presented in sympathetic neurons. Nox activities were 49.9% and 78.6% higher in the sympathetic ganglia of DOCA rats compared to normotensive animals.

Wood et al. [186] showed that endothelial Nox mediated cerebral microvascular dysfunction in sickle-cell transgenic mice. Wilkinson and Landreth [187] proposed that microglial Nox played an important role in the development of oxidative stress in Alzheimer's disease.

In a recent work, Harraz et al. [188] demonstrated that enhanced redox stress in familial amyotrophic lateral sclerosis (ALS) caused by dominant mutations in super-oxide dismutase-1 (SOD1) drastically changed Nox-initiated redox pathway. These authors found that SOD1 regulated Nox-dependent $O_2\cdot^-$ production by binding Rac1 and inhibiting its GTPase activity. Oxidation of Rac1 by H_2O_2 uncoupled SOD1 binding in a reversible fashion, stimulating Nox-derived $O_2\cdot^-$ production. This process of redox-sensitive uncoupling of SOD1 from Rac1 was defective in SOD1 ALS mutants.

2.2.10 CROSS TALK BETWEEN ROS-PRODUCING ENZYMES

The discovery of cross talk between $O_2\cdot^-$-producing enzymes is an important recent event. Desouki et al. [189] demonstrated the existence of cross talk between the mitochondria and Nox1 in cells from breast and ovarian tumors. They found that the inactivation of mitochondrial genes led to downregulation of Nox1, and that the loss of mitochondria control of Nox1 redox signaling contributed to breast and ovarian tumorigenesis. Tephly and Carter [190] found that alveolar macrophages had a high constitutive Nox activity and that the mitochondrial respiratory chain increased $O_2\cdot^-$ production when Nox was inhibited. $O_2\cdot^-$ generation was necessary for macrophage inflammatory protein-2 (MIP-2) gene expression. TNF-α activation of ERK MAPK kinase and ERK activity was an important factor of chemokine gene expression. Collectively, these findings suggest that $O_2\cdot^-$ generation in alveolar macrophages mediates chemokine expression after TNF-α stimulation in an ERK-dependent manner.

2.3 ROS SIGNALING IN REACTIONS CATALYZED BY NITRIC OXIDE SYNTHASES

Nitric oxide synthases (NOS) are important producers of reactive oxygen (ROS) and nitrogen (RNS) species. Discovery and identification of these enzymes capable of producing nitric oxide (NO), the second physiological free radical in biological systems, was the next (after $O_2 \cdot^-$) most important event in free-radical biology. Being unique sources of NO in cells and tissue, NOSs are, at the same time, able to produce $O_2 \cdot^-$ and H_2O_2 under so-called uncoupled conditions. Furthermore, the reaction between $O_2 \cdot^-$ and NO proceeds with a near diffusion rate to form $ONOO^-$, another reactive (but not a free-radical) species. In principle, NOSs are able to catalyze the formation of a great number of different nitrogen-reactive radical and nonradical species, leading to various types of cell injury.

However, first of all, NO produced by NOSs is a well-known important mediator of many physiological processes. It is not by chance that NO was first identified as the endothelium-derived relaxing factor (EDRF), which causes vasodilation of the smooth muscle. Now, other important physiological functions of NO have been developed, among them the interplay between NO and $O_2 \cdot^-$, resulting in the regulation of mitochondrial dioxygen consumption and $O_2 \cdot^-$ signaling in many enzymatic processes. (These functions of NO and $O_2 \cdot^-$ are considered in Chapters 4–7). Collectively, NO and $O_2 \cdot^-$ now outline the principal free-radical-mediated processes in living organisms, including their important signaling functions.

2.3.1 MECHANISMS OF CATALYSIS BY NO SYNTHASES

There are three distinct forms of NOSs expressed in animals and humans that are responsible for NO generation: two constitutive enzymes, neuronal NOS I (nNOS) and endothelial NOS III (eNOS), and one inducible enzyme, NOS II (iNOS). These enzymes catalyze NO and 1-citrulline formation from 1-arginine through the intermediate formation of N-hydroxy-1-arginine and are presented in many cells: neuronal NOS I in neuronal tissue in both the central and peripheral nervous system including neurons, endothelial NOS III in blood vessels, and inducible NOS II in the cardiovascular system and macrophages. NOs is also present in Kupffer cells, hepatocytes, and neutrophils. Recently, the fourth form of NOS, mitochondrial NO synthase (mtNOS) has been discovered. It was found that this NOS isoform is located in the inner mitochondrial membrane in brain, kidney, liver, and muscle [191,192]. Mitochondrial NOS has been characterized as the α-isoform of neuronal NOS, which is phosphorylated at the COOH-terminal end and is different from the endothelial NOS pattern of acylation.

The most important property of NOS is the ability to generate both NO (in the coupled state) and $O_2 \cdot^-$ (in the uncoupled state). The mechanisms of the production of these free radicals by NOSs have been extensively studied. All NOS isoforms contain the N-terminal oxygenase domain with binding sites for heme, $6R$-tetrahydrobiopterin (H_4B), and 1-arginine, and the C-terminal reductase domain with binding sites for FMN, FAD, and NADPH. The enzymes also contain the calmodulin subunit located between oxygenase and reductase domains. The formal scheme of NO production by coupled enzymes can be presented as follows (Figure 2.7).

1-Arginine is converted into N-hydroxy-1-arginine (NOHLA) at the C-terminal reductase domain, which is further oxidized to 1-citrulline and NO by oxygenated ferrous heme at the N-terminal oxygenase domain. Uncoupling of the C-terminal reductase domain and the N-terminal oxygenase domain takes place when the levels of 1-arginine or H_4B are insufficient and the enzymes produce $O_2 \cdot^-$ instead of NO.

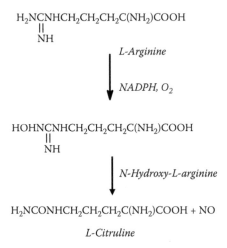

$$H_2NCNHCH_2CH_2CH_2C(NH_2)COOH$$
$$\|$$
$$NH$$

L-Arginine

NADPH, O_2

$$HOHNCNHCH_2CH_2CH_2C(NH_2)COOH$$
$$\|$$
$$NH$$

N-Hydroxy-L-arginine

$$H_2NCONHCH_2CH_2CH_2C(NH_2)COOH + NO$$

L-Citruline

FIGURE 2.7   Formal scheme of NO synthesis by nitric oxide synthase.

Thus, Xia et al. [193] demonstrated that $O_2 \cdot^-$ formation increased while NO generation decreased as the cytosolic arginine levels declined in nNOS-transfected human kidney cells. Similarly, Xia and Zweier [194] showed that the reduced arginine availability resulted in the generation of $O_2 \cdot^-$ and $ONOO^-$ (its reaction product with NO) by iNOS in macrophages.

For a long time, the sites of $O_2 \cdot^-$ production by uncoupled NOS were uncertain. In 1992, Pou et al. [195] showed that brain NOS generated $O_2 \cdot^-$ in a calcium/calmodulin (CaM)-dependent way. It was also found that the CaM binding to nNOS facilitated the transfer of NADPH-derived electrons from the reductase domain to the oxygenase domain, resulting in the conversion of arginine to citrulline and the formation of NO. $O_2 \cdot^-$ was supposedly generated at the heme domain [196]. Later on, Wei et al. [197] demonstrated that the rate of decomposition of the ferrous heme–dioxygen complex corresponded to the rates of H_4B radical formation and arginine hydroxylation by nNOS, eNOS, and iNOS enzymes.

Miller et al. [198] also showed that $O_2 \cdot^-$ production by neuronal NOS is CaM dependent. These authors suggested that $O_2 \cdot^-$ was mostly produced by the reductase domain of the enzyme; however, they believed that both the reductase and oxygenase domains of nNOS can participate in $O_2 \cdot^-$ generation. However, at present, it is concluded that the oxygenase domain is a major source (if not the only one) of $O_2 \cdot^-$ generation. Yoneyama et al. [199] showed that $O_2 \cdot^-$ formation originated from heme of the oxygenase domain of neuronal NOS. It was also found that $O_2 \cdot^-$ generation decreased with decreasing H_4B and 1-arginine. In this regard, authors observed that 1-arginine improved endothelial function in the renal artery and blood pressure in hypertensive rats.

Similarly, the heme moiety is the only source of $O_2 \cdot^-$ production in endothelial NOS [200]. CaM-mediated enzyme regulation affected the production of both NO and $O_2 \cdot^-$. Vasquez-Vivar et al. [201] showed that $O_2 \cdot^-$ was generated from the oxygenase domain of endothelial NOS by dissociation of the ferrous–dioxygen complex. These authors also found that the fully reduced H_4B only was able to diminish $O_2^{\cdot-}$ release from eNOS and that the ratio between oxidized and reduced H_4B metabolites tightly regulated $O_2^{\cdot-}$ formation by this enzyme [202].

Thus, $O_2^{\cdot-}$ formation by uncoupled NOS depends on the oxidation of a heme complex of the oxygenase domain by dioxygen, with subsequent decomposing of the oxygenated complex to $O_2^{\cdot-}$ [203]:

$$nNOS[Fe(II)] + O_2 \leftrightarrow nNOS[Fe(II)O_2] \rightarrow nNOS[Fe(III)] + O_2^- \qquad (2.2)$$

The roles of arginine and tetrahydrobiopterin H_4B in the reactions catalyzed by NOS are rather complicated. It was shown that a decrease in arginine enhanced $O_2^{\cdot-}$ production by NOS, although even in the fully coupled state NOS produced some amount of $O_2^{\cdot-}$. NOS can produce both $O_2^{\cdot-}$ and H_2O_2, and the last is not necessarily formed just by $O_2^{\cdot-}$ dismuting. Rosen et al. [204] suggested that the bound H_4B stimulated direct formation of H_2O_2 at the expense of $O_2^{\cdot-}$ production. The $Fe(II)O_2$ heme complex can either release $O_2^{\cdot-}$ or accept an electron to form H_2O_2 (Reactions 2.3 and 2.4).

$$Fe(II)(Hem)O_2 \rightarrow Fe(III)(Hem) + O_2^{\cdot-} \qquad (2.3)$$

or

$$Fe(II)(Hem)O_2 \xrightarrow{e} Fe(II)(Hem)O_2^{\cdot-} \xrightarrow{2H^+} Fe(III)(Hem) + H_2O_2 \qquad (2.4)$$

These authors also concluded that activated NOS generated $O_2^{\cdot-}$ whether or not enzymes contained the bound H_4B.

Tetrahydrobiopterin apparently takes part in one electron transfer inside the NOS. For example, Schmidt et al. [205] demonstrated the formation of a protonated trihydrobiopterin radical by constitutive NOS. At the same time, tetrahydrobiopterin and L-arginine can affect the production of ROS by eNOS in a different way. Berka et al. [206] point out that there are three distinct radical intermediates that are formed in the reaction between ferrous oxygenase domain (eNOSox) and dioxygen. H_4B-free eNOSox efficiently produced $O_2^{\cdot-}$ in the absence of L-arginine, while L-arginine decreased $O_2^{\cdot-}$ formation. In addition to the H_4B free radical, the formation of a free radical of unknown structure was also observed [207].

There is a difference between the production of ROS by neuronal and inducible NOS. Weaver et al. [208] found that, in the absence of L-arginine, nNOS generated greater amounts of $O_2^{\cdot-}$ and H_2O_2 than iNOS. Masters and her coworkers [209,210] have studied in detail the metabolism of dioxygen by neuronal and endothelial NOS. The addition of CaM to nNOS not only stimulated the dioxygen uptake but also changed the structures of products of uncoupled reactions, confirming the existence of two different sites for electron transfer to dioxygen. The addition of L-arginine initiated the coupled reaction and inhibited dioxygen uptake. Tetrahydrobiopterin affected dioxygen metabolism by decreasing a K_m value of nNOS for dioxygen in the uncoupled reaction. Endothelial NOS is highly coupled in the presence of both L-arginine and tetrahydrobiopterin, but at the same time, L-arginine stimulated dioxygen uptake by this enzyme. 5,6,7,8-Tetrahydrobiopterin suppressed the uncoupled reactions and the formation of O_2·$^-$. These findings demonstrate different mechanisms of dioxygen metabolism for endothelial and neuronal NOS that probably partially explain their functional differences.

Generation of relatively nonreactive physiological free radicals NO and O_2·$^-$ by NOS is accompanied by the formation of more reactive free radicals. Thus, Porasuphatana et al. [211,212] found that neuronal NOS1 oxidized ethanol into hydroxyethyl radical. The formation of this radical was Ca(2+)/CaM-dependent and suppressed by SOD. It was suggested that the perferryl complex of NOS 1 [Fe(V)O](3+)] was responsible for the generation of $CH_3CH(OH)$(.).

For some time, the discovery of mtNOS was questioned. As mentioned earlier, mtNOS is the α-isoform of neuronal NOS. mtNOS activity is associated with the inner mitochondrial membrane, and in intact, succinate-energized mitochondria mtNOS is constitutively active and controls mitochondrial respiration and membrane potential [213,214]. Uptake of Ca^{2+} by mitochondria triggered mtNOS activation and stimulated the release of cytochrome c from mitochondria. Cytochrome c release induced by mtNOS was paralleled by increased lipid peroxidation [215]. It was suggested that Ca^{2+}-induced mtNOS activation resulted in the formation of $ONOO^-$, which led to the initiation of apoptosis. Kanai et al. [216] demonstrated the presence of mtNOS in the cardiomyocytes and urothelial cells; they also showed that it was derived from the neuronal isoform.

2.3.2 ACTIVATORS AND INHIBITORS OF NO SYNTHASES

One group of potential inhibitors/activators of NOSs is antioxidants. Antioxidants can affect the production of NO by NOSs through suppression of the formation of ROS in the uncoupled state. In 1994, Hobbs et al. [217] showed that SOD caused a significant increase in the production of NO by neuronal and inducible NOSs. However, these authors suggested that the effects of SOD were unrelated to the dismutation of O_2·$^-$ anion or activation of NOS. Brady et al. [218] found that extracellular SOD was upregulated together with inducible NOS after NFκB activation. They suggested that it could be a mechanism by which NO production was protected from the reaction with O_2·$^-$ to form dangerous $ONOO^-$. Polytarchou and Papadimitriou [219] have shown that SOD and tempol (membrane-permeable SOD mimetic) as well as Nox inhibitors 4-(2-aminoethyl)-benzenesulfonyl fluoride and apocynin inhibited proliferation and migration of human umbilical vein endothelial cells (HUVEC) and the activity of eNOS.

Another inhibitor of NOS uncoupling is the heat shock protein 90 (hsp90). It was found [220] that hsp90 inhibited strong O_2·$^-$ production by purified rat nNOS in the absence of L-arginine. This inhibition was not due to O_2·$^-$ scavenging because hsp90 did not affect O_2·$^-$ production by XO. Shi et al. [221] confirmed that the association of hsp90 with eNOS is important for increasing NO production and limiting O_2·$^-$ production. Heat shock protein hsp90 apparently plays a critical role in protecting the myocardium against ischemic injury. Pritchard et al. [222] proposed that natural LDL (n-LDL) uncoupled eNOS and increased the formation of O_2·$^-$ by decreasing the association of hsp90 as an initial step in signaling eNOS to

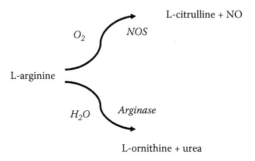

FIGURE 2.8   Competition between NOS and arginase.

generate $O_2\cdot^-$. Association of heat shock protein hsp90 with NOS is an important factor for the protection of the myocardium and cardiomyocytes against ischemic injury. Ilangovan et al. [223] demonstrated that heat shock regulated the respiration of cultured neonatal cardiomyocytes through the activation of NOS. It is of importance that the respiration and the dioxygen concentration in mitochondria of cardiac cells subjected to mild heat shock decreased and NO production increased due to the formation complex of hsp90 with eNOS. In summary, binding of enhanced NO to complexes of mitochondrial respiratory chain downregulated dioxygen consumption in heat-shocked cells. This is an example of the regulatory effects of NO on mitochondrial dioxygen consumption and $O_2\cdot^-$ production, which are considered in detail in Chapter 3.

C-reactive protein (CRP) is a cardiovascular risk marker that induces endothelial dysfunction. Singh et al. [224] have shown that CRP decreased endothelial NOS expression and bioactivity in human aortic endothelial cells (HAECs) through uncoupling of eNOS, which resulted in increased $O_2\cdot^-$ production, decreased NO production, and altered eNOS phosphorylation. ROS formed during uncoupling NOS could be a cause of the inhibition of NOSs. Thus, Kotsonis et al. [225] have shown that ROS and RNS can be responsible for autoinhibition of neuronal NOS.

The activity of NOS is also regulated and suppressed by competition with arginase, an enzyme of the urea cycle, which hydrolyzes L-arginine to urea and L-ornithine. In a recent review, Durante et al. [226] collected and described data concerning the mutual interaction of both enzymes. Arginase is identified in the vasculature where this enzyme regulates NO synthesis and the development of vascular disease. The most important mechanism of the inhibition of NO formation by arginase is competition with NOS for L-arginine (Figure 2.8).

2.3.3   Signaling Mechanisms of Activation of NO Synthases

Numerous signaling processes occur during the activation and catalysis by NOSs. As these enzymes are producers of NO, $O_2\cdot^-$, and H_2O_2—major signaling molecules in many biological processes—the participation of these reactive species in the activation and reactions of NOS should be expected. Important data have been obtained in 1999 by Dimeller et al. [227] and Fulton et al. [228], who demonstrated that the activation of endothelial NOS depended on phosphorylation by serine/threonine protein kinase Akt/PKB. It was found that the phosphatidylinositol-3-OH kinase/Akt pathway is responsible for the direct phosphorylation of eNOS on serine 1179 and enzyme activation.

It is known that endothelial cells release NO in response to increased flow or fluid shear stress (FSS) due to the enhanced phosphorylation and activation of endothelial NOS. Kim et al. [229] showed that the mechanism of eNOS activation by FSS or insulin includes the insulin receptor substrate-1 (IRS-1) tyrosine and serine phosphorylation, an increase in IRS-1-associated phosphatidylinositol 3-kinase activity, phosphorylation of Akt Ser473, and phosphorylation of eNOS Ser1179. Fleming et al. [230] demonstrated that the shear-stress-induced activation of Akt and eNOS in endothelial cells was modulated by the tyrosine phosphorylation of PECAM-1 (platelet endothelial cell adhesion molecule-1). Montagnani et al. [231] showed that insulin regulated eNOS activity using a Ca^{2+}-independent mechanism through the phosphorylation of eNOS by Akt kinase.

It has been suggested that hyperglycemia inhibits endothelial NOS through induced mitochondrial $O_2\cdot^-$ overproduction, and an increase in O-linked N-acetylglucosamine modification and a decrease in O-linked phosphorylation of the transcription factor Sp1. Du et al. [232] have shown that, in bovine aortic endothelial cells, hyperglycemia-associated inhibition of eNOS was accompanied by a twofold increase in O-linked N-acetylglucosamine modification of eNOS and a reciprocal decrease in O-linked serine phosphorylation at residue 1177.

As is well known, insulin resistance and deficit in glucose metabolism promote the development of diabetes. Therefore, the study of the glucose-damaging effects on NO production by eNOS is of real importance. It was demonstrated [233] that prolonged exposure of human aortic endothelial cells to high glucose increased eNOS gene expression, protein expression, and NO release. However, upregulation of eNOS and NO release were accompanied by a sharp enhancement of $O_2\cdot^-$ production. Therefore, chronic exposure to elevated glucose may induce an imbalance between NO and $O_2\cdot^-$, impairing endothelial function and stimulating diabetic vascular disease.

Schnyder et al. [234] have shown that the generation of NO by human endothelial cells after insulin stimulation was inhibited in the presence of high glucose. It was found that insulin-induced endothelial signal transduction was mediated through an immediate complex formation of insulin receptor substrate (IRS) with phosphatidylinositol 3-kinase, which caused serine phosphorylation of the Akt/kinase/eNOS complex. Dysfunction of endothelial IRS cascade and NO generation may further exacerbate the metabolic syndrome of insulin resistance.

Salt et al. [235] also found that high glucose inhibited insulin-stimulated NO production by endothelial NOS. However, high glucose had no effect on insulin-stimulated Akt activation or eNOS phosphorylation at serine Ser1177. These authors therefore suggested that, in the presence of high glucose, Akt-stimulated phosphorylation of eNOS at Ser1177 is not sufficient to stimulate cellular NO synthesis. In a recent work, Chen et al. [236] proposed that high glucose impaired the early and late endothelial progenitor cells by suppression of NO generation through the Akt/eNOS pathway but not by $O_2\cdot^-$-dependent mediated mechanisms.

Aging is an important factor of eNOS inhibition in living organisms. Smith and Hagen [237] suggested that loss of eNOS phosphorylation presents a mechanism that could contribute to the decline in vasomotor function with age. Correspondingly, it was found that aortas from old rats displayed lesser eNOS phosphorylation compared to young animals, supposedly due to a decrease in constitutive Akt activity.

It was suggested that there is at least one more upstream enzyme in the PI3K-Akt-eNOS or Ca-eNOS pathways. Thus, Miller et al. [238] showed that protein tyrosine kinase activity is needed for TNF-α-stimulated induction of inducible NOS in rat hepatocytes. Barsacchi et al. [239] also showed that eNOS activation by TNF-α involved a lipid messenger ceramide. Activation of eNOS in response to TNF-α was correlated with phosphorylation of Akt kinase and eNOS at Ser473 and Ser1179, respectively. This pathway was markedly different from the activation of eNOS through an increase in intracellular calcium concentration and phingomyelinase-independent stimulation of the phosphatidylinositol-3 kinase/Akt pathway.

Matsui et al. [240] have recently proposed that Ca-dependent tyrosine kinase PYK2 is a possible eNOS activator capable of stimulating eNOS phosphorylation. Therefore, the Akt-dependent signaling pathway is not the only mechanism of eNOS activation. Zhang et al. [241] showed that MAPK ERK protein kinase activated inducible NOS through the phosphorylation of Ser745.

Oxidized low-density lipoproteins (ox-LDL) are responsible for NOS activation and the enhancement of $O_2^{\cdot-}$ production by eNOS under some pathophysiological conditions. Fleming et al. [242] have shown that the threonine (Thr) 495 residue was constitutively phosphorylated in eNOS in unstimulated human endothelial cells, but the level of its phosphorylation decreased in the presence of ox-LDL. At the same time, ox-LDL enhanced $O_2^{\cdot-}$ production by endothelial cells, which was partially sensitive to eNOS inhibition. The authors concluded that the effect of ox-LDL was associated with the dephosphorylation of eNOS, the dissociation of the eNOS signaling complex, and the enhanced eNOS-derived $O_2^{\cdot-}$ production. Gharavi et al. [243] demonstrated that oxidized-1-palmitoyl-2-arachidonyl-*sn*-glycero-3-phosphorylcholine (Ox-PAPC), found in atherosclerotic lesions, activated eNOS by phosphorylation of Ser1177 and dephosphorylation of Thr495. Activation of eNOS by Ox-PAPC was regulated through a phosphatidylinositol-3-kinase/Akt-mediated pathway.

These findings suggest that there are two pools of eNOS (the coupled and uncoupled forms), and their activation depends on different activation pathways. Sullivan et al. [244] proposed that the coupled enzyme mostly produced NO after activation, while the uncoupled enzyme produced $O_2^{\cdot-}$ or $O_2^{\cdot-}$ + NO. It is possible that the coupled enzyme is associated with the membrane, whereas the uncoupled enzyme may reside in the cytosol. It has also been shown that the regulation of eNOS depended on the competition between phosphorylation of Thr497 (which probably enhanced the binding of CaM to eNOS, decreasing by this enzymatic activity) and phosphorylation of Ser1179 (which is responsible for the enhancement of NO production). Under pathological conditions, for example, in the presence of oxidized phospholipids in the vasculature, the balance of coupled and uncoupled eNOS might be shifted to the enhancement of $O_2^{\cdot-}$ production and further oxidation of phospholipids, leading to the progression of atherosclerotic lesions [243–245].

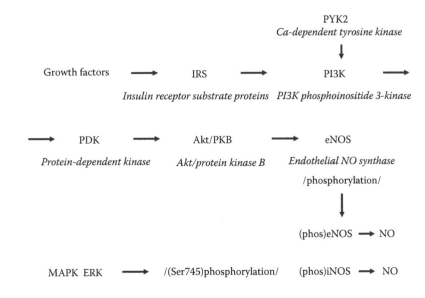

FIGURE 2.9   Principal scheme of ROS signaling in the activation of NOSs [227–243].

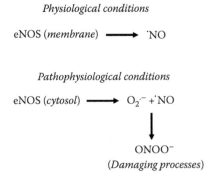

FIGURE 2.10   Two pools of endothelial NOS under physiological and pathological processes (in accord with hypothesis by Sullivan and Pollock [244]).

Signaling processes in the activation of eNOS and iNOS are presented in Figure 2.9 and Figure 2.10, respectively.

2.3.4   SIGNALING PROCESSES CATALYZED BY NOS SYNTHASES

The major signaling pathway by NOSs is well known: NO produced by these enzymes regulates vascular tone and many other physiological functions, activating the soluble isoform of guanylate cyclase (sGC) (Figure 2.11). Activated sGC converts guanosine monophosphate (GMP) to cyclic guanosine monophosphate (cGMP) and pyrophosphate with the subsequent second messenger action of cGMP. The enhancement of calcium release in endothelial cells leads to transient activation of NOS, producing a burst of NO. NO then activates sGC in the adjacent smooth-muscle

$$NOS \longrightarrow NO + sGC \xleftrightarrow{\textit{activation}} NOsGC \longleftrightarrow NOsGC^* \left(\begin{matrix} GTP \\ \\ cGMP \end{matrix}\right.$$

$$NOS \longrightarrow cGMP \uparrow \dashrightarrow src \longrightarrow sGC \textit{ phosphorylation} \qquad [251]$$

$$\underbrace{\qquad\qquad\qquad}_{\text{Endocelial cell}} \qquad\qquad \underbrace{\qquad\qquad\qquad}_{\text{Smooth muscle cell}}$$

ROS *inhibition*

$$ROS \longrightarrow cGC \longrightarrow PKG \textit{ phosphorylation} \longrightarrow VASP \textit{ activation} \qquad [252]$$

Legend: cGC: soluble guanylate cyclase, PKG: protein kinase G, VASP: vasodilator-stimulated phosphoprotein, GMP: guanosine monophosphate, cGMP: cyclic guanosine monophosphate.

FIGURE 2.11   Activation of soluble guanylate cyclase (sGC) by NOS-producing nitric oxide.

cells, resulting in smooth-muscle relaxation (vasodilation). It has been demonstrated that cGMP directly regulates the activities of protein kinase G (PKG) and cyclic nucleotide phosphodiesterases, which in turn regulate the activities of many proteins involved in cellular and physiological processes. Collectively, localization and activity of the NO-cGMP-dependent signaling pathway are regulated by G-protein-coupled receptors, receptor and nonreceptor tyrosine kinases, phosphatases, and other signaling molecules.

Regulation of sGC activity by NO is one of the most important examples of signaling functions of physiological free radicals in enzymatic catalysis. It is one of the clearest examples of cases demonstrating signaling functions of free radical NO through the nucleophilic reaction (addition). Mechanism of sGC activation by NO has been thoroughly studied.

The rate of NO binding to the sGC molecule is very high ($>10^8$ $M^{-1}s^{-1}$) [246]. It is recognized that NO binds to the ferrous ion of a heme prosthetic group attached to sGC at the His105 residue and triggers a conformational change that sharply enhances the catalysis of cGMP synthesis. It was previously thought that the activation of sGC by NO occurs in two steps: binding of NO to the heme to form a biliganded state, and then the rupture of the bond to His105. However, in 1999, Zhao et al. [246] proposed that the regulation of sGC activity by NO depended on the rate of His105 bond cleavage. Both mechanisms have been thoroughly discussed [246–250]. Independently of the credulity of one mechanism or another, both of them demonstrate the nucleophilic mechanism of NO signaling in sGC activation.

Meurer et al. [251] suggested that NOS stimulation of cGMP formation in endothelial cells may trigger the tyrosine phosphorylation of sGC in adjacent smooth-muscle cells through src tyrosine kinase. Russo et al. [252] studied the NO/cGMP/PKG pathway in vascular smooth-muscle cells (VSMC) from the obese Zucker rats. They demonstrated that the activation of sGC by NO activated the vasodilator-stimulated phosphoprotein (VASP) through PKG-stimulated phosphorylation at Ser239. NO/cGMP also activated phosphodiesterase (PDE5) via PKG. ROS caused the inhibition of the NO/cGMP/PKG pathway in obese Zucker rats. Another important physiological signaling function of NOSs is the regulation of dioxygen consumption in the cells through the competition with dioxygen in the mitochondria (see, e.g., Reference 253). This important process with be discussed in detail in Chapter 3.

The ability of NOSs to produce $O_2{}^{.-}$ and H_2O_2 in the uncoupled state makes these enzymes producers of ROS in NOS-catalyzed signaling processes. Wang et al. [254] have shown that eNOS regulated TNF-α, though the mechanism depended on $O_2{}^{.-}$ production and was completely independent of NO. In a subsequent work [255], these authors confirmed that $O_2{}^{.-}$ is a signaling molecule that increased TNF-α production through the activation of p42/44 MAPK protein kinase.

In many cases the nature of signaling-molecules-mediated MAPK activation by NOSs remains obscure. For example, Parenti et al. [256] have shown that the NOS/guanylate cyclase pathway activated the MAPK cascade through the activation of ERK1/2 protein kinase. Gu et al. [257] reported that NO increased the expression of p21gene by a cGMP-dependent pathway through the activation of ERK1/2 and p70(S6k) protein kinases. Callsen et al. [258] found another route of late activation of p42/p44 MAPK kinase by NO in rat mesangial cells, namely, via the inhibition of tyrosine dephosphorylation by cytosolic phosphatases.

Browning et al. [259] demonstrated that exogenously supplied NO stimulated the activation of p38-MAPK in 293T fibroblasts. Phosphorylation of p38 by NO depended on the activation of soluble guanylyl cyclase. Bauer et al. [260] have studied the involvement of the p42/p44 MAPK pathway in the antiproliferative effects of NO on rat aortic smooth muscle cells (RASMC). NO interfered with cell proliferation by inhibition of ornithine decarboxylase (ODC) and polyamine synthesis. Inhibition of ODC contributed to the mechanism by which NO activated the MAPK pathway and promoted the inhibition of cell proliferation. However, apparently NO can be both the activator and the inhibitor of MAPK kinases. For example, Mizuno et al. [261] studied the effect of exogenous NO on cell proliferation and the expression of p53, p21, and phosphorylated p42/44 MAPK in human pulmonary arterial smooth-muscle cells (HPASMC). It was found that NO is able to transiently activate p42/44 MAPK via the induction of p53 and then to suppress its activity via the inactivation of the Ras and Raf cascades.

Contradictory results have been recently received about the regulation by neuronal NOS of the ERK1/2 signaling pathway in human embryonic nNOS-transfected kidney 293 cells [262]. It was found that, in the presence of arginine, nNOS inhibited ERK activity supposedly through the direct attack of NO on the Ras and Raf-1 signaling proteins. However in the absence of arginine, that is, under uncoupled conditions, nNOS-producing $O_2{}^{.-}$ had no effect on EKR activity. A cause of disagreement with the data reported earlier [254–257] is unknown. Raines et al. [262] believe that it may be due to the different rates of NO production in theirs and earlier studies. Zhang et al. [263] demonstrated the involvement of endogenous NOS in Ang II-induced phosphorylation of aortic MAPK (ERK1/2 and p38) in rat vascular smooth muscle cells (RVSMC), which depended on $O_2{}^{.-}$ generation in these cells.

eNOS $\longrightarrow$ $O_2^{\cdot-}$ $\longrightarrow$ p42/44 MAPK $\longrightarrow$ TNFα   [254, 255]

NOS $\longrightarrow$ sGC $\longrightarrow$ ERK1/2   [256]

NOS $\longrightarrow$ phosphatase *inhibition* $\longrightarrow$ p42/p44 *activation*   [258]

NOS $\rightarrow$ NO $\rightarrow$ ODC *inhibition* $\rightarrow$ MAPK *activation* $\rightarrow$ CELL *proliferation*   [260]

p53 *induction* $\longrightarrow$ p42/44 *activation*

NO

Ras/Raf *inactivation* $\longrightarrow$ p42/44 *inactivation*   [261]

NOS $\rightarrow$ $O_2^{\cdot-}$ (?) $\longrightarrow$ Ang II $\longrightarrow$ (ERK1/2 and p38) *phosphorylation*   [263]

FIGURE 2.12   ROS and RNS signaling by NOS (ODC = ornithine decarboxylase).

Recently, White et al. [264] demonstrated the opposing effects of coupled and uncoupled NOS on the Na+-K+ pump in cardiac myocytes. These authors found that coupled NOS stimulated the Na(+)-K(+) pump, while the uncoupling of NOS resulted in $O_2^{\cdot-}$-mediated pump inhibition. Signaling processes catalyzed by NOSs are given in Figure 2.12.

2.3.5   GENERATION OF REACTIVE OXYGEN SPECIES BY NO SYNTHASES IN PATHOPHYSIOLOGICAL PROCESSES

As has been seen, the uncoupled state of NOSs is a producer of ROS and RNS ($O_2^{\cdot-}$, H_2O_2, and ONOO$^-$) and therefore can be a source of pathological disorders. Thus, the uncoupled state was associated with risk factors for atherosclerosis, diabetes, and hypertension. However, it was also hypothesized [244] that uncoupling of eNOS may imply the transfer of an endothelial cell from the quiescent state (that produced NO) into the state that produced $O_2^{\cdot-}$ and H_2O_2, which is adapted for host defense. Correspondingly, the generation of ROS by uncoupled eNOS might be regarded as physiological signaling during injury and infection, and as an essential mechanism in the host defense response. Therefore, in accord with Sullivan and Pollock's hypothesis [244], endothelial NOS might have a Janus face that allows the enzyme to switch from the coupled to the uncoupled state when it encountered infectious agents (Figure 2.13).

Despite this interesting hypothesis, the uncoupled state of NOSs is believed to be an important source of pathological disorders. It is known that diabetes and hypotension are characterized by the enhanced levels of $O_2^{\cdot-}$ production. It has also been shown that there can be different sources of ROS generation in endothelial cells including Nox, mitochondria, and endothelial NOS, and it is quite possible that all of them could be responsible for ROS overproduction.

The importance of uncoupled NOS as a source of $O_2^{\cdot-}$ generation has been demonstrated in various pathological states such as diabetes mellitus, cardiovascular disease, hypotension, and others. Oak and Cai [265] have shown that aortic

Janus eNOS   [244]

Quiescent endothelial cell
Coupled eNOS ⟶ NO

Endothelial cell during injury or infection
Uncoupled eNOS ⟶ $O_2{}^{\cdot-} + H_2O_2$

FIGURE 2.13 Janus eNOS.

$O_2{}^{\cdot-}$ production was significantly increased in streptozotocin-induced diabetic mice, which was inhibited by the inhibitors of eNOS that pointed out at uncoupling of endothelial NOS by Angiotensin II. Thum et al. [266] demonstrated eNOS uncoupling and $O_2{}^{\cdot-}$ overproduction in endothelial progenitor cells (EPCs) from diabetic patients. Glucose-mediated EPC dysfunction depended on PKC and associated with reduced intracellular BH_4 concentrations. Dixon et al. [267] proposed that uncoupled eNOS can reside in patients with hypertension and, to a greater extent, in patients with coexisting hypertension and diabetes, and that it contributed significantly to elevated $O_2{}^{\cdot-}$ production in these pathologic states. In addition to enhanced ROS production, the increased arginase activity in diabetes may contribute to vascular endothelial dysfunction by decreasing L-arginine availability to NOS [268].

It has been shown that neuronal NOS plays important cardiac physiological roles, regulating excitation–contraction coupling and maintaining tissue NO-redox equilibrium. After myocardial infarction (MI), nNOS translocates from the sarcoplasmic reticulum to the cell membrane, where it inhibits β-adrenergic contractility. Saraiva et al. [269] suggested that nNOS exhibited a protective role in the heart after MI despite a potential contribution to left ventricular dysfunction through β-adrenergic hyporesponsiveness. Jones et al. [270] showed that the overexpression of eNOS significantly decreased myocardial infarct size after coronary artery ischemia and reperfusion in rats. These findings support the beneficial role of NO produced by endothelial NOS during myocardial reperfusion injury.

It is believed that the circulating form of folic acid, 5-methyltetrahydrofolate (5-MTHF), might manifest beneficial effects on increased vascular $O_2{}^{\cdot-}$ production in vascular disease states that originated from endothelial NOS uncoupling related to the deficiency of the eNOS cofactor tetrahydrobiopterin. Correspondingly, it was found [271] that 5-MTHF indeed decreased $O_2{}^{\cdot-}$ and $ONOO^-$ production by improving eNOS coupling, mediated by BH_4 availability.

Jin et al. [272] have shown that kallikrein (a member of a subgroup of the serine protease family) increased cardiac endothelial NOS phosphorylation and NO levels, and decreased $O_2{}^{\cdot-}$ formation, TGFβ1 levels, and Smad2 phosphorylation. Furthermore, kallikrein reduced I/R-induced JNK, p38-MAPK, IκBα phosphorylation, and nuclear NFκB activation. In addition, kallikrein improved cardiac performance and reduced infarct size. It is known that myocardial ischemia results in a loss of endothelial NOS activity in the postischemic heart. Dumitrescu et al. [273] showed that BH_4 depletion might be one of the main factors of postischemic eNOS dysfunction.

Cardiovascular disease, diabetes, and hypotension demonstrate the critical role of a decrease in NOS activity and an increase in NOS uncoupling during the development of these pathologies. The importance of deregulation of NOS has also been shown in many other pathological states such as sickle cell disease [274], septic protein leak [275], osteoarthritis [276], and many others. The study of NOSs in pathological states is a big and important problem, but it is far from the tasks and the limits of this book.

REFERENCES

1. JM McCord and I Fridovich. The reduction of cytochrome c by milk xanthine oxidase. *J Biol Chem* 243: 5753–5760, 1968.
2. LM McCord and I Fridovich. The utility of superoxide dismutase in studying free radical reactions. I. Radicals generated by the interaction of sulfite, dimethyl sulfoxide, and oxygen. *J Biol Chem* 244: 6056–6063, 1969.
3. JM McCord and I Fridovich. Superoxide dismutase: An enzymic function for erythrocuprein (hemocuprein). *J Biol Chem* 244: 6049–6055, 1969.
4. PF Knowles, JF Gibson, FM Pick, and RC Bray. Electron-spin-resonance evidence for enzymic reduction of oxygen to a free radical, the superoxide ion. *Biochem J* 111: 53–58, 1969.
5. R Nilsson, FM Pick, and RC Bray. EPR studies on reduction of oxygen to superoxide by some biochemical systems. *Biochim Biophys Acta* 192: 145–148, 1969.
6. RC Bray, FM Pick, and D Samuel. Oxygen-17 hyperfine splitting in the electron paramagnetic resonance spectrum of enzymically generated superoxide. *Eur J Biochem* 15: 352–355, 1970.
7. V Massey, PE Brumby, H Komai, and G Palmer. Studies on milk xanthine oxidase: Some spectral and kinetic properties. *J Biol Chem* 244: 1682–1691, 1969.
8. JS Olson, DP Ballou, G Palmer, and V Massey. The reaction of xanthine oxidase with molecular oxygen. *J Biol Chem* 249: 4350–4362, 1974.
9. T Nishino, T Nishino, LM Schopfer, and V Massey. The reactivity of chicken liver xanthine dehydrogenase with molecular oxygen. *J Biol Chem* 264: 2518–2527, 1989.
10. CM Harris and V Massey. The reaction of reduced xanthine dehydrogenase with molecular oxygen. Reaction kinetics and measurement of superoxide radical. *J Biol Chem* 272: 8370–8379, 1997.
11. R Radi, S Tan, E Prodanov, RA Evans, and DA Park. Inhibition of xanthine oxidase by uric acid and its influence on superoxide radical production. *Biochim Biophys Acta* 1122: 178–182, 1992.
12. S Tan, R Radi, F Gaudier, RA Evans, A Rivera, KA Kirk, and DA Parks. Physiologic levels of uric acid inhibit xanthine oxidase in human plasma. *Pediatr Res* 34: 303–307, 1993.
13. I Fridovich. Quantitative aspects of the production of superoxide anion radical by milk xanthine oxidase. *J Biol Chem* 245: 4053–4057, 1970.
14. P Kuppusamy and JL Zweier. Characterization of free radical generation by xanthine oxidase: Evidence for hydroxyl radical generation. *J Biol Chem* 264: 9880–9884, 1989.
15. RV Lloyd and RP Mason. Evidence against transition metal-independent hydroxyl radical generation by xanthine oxidase. *J Biol Chem* 265: 16733–16736, 1990.
16. BE Britigan, S Pou, GM Rosen, DM Lilleg, and GB Buettner. Hydroxyl radical is not a product of the reaction of xanthine oxidase and xanthine: The confounding problem of adventitious iron bound to xanthine oxidase. *J Biol Chem* 265: 17533–17538, 1990.

17. CA Partridge, FA Blumenstock, and AB Malik. Pulmonary microvascular endothelial cells constitutively release xanthine oxidase. *Arch Biochem Biophys* 294: 184–187, 1992.

18. M Houston, A Estevez, P Chumley, M Aslan, S Marklund, DA Parks, and BA Freeman. Binding of xanthine oxidase to vascular endothelium: Kinetic characterization and oxidative impairment of nitric oxide-dependent signalling. *J Biol Chem* 274: 4985–4994, 1999.

19. PM Hassoun, FS Yu, JJ Zulueta, AC White, and JJ Lanzillo. Effect of nitric oxide and cell redox status on the regulation of endothelial cell xanthine dehydrogenase. *Am J Physiol Lung Cell Mol Physiol* 268: 809–813, 1995.

20. K Ichimori, M Fukahori, H Nakazawa, K Okamoto, and T Nishino. Inhibition of xanthine oxidase and xanthine dehydrogenase by nitric oxide: Nitric oxide converts xanthine-oxidizing enzymes into the desulfo-type inactive form. *J Biol Chem* 274: 7763–7768, 1999.

21. M Houston, P Chumley, R Radi, H Rubbo, and BA Freeman. Xanthine oxidase reaction with nitric oxide and peroxynitrite. *Arch Biochem Biophys* 355: 1–8, 1998.

22. C Lee, X Liu, and JL Zweier. Regulation of xanthine oxidase by nitric oxide and peroxynitrite. *J Biol Chem* 275: 9369–9376, 2000.

23. TM Millar. Peroxynitrite formation from the simultaneous reduction of nitrite and oxygen by xanthine oxidase. *FEBS Lett* 562: 129–133, 2004.

24. M Saleem and H Ohshima. Xanthine oxidase converts nitric oxide to nitroxyl that inactivates the enzyme. *Biochem Biophys Res Commun* 315: 455–462, 2004.

25. JS McNally, A Saxena, H Cai, S Dikalov, and DG Harrison. Regulation of xanthine oxidoreductase protein expression by hydrogen peroxide and calcium. *Arterioscler Thromb Vasc Biol* 25: 1623–1628, 2005.

26. C Galbusera, P Orth, D Fedida, and T Spector. Superoxide radical production by allopurinol and xanthine oxidase. *Biochem Pharmacol* 71: 1747–1752, 2006.

27. XM Wu, M Sawada, and JC Carlson. Stimulation of phospholipase A2 by xanthine oxidase in the rat corpus luteum. *Biol Reprod* 47: 1053–1058, 1992.

28. C Gustafson, M Lindahi, and C Tagesson. Hydrogen-peroxide stimulates phospholipase-A-mediated arachidonic-acid release in cultured intestinal epithelial cells. *Scand J Gastroenterol* 26: 237–247, 1991.

29. C Zhang, TW Hein, W Wang, Y Ren, RD Shipley, and L Kuo. Activation of JNK and xanthine oxidase by TNF-alpha impairs nitric oxide-mediated dilation of coronary arterioles. *J Mol Cell Cardiol* 40: 247–257, 2006.

30. J Ross and WM Armstead. Differential role of PTK and ERK MAPK in superoxide impairment of KATP and KCa channel cerebrovasodilation. *Am J Physiol Regul Integr Comp Physiol* 285: R149–R154, 2003.

31. Q Wang and CM Doerschuk. The p38 mitogen-activated protein kinase mediates cytoskeletal remodeling in pulmonary microvascular endothelial cells upon intracellular adhesion molecule-1 ligation. *J Immunol* 166: 6877–6884, 2001.

32. AK Brzezinska, N Lohr, and WM Chilian. Electrophysiological effects of superoxide (O2·⁻) on the plasma membrane in vascular endothelial cells. *Am J Physiol Heart Circ Physiol* 289: 2379–2386, 2005.

33. C Gaitanaki, M Papatriantafyllou, K Atathopoulou, and I Beis. Effects of various oxidants and antioxidants on the p38-MAPK signalling pathway in the perfused amphibian heart. *Mol Cell Biochem* 291: 107–117, 2006.

34. N Matesanz, N Lafuente, V Azcutia, D Martin, A Cuadrado, J Nevado, L Rodriguez-Manas, CF Sanchez-Ferrer, and C Peiro. Xanthine oxidase-derived extracellular superoxide anions stimulate activator protein 1 activity and hypertrophy in human vascular smooth muscle via c-Jun N-terminal kinase and p38 mitogen-activated protein kinases. *J Hypertens* 25: 609–618, 2007.

35. R Larsson and P Cerutti. Translocation and enhancement of phosphotransferase activity of protein kinase c following exposure in mouse epidermal cells to oxidants. *Cancer Res* 49: 5627–5632, 1989.

36. LT Knapp and E Klann. Superoxide-induced stimulation of protein kinase C via thiol modification and modulation of zinc content. *J Biol Chem* 275: 24136–24145, 2000.

37. GB Silva, PA Ortiz, NJ Hong, and JL Garvin. Superoxide stimulates NaCl absorption in the thick ascending limb via activation of protein kinase C. *Hypertension* 48: 467–472, 2006.

38. J Zhen, H Lu, XQ Wang, ND Vaziri, and XJ Zhou. Upregulation of endothelial and inducible nitric oxide synthase expression by reactive oxygen species. *Am J Hypertens* 21: 28–34, 2008.

39. JW de Jong, RG Schoemaker, R de Jonge, P Bernocchi, E Keijzer, R Harrison, HS Sharma, and C Ceconi. Enhanced expression and activity of xanthine oxidoreductase in the failing heart. *J Mol Cell Cardiol* 32: 2083–2089, 2000.

40. WF Saavedra, N Paolocci, ME St. John, MW Skaf, GC Stewart, J-S Xie, RW Harrison, J Zeichner, D Mudrick, E Marbán, DA Kass, and JM Hare. Imbalance between xanthine oxidase and nitric oxide synthase signaling pathways underlies mechanoenergetic uncoupling in the failing heart. *Circ Res* 90: 297–304, 2002.

41. U Landmesser, S Spiekermann, S Dikalov, H Tatge, R Wilke, C Kohler, DG Harrison, B Hornig, and H Drexler. Vascular oxidative stress and endothelial dysfunction in patients with chronic heart failure: Role of xanthine-oxidase and extracellular superoxide dismutase. *Circulation* 106: 3073–3078, 2002.

42. JG Duncan, R Ravi, LB Stull, and AM Murphy. Chronic xanthine oxidase inhibition prevents myofibrillar protein oxidation and preserves cardiac function in a transgenic mouse model of cardiomyopathy. *Am J Physiol Heart Circ Physiother* 289: H1512–H1518, 2005.

43. SA Khan, K. Lee, KM Minhas, DR Gonzalez, SV Raju, AD Tejani, D Li, DE Berkowitz, and JM Hare. Neuronal nitric oxide synthase negatively regulates xanthine oxidoreductase inhibition of cardiac excitation-contraction coupling. *Proc Natl Acad Sci USA* 101: 15944–15948, 2004.

44. U Landmesser, S Spiekermann, C Preuss, S Sorrentino, D Fischer, C Manes, M Muller, and H Drexler. Angiotensin II induces endothelial xanthine oxidase activation: role for endothelial dysfunction in patients with coronary disease. *Arterioscler Thromb Vasc Biol* 27, 943–948, 2007.

45. MA Newaz, Z Yousefipour, and A Oyekan. Oxidative stress-associated vascular aging is xanthine oxidase-dependent but not NAD(P)H oxidase-dependent. *J Cardiovasc Pharmacol* 48: 88–94, 2006.

46. R Aranda, E Domenech, AD Rus, JT Real, J Sastre, J Vina, and FV Pallardo. Age-related increase in xanthine oxidase activity in human plasma and rat tissues. *Free Radic Res* 41: 1195–1200, 2007.

47. TG Gabig and BM Babior. The $O_2(-)$-forming oxidase responsible for the respiratory burst in human neutrophils: Properties of the solubilized enzyme. *J Biol Chem* 254: 9070–9074, 1979.

48. BM Babior. NADPH oxidase: An update. *Blood* 93: 1464–1476, 1999.

49. KK Griendling, D Sorescu, and M Ushio-Fukai. NAD(P)H oxidase: Role in cardiovascular biology and disease. *Circ Res* 86: 494–501, 2000.

50. FR DeLeo, L-AH Allen, M Apicella, and WM Nauseef. NADPH oxidase activation and assembly during phagocytosis. *J Immunol* 163: 6732–6740, 1999.

51. MT Quinn and KA Gauss. Structure and regulation of the neutrophil respiratory burst oxidase: Comparison with nonphagocyte oxidases. *J Leukoc Biol* 76: 760–781, 2004.

52. WM Nauseef, BD Volpp, S McCormick, KG Leidal, and RA Clark. Assembly of the neutrophil respiratory burst oxidase. Protein kinase C promotes cytoskeletal and membrane association of cytosolic oxidase components. *J Biol Chem* 266: 5911–5917, 1991.

53. J El Benna, LRP Faust, JL Johnson, and BM Babior. Phosphorylation of the respiratory burst oxidase subunit p47 as determined by two-dimensional phosphopeptide mapping. *J Biol Chem* 271: 6374–6378, 1996.

54. I Beck-Speier, JG Liese, BH Belohradsky, and JJ Godleski. Sulfite stimulates NADPH oxidase of human neutrophils to produce active oxygen radicals via protein kinase C and calcium/calmodulin pathways. *Free Radic Biol Med* 14: 661–668, 1993.

55. PM Dang, A Fontayne, J Hakim, J El Benna, and A Perianin. Protein kinase C zeta phosphorylates a subset of selective sites of the NADPH oxidase component p47(phox) and participates in formyl peptide-mediated neutrophil respiratory burst. *J Immunol* 166: 1206–1213, 2001.

56. F Watson, J Robinson, and SW Edwards. Protein kinase-C-dependent and C-independent activation of NADPH oxidase of human neutrophils. *J Biol Chem* 266: 7432–7440, 1991.

57. JT Curnutte, RW Erickson, J Ding, and JA Badwey. Reciprocal interactions between protein kinase C and components of the NADPH oxidase complex may regulate superoxide production by neutrophils stimulated with a phorbol ester. *J Biol Chem* 269: 10813–10819, 1994.

58. RL Miller, GY Sun, and AY Sun. Cytotoxicity of paraquat in microglial cells: Involvement of PKCdelta- and ERK1/2-dependent NADPH oxidase. *Brain Res* 2007 Jul 10 [Epub ahead of print].

59. YL Siow, KK Au-Yeung, CW Woo, and K O. Homocysteine stimulates phosphorylation of NADPH oxidase p47 phox and p67 phox subunits in monocytes via protein kinase C-beta activation. *Biochem J* 398: 73–82, 2006.

60. N Cheng, R He, J Tian, MC Dinauer, and RD Ye. A critical role of protein kinase Cδ activation loop phosphorylation in formyl-methionyl-leucyl-phenylalanine-induced phosphorylation of p47[phox] and rapid activation of nicotinamide adenine dinucleotide phosphate oxidase. *J Immunol* 179: 7720–7728, 2007.

61. RA Clark, KG Leidal, DW Pearson, and WM Nauseef. NADPH oxidase of human eutrophils. Subcellular localization and characterization of an arachidonate-activatable superoxide-generating system. *J Biol Chem* 262: 4065–4074, 1987.

62. LM Henderson, JB Chappell, and OT Jones. Superoxide generation is inhibited by phospholipase A2 inhibitors: Role for phospholipase A2 in the activation of the NADPH oxidase. *Biochem J* 264: 249–255, 1989.

63. V Cherny, LM Henderson, W Xu, L Thomas, and TE DeCoursey. Activation of NADPH oxidase-related proton and electron currents in human eosinophils by arachidonic acid. *J Physiol* 535: 783–794, 2001.

64. I Hazan, R Dana, Y Granot, and R Levy. Cytosolic phospholipase A2 and its mode of activation in human neutrophils by opsonized zymosan. Correlation between 42/44 kDa mitogen-activated protein kinase, cytosolic phospholipase A2 and NADPH oxidase. *Biochem J* 326: 867–876, 1997.

65. X Zhao, EA Bey, FB Wientjes, and MK Cathcart. Cytosolic phospholipase a2 (cpla2) regulation of human monocyte NADPH oxidase activity: cPLA2 affects translocation but not phosphorylation of p67phox and p47phox. *J Biol Chem* 277: 25385–25392, 2002.

66. R Dana, HL Malech, and R Levy. The requirement for phospholipase A2 for activation of the assembled NADPH oxidase in human neutrophils. *Biochem J* 297: 217–223, 1994.

67. YM O'Dowd, J El-Benna, A Perianin, and P Newsholme. Inhibition of formyl-methionyl-leucyl-phenylalanine-stimulated respiratory burst in human neutrophils by adrenaline: inhibition of phospholipase A(2) activity but not p47phox phosphorylation and trans-location. *Biochem Pharmacol* 67: 183–190, 2004.

68. W Zhang, Y Wang, CW Chen, K Xing, S Vivekanandan, and MF Lou. The positive feed-back role of arachidonic acid in the platelet-derived growth factor-induced signaling in lens epithelial cells. *Mol Vis* 12: 821–831, 2006.

69. C Chenevier-Gobeaux, C Simonneau, P Therond, D Bonnefont-Roussellot, S Poiraudeau, OG Ekindjian, and D Borderie. Implication of cytosolic phospholipase A(2) (cPLA(2)) in the regulation of human synoviocyte NADPH oxidase (Nox2) activity. *Life Sci* 81: 1050–1058, 2007.

70. J Ding, CJ Vlahos, R Liu, RF Brown, and JA Badwey. Antagonists of phosphatidylinositol 3-kinase block activation of several novel protein kinases in neutrophils. *J Biol Chem* 270: 11684–11694, 1995.

71. CJ Vlahos, WF Matter, RF Brown, AE Traynor-Kaplan, PG Heyworth, ER Prossnitz, RD Ye, P Marder, JA Schelm, and KJ Rothfuss. Investigation of neutrophil signal transduction using a specific inhibitor of phosphatidylinositol 3-kinase. *J Immunol* 154: 2413–2422, 1995.

72. Q Chen, DW Powell, MJ Rane, S Singh, W Butt, JB Klein, and KR McLeish. Akt phosphorylates p47phox and mediates respiratory burst activity in human neutrophils. *J Immunol* 170: 5302–5308, 2003.

73. T Yamamori, O Inanami, H Sumimoto, T Akasaki, H Nagahata, and M Kuwabara. Relationship between p38 mitogen-activated protein kinase and small GTPase Rac for the activation of NADPH oxidase in bovine neutrophils. *Biochem Biophys Res Commun* 293: 1571–1578, 2002.

74. CI Suh, ND Stull, HJ Li, W Tian, MO Price, S Grinstein, MB Yaffe, S Atkinson, and MC Dinauer. The phosphoinositide-binding protein p40phox activates the NADPH oxidase during Fc{gamma}IIA receptor-induced phagocytosis. *J Exp Med* 203: 1915–1925, 2006.

75. TM Poolman, LL Ng, PB Farmer, and MM Manson. Inhibition of the respiratory burst by resveratrol in human monocytes: Correlation with inhibition of PI3K signaling. *Free Radic Biol Med* 39: 118–132, 2005.

76. XP Gao, X Zhu, J Fu, Q Liu, RS Frei, and AB Malik. Blockade of class IA phospho-inositide 3-kinase in neutrophils prevents NADPH oxidase activation- and adhesion-dependent inflammation. *J Biol Chem* 282: 6116–6125, 2007.

77. CR Hoyal, A Gutierrez, BM Young, SD Catz, J-H Lin, PN Tsichlis, and BM Babior. Modulation of p47[phos] activity by site-specific phosphorylation: Akt-dependent activa-tion of the NADPH oxidase. *Proc Natl Acad Sci USA* 100: 5130–5135, 2003.

78. KR McLeish, C Knall, RA Ward, P Gerwins, PY Coxon, JB Klein, and GL Johnson. Activation of mitogen-activated protein kinase cascades during priming of human neutrophils by TNF-alpha and GM-CSF. *J Leukocyte Biol* 64: 537–545, 1998.

79. Y-L Zu, J Qi, A Gilchrist, GA Fernandez, D Vazquez-Abad, DL Kreutzer, C-K Huang, and RI Sha'afi. p38 Mitogen-activated protein kinase activation is required for human neutrophil function triggered by TNF-α or FMLP stimulation. *J Immunol* 160: 1982–1989, 1998.

80. GE Brown, MQ Stewart, SA Bissonnette, AE Elia, E Wilker, and MB Yaffe. Distinct ligand-dependent roles for p38 MAPK in priming and activation of the neutrophil NADPH oxidase. *J Biol Chem* 279: 27059–27068, 2004.

81. C Dewas, M Fay, M-A Gougerot-Pocidalo, and J El-Benna. The mitogen-activated protein kinase extracellular signal-regulated kinase 1/2 pathway is involved in formyl-methionyl-leucyl-phenylalanine-induced p47phox phosphorylation in human neutro-phils. *J Immunol* 165: 5238–5244, 2000.

82. PM Dang, A Stensballe, T Boussetta, H Raad, C Dewas, Y Kroviarski, G Havem, ON Jensen, MA Gougerot-Pocidalo, and J El-Benna. A specific p47-serine phosphorylated by convergent MAPKs mediates neutrophil NADPH oxidase priming at inflammatory sites. *J Clin Invest* 116, 2033–2043, 2006.

83. R Seru, P Mondola, S Damiano, S Svegliati, S Agnese, EV Avvedimento, and M Santillo. HaRas activates the NADPH oxidase complex in human neuroblastoma cells via extracellular signal-regulated kinase 1/2 pathway. *J Neurochem* 91: 613–622, 2004.

84. JS Kim, JG Kim, CY Jeon, HY Won, MY Moon, JY Seo, JI Kim, J Kim, JY Lee, SY Choi, J Park, JH Yoon Park, KS Ha, PH Kim, and JB Park. Downstream components of RhoA required for signal pathway of superoxide formation during phagocytosis of serum opsonized zymosans in macrophages. *Exp Mol Med* 37: 575–587, 2005.

85. G Alba, R El Bekay, M Alvarez-Maqueda, P Chacon, A Vega, J Monteseirin, C Santa Maria, E Pintado, FJ Bedoya, R Bartrons, and F Sobrino. Stimulators of AMP-activated protein kinase inhibit the respiratory burst in human neutrophils. *FEBS Lett* 573: 219–225, 2004.

86. KD Martyn, MJ Kim, MT Quinn, MC Dinauer, and UG Knaus. P-21 activated kinase (Pak) regulates NADPH oxidase activation in human neutrophils. *Blood* 106: 3962–3969, 2005.

87. K Roepstorff, I Rasmussen, M Sawada, C Cudre-Maroux, P Salmon, G Bokoch, B van Deur, and F Vilhardt. Stimulus dependent regulation of the phagocyte NADPH oxidase by a VAV1, rac1, and PAK1 signaling axis. *J Biol Chem* 283: 7983–7993, 2008.

88. R El Bekay, G Alba, ME Reyes, P Chacon, A Vega, J Martin-Nieto, J Jimenez, E Ramos, J Olivan, E Pintado, and F Sobrino. Rac2 GTPase activation by angiotensin II is modulated by Ca2+/calcineurin and mitogen-activated protein kinases in human neutrophils. *J Mol Endocrinol* 39: 351–363, 2007.

89. P Lin, EJ Welch, XP Gao, AB Malik, and RD Ye. Lysophosphatidylcholine modulates neutrophil oxidant production through elevation of cyclic AMP. *J Immunol* 174: 2981–2989, 2005.

90. AJ Fay, X Qian, YN Jan, and LY Jan. SK channels mediate NADPH oxidase-independent reactive oxygen species production and apoptosis in granulocytes. *Proc Natl Acad Sci USA* 103: 17548–17553, 2006.

91. E Shima, M Katsube, T Kato, M Kitagawa, F Hato, M Hino, T Takahashi, H Fujita, and S Kitagawa. Calcium channel blockers suppress cytokine-induced activation of human neutrophils. *Am J Hypertens* 21: 78-84, 2008.

92. LM Henderson, JB Chappell, and OT Jones. The superoxide-generating NADPH oxidase of human neutrophils is electrogenic and associated with an H+ channel. *Biochem J* 246: 325–329, 1987.

93. J Schrenzel, L Serrander, B Banfi, O Nusse, R Fouyouzi, DP Lew, N Demaurex, and KH Krause. Electron currents generated by the human phagocyte NADPH oxidase. *Nature* 392(6677): 734–737, 1998.

94. TE DeCoursey, V Cherny, AG DeCoursey, W Xu, and L Thomas. Interactions between NADPH oxidase-related proton and electron currents in human eosinophils. *J Physiol* 535: 767–781, 2001.

95. TE DeCoursey, D Morgan, and VV Cherny. The voltage dependence of NADPH oxidase reveals why phagocytes need proton channels. *Nature* 422(6931): 531–534, 2003.

96. KA Gauss, LK Nelson-Overton, DW Siemsen, Y Gao, FR Deleo, and MT Quinn. Role of NF-{kappa}B in transcriptional regulation of the phagocyte NADPH oxidase by tumor necrosis factor-{alpha}. *J Leukocyte Biol* 82: 729–741, 2007.

97. M Alvarez-Maqueda, R El Bekay, J Monteseirin, G Alba, P Chacon, A Vega, C Santa Maria, JR Tejedo, J Martin-Nieto, FJ Bedoya, E Pintado, and F Sobrino. Homocysteine enhances superoxide anion release and NADPH oxidase assembly by human neutrophils: Effects on APK activation and neutrophil migration. *Atherosclerosis* 172: 229–238, 2004.
98. YL Siow, KK Au-Yeung, CW Woo, and K O. Homocysteine stimulates phosphorylation of NADPH oxidase p47 phox and p67 phox subunits in monocytes via protein kinase C-beta activation. *Biochem J* 398: 73–82, 2006.
99. LE Marcal, PM Dias-da-Motta, J Rehder, RL Mamoni, MH Blotta, CB Whitney, PE Newburger, FF Costa, ST Saad, and A Condino-Neto. Up-regulation of NADPH oxidase components and increased production of interferon-gamma by leukocytes from sickle cell disease patients. *Am J Hematol* 83: 41–45, 2007.
100. J Anrather, G Racchumi, and C Iadecola. NF-kappa B regulates phagocytic NADPH oxidase by inducing the expression of gp91phox. *J Biol Chem* 281: 5657–5667, 2006.
101. MJ Coffey, CH Serezani, SM Phare, N Flamand, and M Peters-Golden. NADPH oxidase deficiency results in reduced alveolar macrophage 5-lipoxygenase expression and decreased leukotriene synthesis. *J Leukoc Biol* 82: 1585–1591, 2007.
102. G Zalba, A Fortuño, J Orbe, G San José, MU Moreno, M Belzunce, JA Rodríguez, O Beloqui, JA Páramo, and J Díez. Phagocytic NADPH oxidase-dependent superoxide production stimulates matrix metalloproteinase-9: Implications for human atherosclerosis. *Arterioscler Thromb Vasc Biol* 27: 587–593, 2007.
103. A Vega, P Chacon, G Alba, R El Bekay, J Martin-Nieto, and F Sobrino. Modulation of IgE-dependent COX-2 gene expression by reactive oxygen species in human neutrophils. *J Leukoc Biol* 80: 152–163, 2006.
104. B Lassègue and RE Clempus. Vascular NAD(P)H oxidases: Specific features, expression, and regulation. *Am J Physiol Reg Integr Comp Physiol* 285: R277–R297, 2003.
105. J-M Li and AM Shah. Intracellular localization and preassembly of the NADPH oxidase complex in cultured endothelial cells. *J Biol Chem* 277: 19952–19960, 2002.
106. A Gorlach, RP Brandes, K Nguyen, M Amidi, F Dehghani, and R Busse. A gp91phox containing NADPH oxidase selectively expressed in endothelial cells is a major source of oxygen radical generation in the arterial wall. *Circ Res* 87: 26–32, 2000.
107. SH Ellmark, GJ Dusting, M Ng Tang Fui, N Guzzo-Pernell, and GR Drummond. The contribution of Nox4 to NADPH oxidase activity in mouse vascular smooth muscle. *Cardiovasc Res* 65: 495–504, 2005.
108. A San Martin, R Foncea, FR Laurindo, R Ebensperger, KK Griendling, and F Leighton. Nox1-based NADPH oxidase-derived superoxide is required for VSMC activation by advanced glycation end-products. *Free Radic Biol Med* 42: 1671–1679, 2007.
109. SA Jones, VB O'Donnell, JD Wood, JP Broughton, EJ Hughes, and OT Jones. Expression of phagocyte NADPH oxidase components in human endothelial cells. *Am J Physiol* 271: H1626–H1634, 1996.
110. B Yang and V Rizzo. TNF{alpha} potentiates protein-tyrosine nitration through activation of NADPH oxidase and eNOS localized in membrane rafts and caveolae of bovine aortic endothelial cells. *Am J Physiol Heart Circ Physiol* 292: H954–H962, 2007.
111. FL Miller, Jr, M Filali, GJ Huss, B Stanic, A Chamseddine, TJ Barna, and FS Lamb. Cytokine activation of nuclear factor kappaB in vascular smooth muscle cells requires signaling endosomes containing Nox1 and ClC-3. *Circ Res* 101: 663–671, 2007.
112. B Lassegue. How does the chloride/proton antiporter ClC-3 control NADPH oxidase? *Circ Res* 101: 648–650, 2007.
113. KK Griendling, CA Minieri, JD Ollerenshaw, and RW Alexander. Angiotensin II stimulates NADH and NADPH oxidase activity in cultured vascular smooth muscle cells. *Circ Res* 74: 1141–1148, 1994.

114. PJ Pagano, SJ Chanock, DA Siwik, WS Colucci, and JK Clark. Angiotensin II induces p67phox mRNA expression and NADPH oxidase superoxide generation in rabbit aortic adventitial fibroblasts. *Hypertension* 32: 331–337, 1998.

115. ME Cifuentes, FE Rey, OA Carretero, and PJ Pagano. Upregulation of p67*phox* and gp91*phox* in aortas from angiotensin II-infused mice. *Am J Physiol* 279: H2234–H2240, 2000.

116. D Lang, SI Mosfer, A Shakesby, F Donaldson, and MJ Lewis. Coronary microvascular endothelial cell redox state in left ventricular hypertrophy: The role of angiotensin II. *Circ Res* 86: 463–469, 2000.

117. H Zhang, A Schmeisser, CD Garlichs, K Plotze, U Damme, A Mugge, and WG Daniel. Angiotensin II-induced superoxide anion generation in human vascular endothelial cells: Role of membrane-bound NADH-/NADPH-oxidases. *Cardiovasc Res* 44: 215–222, 1999.

118. JK Bendall, AC Cave, C Heymes, N Gall, and AM Shah. Pivotal role of a gp91[phox]-containing NADPH oxidase in angiotensin II-induced cardiac hypertrophy in mice. *Circulation* 105, 293–296, 2002

119. J-M Li and AM Shah. Mechanism of endothelial cell NADPH oxidase activation by angiotensin II: Role of the p47[phox] subunit. *J Biol Chem* 278: 12094–12100, 2003.

120. L Li, GD Fink, SW Watts, CA Northcott, JJ Galligan, PJ Pagano, and AF Chen. Endothelin-1 increases vascular superoxide via endothelin$_A$-NADPH oxidase pathway in low-renin hypertension. *Circulation* 107: 1053–1058, 2003.

121. SJ An, R Boyd, M Zhu, S Chapman, DR Pimental, and HD Wang. NADPH oxidase mediates angiotensin II-induced endothelin-1 expression in vascular adventitial fibroblasts. *Cardiovasc Res* 75: 702–709, 2007.

122. ED Loomis, JC Sullivan, DA Osmond, DM Pollock, and JS Poll. Endothelin mediates superoxide production and vasoconstriction through activation of NADPH oxidase and uncoupled NOS in the rat aorta. *J Pharmacol Exp Ther* 315: 1058–1064, 2005.

123. LM Li, AM Mullen, S Yun, F Wientjes, GY Brouns, AJ Thrasher, and AM Shah. Essential role of the NADPH oxidase subunit p47(phox) in endothelial cell superoxide production in response to phorbol ester and tumor necrosis factor-alpha. *Circ Res* 90: 123–124, 2002.

124. YS Kim, MJ Morgan, S Choksi, and ZG Liu. TNF-induced activation of the nox1 NADPH oxidase and its role in the induction of necrotic cell death. *Mol Cell* 26, 675–687, 2007.

125. LM Li and AM Shah. Differential NADPH- versus NADH-dependent superoxide production by phagocyte-type endothelial cell NADPH oxidase. *Cardiovasc Res* 52: 477–486, 2001.

126. T Munzel, IB Afanas'ev, AL Kleschev, and DG Harrison. Detection of superoxide in vascular tissue. *Arterioscler Thromb Vasc Biol* 22: 1761–1768, 2002.

127. A Manea, SA Manea, AV Gafencu, and M Raicu. Regulation of NADPH oxidase subunit p22(phox) by NF-kB in human aortic smooth muscle cells. *Arch Physiol Biochem* 113: 163–172, 2007.

128. VE Edirimanne, CW Woo, YL Siow, GN Pierce, JY Xie, and K O. Homocysteine stimulates NADPH oxidase-mediated superoxide production leading to endothelial dysfunction in rats. *Can J Physiol Pharmacol* 85: 1236–1247, 2007.

129. F Yi, QZ Chen, S Jin, and PL Li. Mechanism of homocysteine-induced rac1/NADPH oxidase activation in mesangial cells: Role of guanine nucleotide exchange factor VAV2. *Cell Physiol Biochem* 20: 909–918, 2007.

130. JS Becker, A Adler, A Schneeberger, H Huang, Z Wang, E Walsh, A Koller, and TH Hintze. Hyperhomocysteinemia, a cardiac metabolic disease: Role of nitric oxide and the p22phox subunit of NADPH oxidase. *Circulation* 111: 2112–2118, 2005.

131. CH Coyle, LJ Martinez, MC Coleman, DR Spitz, NL Weintraub, and KN Kader. Mechanisms of H(2)O(2)-induced oxidative stress in endothelial cells. *Free Radic Biol Med* 40: 2206–2213, 2006.

132. F Jiang, SJ Roberts, SR Datla, and GJ Dusting. NO modulates NADPH oxidase function via heme oxygenase-1 in human endothelial cells. *Hypertension* 48, 950–957, 2006.

133. S Selemidis, GJ Dusting, H Peshavariva, BK Kemp-Harper, and GR Drummond. Nitric oxide suppresses NADPH oxidase-dependent superoxide production by S-nitrosylation in human endothelial cells. *Cradiovasc Res* 75, 349–358, 2007.

134. M Pleskova, KF Beck, MH Behrens, A Huwiler, B Fichtlscherer, O Wingerter, RP Brandes, A Mulsch, and J Pfeilschifter. Nitric oxide down-regulates the expression of the catalytic NADPH oxidase subunit Nox1 in rat renal mesangial cells. *FASEB J* 10.1096/fj.05-3791fje, 2005.

135. S Heumuller, S Wind, E Barbosa-Sicard, HH Schmidt, R Busse, K Schroder, and RP Brandes. Apocynin is not an inhibitor of vascular reduced nicotinamide-adenine dinucleotide phosphate oxidases but an antioxidant. *Hypertension* 51: 211–217, 2008.

136. G Hu, G Zheng, JI Zweier, S Deshpande, K Irani, and RC Ziegelstein. NADPH oxidase activation increases the sensitivity of intracellular Ca2+ stores to inositol 1,4,5-triphosphate in human endothelial cells. *J Biol Chem* 275: 15749–15757, 2000.

137. C Fan, M Katsuyama, T Nishinaka, and C Yabe-Nishimura. Transactivation of the EGF receptor and a PI3 kinase-ATF-1 pathway is involved in the upregulation of NOX1, a catalytic subunit of NADPH oxidase. *FEBS Lett* 579: 1301–1305, 2005.

138. A Ruhul Abid, KC Spokes, SC Shih, and WC Aird. NADPH oxidase activity selectively modulates vascular endothelial growth factor signaling pathways. *J Biol Chem* 282: 35373–35385, 2007.

139. A Sadok, V Bourgarel-Rey, F Gattacceca, C Penel, M Lehmann, and H Kovacic. Nox1-dependent superoxide production controls colon adenocarcinoma cell migration. *Biochim Biophys Acta* 1783: 23–33, 2008.

140. MM Harraz, A Park, D Abbott, W Zhou, Y Zhang, and JF Engelhardt. MKK6 phosphorylation regulates production of superoxide by enhancing Rac GTPase activity. *Antioxid Redox Signal* 9: 1803–1813, 2007.

141. T Inoue, Y Suzuki, T Yoshimaru, and C Ra. Reactive oxygen species produced up- or downstream of calcium influx regulate proinflammatory mediator release from mast cells: Role of NADPH oxidase and mitochondria. *Biochim Biophys Acta* 1783: 789–802, 2008.

142. JK Bendall, R Rinze, D Adlam, AL Tatham, J de Bono, and KM Channon. Endothelial Nox2 overexpression potentiates vascular oxidative stress and hemodynamic response to angiotensin II: Studies in endothelial-targeted Nox2 transgenic mice. *Circ Res* 100: 1016–1025, 2007.

143. J Li, M Stouffs, L Serrander, B Banfi, E Bettiol, Y Charnay, K Steger, K-H Krause, and ME Jaconi. The NADPH oxidase nox4 drives cardiac differentiation: Role in regulating cardiac transcription factors and MAP kinase activation. *Mol Biol Cell* 17: 3978–3988, 2006.

144. M Ushio-Fukai, RW Alexander, M Akers, and KK Griendling. p38 Mitogen-activated protein kinase is a critical component of the redox-sensitive signaling pathways activated by angiotensin II: Role in vascular smooth muscle cell hypertrophy. *J Biol Chem* 273: 15022–15029, 1998.

145. B Schieffer, M Luchtefeld, S Braun, A Hilfiker, D Hilfiker-Kleiner, and H Drexler. Role of NAD(P)H oxidase in angiotensin II-induced JAK/STAT signaling and cytokine induction. *Circ Res* 87: 1195–1201, 2000.

146. SH Chan, LL Wang, HL Tseng, and JY Chan. Upregulation of AT1 receptor gene on activation of protein kinase Cbeta/nicotinamide adenine dinucleotide diphosphate oxidase/ERK1/2/c-fos signaling cascade mediates long-term pressor effect of angiotensin II in rostral ventrolateral medulla. *J Hypertens* 25: 1845–1861, 2007.

147. G Ding, A Zhang, S Huang, X Pan, R Chen, and T Yang. Angiotensin II induces c-Jun NH2-terminal kinase activation and proliferation of human mesangial cells via redox-sensitive transactivation of the EGF receptor. *Am J Physiol Renal Physiol* 293: F1889–F1897, 2007.

148. SD Hingtgen, X Tian, J Yang, SM Dunlay, AS Peek, Y Wu, RV Sharma, JF Engelhardt, and RL Davisson. Nox2-containing NADPH oxidase and Akt activation play a key role in angiotensin II-induced cardiomyocyte hypertrophy. *Physiol Genomics* 26: 180–191, 2006.

149. D Feliers, Y Gorin, G Ghosh-Choudhury, HE Abboud, and BS Kasinath. Angiotensin II stimulation of VEGF mRNA translation requires production of reactive oxygen species. *Am J Physiol Renal Physiol* 290: F927–F936, 2006.

150. H Nakagami, M Takemoto, and JK Liao. NADPH oxidase-derived superoxide anion mediates angiotensin II-induced cardiac hypertrophy. *J Mol Cell Cardiol* 35: 851–859, 2003.

151. H Hua, S Munk, H Goldberg, IG Fantus, and CI Whiteside. High glucose-suppressed endothelin-1 Ca2+ signaling via NADPH oxidase and diacylglycerol-sensitive protein kinase C isozymes in mesangial cells. *J Biol Chem* 278: 33951–33962, 2003.

152. L Xia, H Wang, S Munk, H Frecker, HJ Goldberg, IG Fantus, and CI Whiteside. Reactive oxygen species, PKC-{beta}1, and PKC-{zeta} mediate high glucose-induced vascular endothelial growth factor expression in mesangial cells. *Am J Physiol Endocrinol Metab* 293: E1280–E1288, 2007.

153. S Liu, X Ma, M Gong, L Shi, T Lincoln, and S Wang. Glucose down-regulation of cGMP-dependent protein kinase I expression in vascular smooth muscle cells involves NAD(P)H oxidase-derived reactive oxygen species. *Free Radic Biol Med* 42: 852–863, 2007.

154. F Iwashima, T Yoshimoto, I Minami, M Sakurada, Y Hirono, and Y Hirata. Aldosterone induces superoxide generation via Rac1 activation in endothelial cells. *Endocrinology* 149: 1009–1014, 2008.

155. R Harfouche, NA Abdel-Malak, RP Brandes, A Karsan, K Irani, and SN Hussain. Roles of reactive oxygen species in angiopoietin-1/tie-2 receptor signaling. *FASEB J* 2005, 10.1096/fj.04-3621fje.

156. J-X Chen, H Zeng, ML Lawrence, TS Blackwell, and B Meyrick. Angiopoietin-1-induced angiogenesis is modulated by endothelial NADPH oxidase. *Am J Physiol Heart Circ Physiol* 291: H1563–H1572, 2006.

157. SR Datla, H Peshavariya, GJ Dusting, and F Jiang. Important role of Nox4 type NADPH oxidase in angiogenic responses in human microvascular endothelial cells in vitro. *Arterioscler Thromb Vasc Biol* 27: 2319–2324, 2007.

158. J-M Li, LM Fan, MR Christie, and AM Shah. Acute tumor necrosis factor alpha signaling via NADPH oxidase in microvascular endothelial cells: Role of p47phox phosphorylation and binding to TRAF4. *Mol Cell Biol* 25: 2320–2330, 2005.

159. JX Chen, H Zeng, QH Tuo, H Yu, B Meyrick, and JL Aschner. NADPH oxidase modulates myocardial Akt, ERK1/2 activation, and angiogenesis after hypoxia-reoxygenation. *Am J Physiol Heart Circ Physiol* 292: H1664–H1674, 2007.

160. XL Cui and JG Douglas. Arachidonic acid activates c-jun N-terminal kinase through NADPH oxidase in rabbit proximal tubular epithelial cells. *Proc Natl Acad Sci USA* 94: 3771–3776, 1997.

161. J Mitsushita, JD Lambeth, and T Kamata. The superoxide-generating oxidase Nox1 is functionally required for Ras oncogene transformation. *Cancer Res* 64: 3580–3585, 2004.

162. T Mochizuki, S Furuta, J Mitsushita, WH Shang, M Ito, Y Yokoo, M Yamaura, S Ishizone, J Nakayama, A Konagai, K Hirose, K Kiyosawa, and T Kamata. Inhibition of NADPH oxidase 4 activates apoptosis via the AKT/apoptosis signal-regulating kinase 1 pathway in pancreatic cancer PANC-1 cells. *Oncogene* 25: 3699–3707, 2006.

163. A El Jamali, AJ Valente, JD Lechleiter, MJ Gamez, DW Pearson, WM Mauseef, and RA Clark. Novel redox-dependent regulation of NOX5 by the tyrosine kinase c-Abl. *Free Radic Biol Med* 44(5): 868–881, 2008.

164. JD Lamberth. Nox enzymes, ROS, and chronic disease: An example of antagonistic pleiotropy. *Free Radic Biol Med* 43, 332–347, 2007.

165. BJ Goldstein, M Kalyankar, and X Wu. Redox paradox: Insulin action is facilitated by insulin-stimulated reactive oxygen species with multiple potential signaling targets. *Diabetes* 54: 311–321, 2005.

166. Y Wei, JR Sowers, SE Clark, W Li, CM Ferrario, and CS Stump. Angiotensin II-induced skeletal muscle insulin resistance is mediated by NF-{kappa}B activation via NADPH oxidase. *Am J Physiol Endocrinol Metab* 294: E345–E351, 2008.

167. V Thallas-Bonke, SR Thorpe, MT Coughlan, K Fukami, FY Yap, K Sourris, S Penfold, LA Bach, ME Cooper, and JM Forbes. Inhibition of NADPH oxidase prevents AGE mediated damage in diabetic nephropathy through a protein kinase C-{alpha} dependent pathway. *Diabetes* 57: 460–469, 2008.

168. HY Shi, FF Hou, HX Niu, GB Wang, D Xie, ZJ Guo, ZM Zhou, F Yang, JW Tian, and X Zhang. Advanced oxidation protein products promote inflammation in diabetic kidney through activation of renal NADPH oxidase. *Endocrinology* 149: 1829–1839, 2008.

169. F Qin, M Simeone, and R Patel. Inhibition of NADPH oxidase reduces myocardial oxidative stress and apoptosis and improves cardiac function in heart failure after myocardial infarction. *Free Radic Biol Med* 43: 271–281, 2007.

170. H Hong, JS Zeng, DL Kreulen, DI Kaufman, and AF Chen. Atorvastatin protects against cerebral infarction via inhibiting NADPH oxidase-derived superoxide in ischemic stroke. *Am J Physiol Heart Circ Physiol* 291: H2210–H2215, 2006.

171. TJ Guzik, J Sadowski, B Guzik, A Jopek, B Kapelak, P Przybylowski, K Wierzbicki, R Korbut, DG Harrison, and KM Channon. Coronary artery superoxide production and Nox isoform expression in human coronary artery disease. *Arterioscler Thromb Vasc Biol* 26: 333–339, 2006.

172. C Nediani, E Borchi, C Giordano, S Baruzzo, V Ponziani, M Sebastiani, P Nassi, A Mugelli, G d'Amani, and E Cerbai. NADPH oxidase-dependent redox signaling in human heart failure: Relationship between the left and right ventricle. *J Mol Cell Cardiol* 42: 826–834, 2007.

173. A Oudot, C Martin, D Busseuil, C Vergely, L Demaion, and L Rochette. NADPH oxidases are in part responsible for increased cardiovascular superoxide production during aging. *Free Radic Biol Med* 40: 2214–2222, 2006.

174. RS Gupte, V Vijay, B Marks, RJ Levine, HN Sabbah, MS Wolin, FA Recchia, and SA Gupte. Upregulation of glucose-6-phosphate dehydrogenase and NAD(P)H oxidase activity increases oxidative stress in failing human heart. *J Card Fail* 13: 497–506, 2007.

175. YA Suh, RS Arnold, B Lassegue, J Shi, X Xu, D Sorescu, AB Chung, KK Riendling, and D Lambeth. Cell transformation by the superoxide-generating oxidase Mox1. *Nature* 401(6748): 79–82, 1999.

176. M Edderkaoui, P Hong, EC Vaquero, JK Lee, L Fischer, H Friess, MW Buchler, MM Lerch, SJ Pandol, and AS Gukovskaya. Extracellular matrix stimulates reactive oxygen species production and increases pancreatic cancer cell survival through 5-lipoxygenase and NADPH oxidase. *Am J Physiol Gastrointest Liver Physiol* 289: G1137–G1147, 2005.

177. A Laurent, C Nicco, C Chéreau, C Goulvestre, J Alexandre, A Alves, E Lévy, F Goldwasser, Y Panis, O Soubrane, B Weill, and F Batteux. Controlling tumor growth by modulating endogenous production of reactive oxygen species. *Cancer Res* 65: 948–956, 2005.

178. A Gupta, SF Rosenberger, and GT Bowden. Increased ROS levels contribute to elevated transcription factor and MAP kinase activities in malignantly progressed mouse keratinocyte cell lines. *Carcinogenesis* 20: 2063–2073, 1999.

179. H-J Cho, HG Jeong, J-S Lee, E-R Woo, J-W Hyun, M-H Chung, and HJ You. Oncogenic H-ras enhances DNA repair through the ras/phosphatidylinositol 3-kinase/rac1 pathway in NIH3T3 cells. Evidence for association with reactive oxygen species. *J Biol Chem* 277: 19358–19366, 2002.

180. RS Arnold, J He, A Remo, D Ritsick, Q Yin-Goen, JD Lambeth, MW Datta, AN Young, and JA Petros. Nox1 expression determines cellular reactive oxygen and modulates c-fos-induced growth factor, interleukin-8, and cav-1. *Am J Pathol* 171: 2021–2032, 2007.

181. Y Chen, PS Gill, and WJ Welch. Oxygen availability limits renal NADPH-dependent superoxide production. *Am J Physiol Renal Physiol* 289: 749–753, 2005.

182. JJ Marden, Y Zhang, FD Oakley, W Zhou, M Luo, HP Jia, PB McCray, Jr, M Yaniv, JB Weitzman, and JF Engelhardt. JunD protects the liver from ischemia/reperfusion injury by dampening AP-1 transcriptional activation. *J Biol Chem* 283: 6687–6695, 2008.

183. JQ Liu, IN Zelko, EM Erbynn, JS Sham, and RJ Folz. Hypoxic pulmonary hypertension: Role of superoxide and NADPH oxidase (gp91phox). *Am J Physiol Lung Cell Mol Physiol* 290: L2–L10, 2006,

184. JX Chen, H Zeng, QH Tuo, H Yu, B Mevrick, and JL Ascher. NADPH oxidase modulates myocardial Akt, ERK1/2 activation, and angiogenesis after hypoxia-reoxygenation. *Am J Physiol Heart Circ Physiol* 292: H1664–H1674, 2007.

185. X Dai, X Cao, and DL Kreulen. Superoxide anion is elevated in sympathetic neurons in doca-salt hypertension via activation of nadph oxidase. *Am J Physiol Heart Circ Physiol* 290: H1019–H1026, 2006.

186. KC Wood, RP Hebbel, and DN Granger. Endothelial cell NADPH oxidase mediates the cerebral microvascular dysfunction in sickle cell transgenic mice. *FASEB J* 19: 989–991, 2005.

187. BL Wilkinson and GE Landreth. The microglial NADPH oxidase complex as a source of oxidative stress in Alzheimer's disease. *J Neuroinflammation* 3: 30–34, 2006.

188. MM Harraz, JJ Marden, W Zhou, Y Zhang, A Williams, VS Sharov, K Nelson, LM Paulson, H Schoneich, and JF Engelgardt. SOD1 mutations disrupt redox-sensitive Rac regulation of NADPH oxidase in a familial ALS model. *J Clin Invest* 118: 659–670, 2008.

189. MM Desouki, M Kulawiec, S Bansal, GM Das, and KK Singh. Cross talk between mitochondria and superoxide generating NADPH oxidase in breast and ovarian tumors. *Cancer Biol Ther* 4: 1367–1373, 2005.

190. LA Tephly and AB Carter. Constitutive NADPH oxidase and increased mitochondrial respiratory chain activity regulate chemokine gene expression. *Am J Physiol Lung Cell Mol Physiol* 293: L1143–L1155, 2007.

191. L Kobzik, B Stringer, JL Balligand, MB Reid, and JS Stamler. Endothelial type nitric oxide synthase in skeletal muscle fibers: Mitochondrial relationships. *Biochem Biophys Res Commun* 211: 375–381, 1995.

192. TE Bates, A Loesch, G Burnstock, and JB Clark. Immunocytochemical evidence for a mitochondrially located nitric oxide synthase in brain and liver. *Biochem Biophys Res Commun* 213: 896–900,1995.

193. Y Xia, VL Dawson, TM Dawson, SH Snyder, and JL Zweier. Nitric oxide synthase generates superoxide and nitric oxide in arginine-depleted cells leading to peroxynitrite-mediated cellular injury. *Proc Natl Acad Sci USA* 93: 6770–6774, 1996.

194. Y Xia and JL Zweier. Superoxide and peroxynitrite generation from inducible nitric oxide synthase in macrophages. *Proc Natl Acad Sci USA* 94: 6954–6958, 1997.

195. S Pou, WS Pou, DS Bredt, SH Snyder, and GM Rosen. Generation of superoxide by purified brain nitric oxide synthase. *J Biol Chem* 267: 24173–24176, 1992.

196. S Pou, L Keaton, W Surichamorn, and GM Rosen. Mechanism of superoxide generation by neuronal nitric-oxide synthase. *J Biol Chem* 274: 9573–9580, 1999.

197. C-C Wei, Z-Q Wang, D Durra, C Hemann, R Hille, ED Garcin, ED Getzoff, and DJ Stuehr. The three nitric-oxide synthases differ in their kinetics of tetrahydrobiopterin radical formation, heme-dioxy reduction, and arginine hydroxylation. *J Biol Chem* 280: 8929–8935, 2005.

198. RT Miller, P Martasek, LJ Roman, JS Nishimura, and BSS Masters. Involvement of the reductase domain of neuronal nitric oxide synthase in superoxide anion production. *Biochemistry* 36: 15277–15284, 1997.

199. H Yoneyama, A Yamamoto, and H Kosaka. Neuronal nitric oxide synthase generates superoxide from the oxygenase domain. *Biochem J* 360: 247–253, 2001.

200. E Stroes, M Hijmering, M van Zandvoort, R Wever, TJ Rabelink, and EE van Faassen. Origin of superoxide production by endothelial nitric oxide synthase. *FEBS Lett* 438: 161–164, 1998.

201. J Vasquez-Vivar, B Kalyanaraman, P Martasek, N Hogg, BSS Masters, H Karoui, P Tordo, and KA Pritchard, Jr. Superoxide generation by endothelial nitric oxide synthase: The influence of cofactors. *Proc Natl Acad Sci USA* 95: 9220–9225, 1998.

202. J Vasquez-Vivar, P Martasek, J Whitsett, J Joseph, and B Kalyanaraman. The ratio between tetrahydrobiopterin and oxidized tetrahydrobiopterin analogues controls superoxide release from endothelial nitric oxide synthase: An EPR spin trapping study. *Biochem J* 362: 733–739, 2002.

203. HM Abu-Soud, R Gachhui, FM Raushel, and DJ Stuehr. The ferrous-dioxy complex of neuronal nitric oxide synthase. *J Biol Chem* 272: 17349–17353, 1997.

204. GM Rosen, P Tsai, J Weaver, S Porasuphatana, LJ Roman, AA Starkov, G Fiskum, and S Pou. The role of tetrahydrobiopterin in the regulation of neuronal nitric-oxide synthase-generated superoxide. *J Biol Chem* 277: 40275–40280, 2002.

205. PP Schmidt, R Lange, AC Gorren, ER Werner, B Mayer, and KK Andersson. Formation of a protonated trihydrobiopterin radical cation in the first reaction cycle of neuronal and endothelial nitric oxide synthase detected by electron paramagnetic resonance spectroscopy. *J Biol Inorg Chem* 6: 151–158, 2001.

206. V Berka, G Wu, H-C Yeh, G Palmer, and A Tsai. Three different oxygen-induced radical species in endothelial nitric-oxide synthase oxygenase domain under regulation by L-arginine and tetrahydrobiopterin. *J Biol Chem* 79: 32243–32251, 2004.

207. P Tsai, J Weaver, GL Cao, S Pou, LJ Roman, AA Starkov, and GM Rosen. L-arginine regulates neuronal nitric oxide synthase production of superoxide and hydrogen peroxide. *Biochem Pharmacol* 69: 971–979, 2005.

208. J Weaver, S Porasuphatana, P Tsai, S Pou, LJ Roman, and GM Rosen. A comparative study of neuronal and inducible nitric oxide synthases: Generation of nitric oxide, super-oxide, and hydrogen peroxide. *Biochim Biophys Acta* 1726: 302–308, 2005.

209. YT Gao, SP Panda, LJ Roman, P Martasek, Y Ishimura, and BS Masters. Oxygen metabolism by neuronal nitric-oxide synthase. *J Biol Chem* 282: 7921–7929, 2007.

210. YT Gao, LJ Roman, P Martásek, SP Panda, Y Ishimura, and BS Masters. Oxygen metabolism by endothelial nitric-oxide synthase. *J Biol Chem* 282: 28557–28565, 2007.
211. S Porasuphatana, P Tsai, S Pou, and GM Rosen. Involvement of the perferryl complex of nitric oxide synthase in the catalysis of secondary free radical formation. *Biochim Biophys Acta* 1526: 95–104, 2001.
212. S Porasuphatana, P Tsai, S Pou, and GM Rosen. Perferryl complex of nitric oxide synthase: Role in secondary free radical formation. *Biochem Biophys Acta* 1569: 111–116, 2002.
213. P Ghafourifar and C Richter. Nitric oxide synthase activity in mitochondria. *FEBS Lett* 418: 291–296, 1997.
214. C Giulivi, JJ Poderoso, and A Boveris. Production of nitric oxide by mitochondria. *J Biol Chem* 273: 11038–11043, 1998.
215. P Ghafourifar, U Schenk, SD Klein, and C Richter. Mitochondrial nitric-oxide synthase stimulation causes cytochrome *c* release from isolated mitochondria: Evidence for intramitochondrial peroxynitrite formation. *J Biol Chem* 274: 31185–31188, 1999.
216. A Kanai, M Epperly, L Pearce, L Birder, M Zeidel, S Meyers, J Greenberger, W De Groat, G Apodaca, and J Peterson. Differing roles of mitochondrial nitric oxide synthase in cardiomyocytes and urothelial cells. *Am J Physiol Heart Circ Physiol* 286: H13–H21, 2004.
217. AJ Hobbs, JM Fukuto, and LJ Ignarro. Formation of free nitric oxide from l-arginine by nitric oxide synthase: Direct enhancement of generation by superoxide dismutase. *Proc Natl Acad Sci USA* 91: 10992–10996, 1994.
218. TC Brady, LY Chang, BJ Day, and JD Crapo. Extracellular superoxide dismutase is upregulated with inducible nitric oxide synthase after NF-kappa B activation. *Am J Physiol* 273: L1002–L1006, 1997.
219. C Polytarchou and E Papadimitriou. Antioxidants inhibit human endothelial cell functions through down-regulation of endothelial nitric oxide synthase activity. *Eur J Pharmacol* 510: 31–38, 2005.
220. Y Song, AJ Cardounel, JL Zweier, and Y Xia. Inhibition of superoxide generation from neuronal nitric oxide synthase by heat shock protein 90: Implications in NOS regulation. *Biochemistry* 41: 10616–10622, 2002.
221. Y Shi, JE Baker, C Zhang, JS Tweddell, J Su, and KA Pritchard, Jr. Chronic hypoxia increases endothelial nitric oxide synthase generation of nitric oxide by increasing heat shock protein 90 association and serine phosphorylation. *Circ Res* 91: 300–306, 2002.
222. KA Pritchard, AW Ackerman, J Ou, M Curtis, DM Smalley, JT Fontana, MB Stemerman, and WC Sessa. Native low-density lipoprotein induces endothelial nitric oxide synthase dysfunction: Role of heat shock protein 90 and caveolin-1. *Free Radic Biol Med* 33: 52–62, 2002.
223. G Ilangovan, S Osinbowale, A Bratasz, M Bonar, AJ Cardounel, JL Zweier, and P Kuppusamy. Heat shock regulates the respiration of cardiac H9c2 cells through upregulation of nitric oxide synthase. *Am J Physiol Cell Physiol* 287: C1472–C1481, 2004.
224. U Singh, S Devaraj, J Vasquez-Vivar, and I Jialal. C-reactive protein decreases endothelial nitric oxide synthase activity via uncoupling. *J Mol Cell Cardiol* 43: 780–791, 2007.
225. P Kotsonis, A Frey, LG Frohlich, H Hofmann, A Reif, DA Wink, M Feelisch, and HH Schmidt. Autoinhibition of neuronal nitric oxide synthase: Distinct effects of reactive nitrogen and oxygen species on enzyme activity. *Biochem* 340: 745–752, 1999.
226. W Durante, FK Johnson, and RA Johnson. Arginase: A critical regulator of nitric oxide synthesis and vascular function. *Clin Exp Pharmacol Physiol* 34: 906–911, 2007.
227. S Dimmeler, I Fleming, B Fisslthaler, C Hermann, R Busse, and AM Zeiher. Activation of nitric oxide synthase in endothelial cells by Akt-dependent phosphorylation. *Nature* 399(6736): 601–605, 1999.

228. D Fulton, JP Gratton, TJ McCabe, J Fontana, Y Fujio, K Walsh, and TF Franke, A Papapetropoulos, and WC Sessa. Regulation of endothelium-derived nitric oxide production by the protein kinase Akt. *Nature* 399 (6736): 597–601, 1999.

229. F Kim, B Gallis, and MA Corson. TNF-α inhibits flow and insulin signaling leading to NO production in aortic endothelial cells. *Am J Physiol Cell Physiol* 280: C1057–C1065, 2001.

230. I Fleming, B Fisslthaler, M Dixit, and R Busse. Role of PECAM-1 in the shear-stress-induced activation of Akt and the endothelial nitric oxide synthase (eNOS) in endothelial cells. *J Cell Sci* 118: 4103–4111, 2005.

231. M Montagnani, H Chen, VA Barr, and MJ Quon. Insulin-stimulated activation of eNOS is independent of Ca2+ but requires phosphorylation by Akt at Ser1179. *J Biol Chem* 276: 30392–30398, 2001.

232. XL Du, D Edelstein, S Dimmeler, Q Ju, C Sui, and M Brownlee. Hyperglycemia inhibits endothelial nitric oxide synthase activity by posttranslational modification at the Akt site. *J Clin Invest* 108: 1341–1348, 2001.

233. F Cosentino, K Hishikawa, ZS Katusic, and TF Luscher. High glucose increases nitric oxide synthase expression and superoxide anion generation in human aortic endothelial cells. *Circulation* 96: 25–28, 1997.

234. B Schnyder, M Pittet, J Durand, and S Schnyder-Candrian. Rapid effects of glucose on the insulin signaling of endothelial NO generation and epithelial Na transport. *Am J Physiol Endocrinol Metab* 282: E87–E94, 2002.

235. P Salt, VA Morrow, FM Brandie, JMC Connell, and JR Petrie. High glucose inhibits insulin-stimulated nitric oxide production without reducing endothelial nitric-oxide synthase Ser1177 phosphorylation in human aortic endothelial cells. *J Biol Chem* 278: 18791–18797, 2003.

236. YH Chen, SJ Lin, FY Lin, TC Wu, CR Tsao, PH Huang, PL Liu, YL Chen, and JW Chen. High glucose impairs early and late endothelial progenitor cells by modifying nitric oxide-related but not oxidative stress-mediated mechanisms. *Diabetes* 56: 1559–1568, 2007.

237. AR Smith and TM Hagen. Vascular endothelial dysfunction in aging: Loss of Akt-dependent endothelial nitric oxide synthase phosphorylation and partial restoration by (R)-alpha-lipoic acid. *Biochem Soc Trans* 31: 1447–1449, 2003.

238. R Miller, JM Collier, and RE Billings. Protein tyrosine kinase activity regulates nitric oxide synthase induction in rat hepatocytes. *Am J Physiol* 272: G207–G214, 1997.

239. R Barsacchi, C Perrotta, S Bulotta, S Moncada, N Borgese, and E Clementi. Activation of endothelial nitric-oxide synthase by tumor necrosis factor-alpha: A novel pathway involving sequential activation of neutral sphingomyelinase, phosphatidylinositol-3' kinase, and Akt. *Mol Pharmacol* 63: 886–895, 2003.

240. A Matsui, M Okigaki, K Amano, Y Adachi, D Jin, S Takai, T Yamashita, S Kawashima, T Kurihara, M Miyazaki, K Tateishi, S Matsunaga, A Katsume, S Honshou, T Takahashi, S Matoba, T Kusaba, T Tatsumi, and H Matsubara. Central role of calcium-dependent tyrosine kinase PYK2 in endothelial nitric oxide synthase–mediated angiogenic response and vascular function. *Circulation* 116: 1041–1051, 2007.

241. Y Zhang, V Brovkovych, S Brovkovych, F Tan, B-S Lee, T Sharma, and RA Skidgel. Dynamic receptor-dependent activation of inducible nitric-oxide synthase by ERK-mediated phosphorylation of Ser745. *J Biol Chem* 282: 32453–32461, 2007.

242. I Fleming, A Mohamed, J Galle, L Turchanowa, RP Brandes, B Fisslthaler, and R Busse. Oxidized low-density lipoprotein increases superoxide production by endothelial nitric oxide synthase by inhibiting PKCalpha. *Cardiovasc Res* 65: 897–906, 2005.

243. NM Gharavi, NA Baker, KP Mouillesseaux, W Yeung, HM Honda, X Hsieh, M Yeh, EJ Smart, and JA Berliner. Role of endothelial nitric oxide synthase in the regulation of SREBP activation by oxidized phospholipids. *Circ. Res* 98: 768–776, 2006.

244. JC Sullivan and JS Pollock. Coupled and uncoupled NOS: Separate but equal?: Uncoupled NOS in endothelial cells is a critical pathway for intracellular signaling. *Circ Res* 98: 717–719, 2006.
245. MI Lin, D Fulton, R Babbitt, I Fleming, R Busse, KA Pritchard, Jr, and WC Sessa. Phosphorylation of threonine 497 in endothelial nitric-oxide synthase coordinates the coupling of L-arginine metabolism to efficient nitric oxide production. *J Biol Chem* 278: 44719–44726, 2003.
246. Y Zhao, PE Brandish, DP Ballou, and MA Marletta. A molecular basis for nitric oxide sensing by soluble guanylate cyclase. *Proc Natl Acad Sci USA* 96: 14753–14758, 1999.
247. DP Ballou, Y Zhao, PE Brandish, and MA Marletta. Revisiting the kinetics of nitric oxide (NO) binding to soluble guanylate cyclase: The simple NO-binding model is incorrect. *Proc Natl Acad Sci USA* 99: 12097–12101, 2002.
248. TC Bellamy, J Wood, and J Garthwaite. On the activation of soluble guanylyl cyclase by nitric oxide. *Proc Natl Acad Sci USA* 99: 507–510, 2002.
249. B Roy and J Garthwaite. Nitric oxide activation of guanylyl cyclase in cells revisited. *Proc Natl Acad Sci USA* 103: 12185–12190, 2006.
250. JA Winger, ER Derbyshire, and MA Marletta. Dissociation of nitric oxide from soluble guanylate cyclase and heme-nitric oxide/oxygen binding domain constructs. *J Biol Chem* 282: 897–907, 2007.
251. S Meurer, S Pioch, S Gross, and W Muller-Esterl. Reactive oxygen species induce tyrosine phosphorylation of and src kinase recruitment to NO-sensitive guanylyl cyclase. *J Biol Chem* 280: 33149–33156, 2005.
252. I Russo, P Del Mese, G Doronzo, L Mattiello, M Viretto, A Bosia, G Anfossi, and M Trovatti. Resistance to the nitric oxide/ cyclic GMP/ protein kinase G pathway in vascular smooth muscle cells from the obese Zucker rat, a classical animal model of insulin resistance: Role of oxidative stress. *Endocrinology* 149: 1480–1489, 2008.
253. KE Loke, PI McConnell, JM Tuzman, EG Shesely, CJ Smith, CJ Stackpole, CI Thompson, G Kaley, MS Wolin, and TH Hintze. Endogenous endothelial nitric oxide synthase–derived nitric oxide is a physiological regulator of myocardial oxygen consumption. *Circ Res* 84: 840–845, 1999.
254. W Wang, S Wang, L Yan, P Madara, AD Cintron, RA Wesley, and RL Danner. Superoxide production and signaling by endothelial nitric oxide synthase. *J Biol Chem* 275: 16899–16903, 2000.
255. W Wang, S Wang, EV Nishanian, AD Cintron, RA Wesley, and RL Danner. Signaling by eNOS through a superoxide-dependent p42/44 mitogen-activated protein kinase pathway. *Am J Physiol Cell Physiol* 281: C544–C554, 2001.
256. A Parenti, L Morbidelli, X-L Cui, JG Douglas, JD Hood, HJ Granger, F Ledda, and M Ziche. Nitric oxide is an upstream signal of vascular endothelial growth factor-induced extracellular signal-regulated kinase1/2 activation in postcapillary endothelium. *J Biol Chem* 273: 4220–4226, 1998.
257. M Gu, J Lynch, and P Brecher. Nitric oxide increases p21(Waf1/Cip1) expression by a cGMP-dependent pathway that includes activation of extracellular signal-regulated kinase and p70(S6k). *J Biol Chem* 275: 11389–11396, 2000.
258. D Callsen, J Pfeilschifter, and B Brune. Rapid and delayed p42/p44 mitogen-activated protein kinase activation by nitric oxide: The role of cyclic GMP and tyrosine phosphatase inhibition. *J Immunol* 161: 4852–4858, 1998.
259. DD Browning, MP McShane, C Marty, and RD Ye. Nitric oxide activation of p38 mitogen-activated protein kinase in 293T fibroblasts requires cGMP-dependent protein kinase. *J Biol Chem* 275: 2811–2816, 2007.
260. PM Bauer, GM Buga, and LJ Ignarro. Role of p42/p44 mitogen-activated-protein kinase and p21[waf1/cip1] in the regulation of vascular smooth muscle cell proliferation by nitric oxide. *Proc Natl Acad Sci USA* 98: 12802–12807, 2001.

261. S Mizuno, M Kadowaki, Y Demura, S Ameshima, I Miyamori, and T Ishizaki. p42/44 mitogen-activated protein kinase regulated by p53 and nitric oxide in human pulmonary arterial smooth muscle cells. *Am J Respir Cell Mol Biol* 31: 184–192, 2004.
262. KW Raines, G-L Cao, S Porsuphatana, P Tsai, GM Rosen, and P Shapiro. Nitric oxide inhibition of ERK1/2 activity in cells expressing neuronal nitric-oxide synthase. *J Biol Chem* 279: 3933–3940, 2004.
263. GX Zhang, Y Nagai, T Nakagawa, H Miyanaka, Y Fujisawa, A Nishiyama, K Izuishi, K Ohmori, and S Kimura. Involvement of endogenous nitric oxide in angiotensin II-induced activation of vascular mitogen-activated protein kinases. *Am J Physiol Heart Circ Physiol* 293: H2403–H2408, 2007.
264. CN White, EJ Hamilton, A Garcia, D Wang, KK Chia, GA Figtree, and HH Rasmussen. Opposing effects of coupled and uncoupled NOS activity on the Na+-K+ pump in cardiac myocytes. *Am J Physiol Cell Physiol* 10.1152/ajpcell.00242, 2007.
265. JH Oak and H Cai. Attenuation of angiotensin II signaling recouples eNOS and inhibits nonendothelial NOX activity in diabetic mice. *Diabetes* 56: 118–126, 2007.
266. T Thum, D Fraccarollo, M Schultheiss, S Froese, P Galuppo, JD Widder, D Tsikas, G Ertl, and J Bauersachs. Endothelial nitric oxide synthase uncoupling impairs endothelial progenitor cell mobilization and function in diabetes. *Diabetes* 56: 666–674, 2007.
267. LJ Dixon, SM Hughes, K Rooney, A Madden, A Devine, W Leahey, W Henry, GD Johnston, and GE McVeigh. Increased superoxide production in hypertensive patients with diabetes mellitus: Role of nitric oxide synthase. *Am J Hypertens* 18: 839–843, 2005.
268. MJ Romero, DH Platt, HE Tawfik, M Labazi, AB El-Remessy, M Bartoli, RB Caldwell, and RW Caldwell. Diabetes-induced coronary vascular dysfunction involves increased arginase activity. *Circ Res* 102: 95–102, 2008.
269. RM Saraiva, RM Minhas, CV Raju, LA Barouch, E Pitz, KH Schuleri, K Vandegaer, D Li, and LM Hare. Deficiency of neuronal nitric oxide synthase increases mortality and cardiac remodeling after myocardial infarction: Role of nitroso-redox equilibrium. *Circulation* 112: 3415–3422, 2005.
270. SP Jones, JJM Greer, AK Kakkar, PD Ware, RH Turnage, M Hicks, R van Haperen, R de Crom, S Kawashima, MI Yokoyama, and DJ Lefer. Endothelial nitric oxide synthase overexpression attenuates myocardial reperfusion injury. *Am J Physiol Heart Circ Physiol* 286: H276–H282, 2004.
271. C Antoniades, C Shirodaria, N Warrick, S Cai, J de Bono, J Lee, P Leeson, S Neubauer, C Ratnatunga, R Pillai, H Refsum, and KM Channon. 5-Methyltetrahydrofolate rapidly improves endothelial function and decreases superoxide production in human vessels: Effects on vascular tetrahydrobiopterin availability and endothelial nitric oxide synthase coupling. *Circulation* 114: 1193–1201, 2006.
272. H Yin, L Chao, and J Chao. Nitric oxide mediates cardiac protection of tissue kallikrein by reducing inflammation and ventricular remodeling after myocardial ischemia/reperfusion. *Life Sci* 82: 156–165, 2008.
273. C Dumitrescu, R Biondi, Y Xia, AJ Cardounel, LJ Druhan, G Ambrosio, and JL Zweier. Myocardial ischemia results in tetrahydrobiopterin (BH_4) oxidation with impaired endothelial function ameliorated by BH_4. *Proc Natl Acad Sci USA* 104: 15081–15086, 2007.
274. KC Wood, RP Hebbel, DJ Lefer, and DN Granger. Critical role of endothelial cell-derived nitric oxide synthase in sickle cell disease-induced microvascular dysfunction. *Free Radic Biol Med* 40: 1443–1453, 2006.
275. JL Shelton, L Wang, G Cepinskas, M Sandig, JA Scott, ML North, R Inculet, and S Mehta. Inducible NO synthase (iNOS) in human neutrophils but not pulmonary microvascular endothelial cells (PMVEC) mediates septic protein leak in vitro. *Microvasc Res* 74: 23–31, 2007.
276. JU Scher, MH Pillinger, and SB Abramson. Nitric oxide synthases and osteoarthritis. *Curr Rheumatol Rep* 9: 9–15, 2007.

3 Reactive Oxygen and Nitrogen Species Signaling in Mitochondria

Together with xanthine oxidase (XO), NADPH oxidases (Noxs), and nitric oxide synthases (NOSs), mitochondria is one of the major producers of reactive oxygen (ROS) and nitrogen (RNS) species. These enzymes and mitochondria are not only major producers of free radicals in cells and tissues but generate them as signaling species in many enzymatic reactions, and not only as the initiators of numerous damaging processes. The major distinction of mitochondria from the enzymes just mentioned is uncertainty of mode and means of production of oxygen and nitrogen radicals. At present, there are seemingly no doubts that mitochondrial superoxide ($O_2 \cdot^-$) production is not an in vitro artifact but a phenomenon observed in in vivo systems. (However, it should be noted that there are some objections concerning the actual levels of $O_2 \cdot^-$ production by mitochondria in vivo [1]). $O_2 \cdot^-$ is a major ROS produced in mitochondria as a result of a leak of electrons from the two-electron transport respiratory mitochondrial chain to the one-electron reduction of dioxygen, and may be the source of damage in normal biological processes. This is a very important function of $O_2 \cdot^-$, but up-to-date studies also highlight the important signaling activity of mitochondria-producing $O_2 \cdot^-$ in many enzymatic processes. We will now consider the mechanisms of $O_2 \cdot^-$ production by mitochondria.

3.1 ROS SIGNALING IN MITOCHONDRIA

3.1.1 Mechanisms of Superoxide Production by Mitochondria

The mitochondrial electron transfer chain consists of several electron transfer carriers (Figure 3.1). The sites of electron leak from the two-electron transfer chain and $O_2 \cdot^-$ formation were (and are) the subject of long discussion. On a thermodynamic basis, electron transfer from any reduced form of biomolecule (X) to dioxygen is possible if the difference between their reduction potentials is negative: $(E^o(X) - E^o(O_2)) < 0$. It is obvious that, with the one-electron reduction potential of dioxygen $E^o(O_2)$ being equal to -0.16 V, several electron carriers of mitochondria with the electron reduction potential ranging from -0.320 V to $+0.380$ V can reduce O_2 to $O_2 \cdot^-$. (This question is not so simple because only the reduction potentials of cytochromes and ubiquinone undoubtedly correspond to the one-electron reduction potential. Unfortunately, the comparison of two- and one-reduction potentials for the estimation of the direction of electron transfer is impossible.)

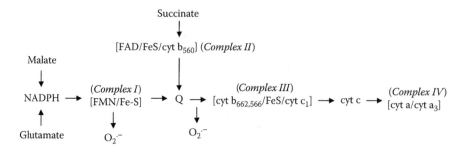

FIGURE 3.1 Mitochondrial electron transfer chain.

$$O_2^{\cdot-} \qquad O_2^{\cdot-}$$

$$\uparrow e \qquad \uparrow e$$

$$NADH \xrightarrow{2e} FMN \longrightarrow FMNH_2 \longrightarrow FMNH^{\cdot} \longrightarrow [4Fe\text{-}4S] \longrightarrow [2Fe\text{-}2S]$$

E_{m7} (mV) -320 -340 -250 -370

FIGURE 3.2 Superoxide formation by mitochondrial Complex I. (Adapted from *J Biol Chem* 280: 37339–37348, 2005.)

Complex I (containing the flavin mononucleotide [FMN] and iron sulfide [FeS] potential $O_2^{\cdot-}$-producing sites) was, for a long time, considered one of two major sites of $O_2^{\cdot-}$ production [2,3]. It is seen from Figure 3.2 [4] that the difference between reduction potentials of $O_2^{\cdot-}$-producing sites of Complex I and dioxygen is <0, and, therefore, the generation of $O_2^{\cdot-}$ by FMS or FeS is a thermodynamically possible process. Turrens and Boveris [3] concluded that NADH dehydrogenase (Complex I) is responsible for about 50% $O_2^{\cdot-}$ production by bovine heart mitochondria. Genova et al. [5] suggested that the one-electron donor for dioxygen in the Complex I is a redox center located prior to the binding sites of three different types of coenzyme quinone (CoQ) antagonists, presumably an iron–sulfur (Fe-NS) cluster N2. The same conclusion was achieved by Herrero and Barja [6] for Complex I in rodent mitochondria. Kushnareva et al. [7] proposed that the $O_2^{\cdot-}$-generating site of Complex I is the Fe–S complex N-1a, which is 50 mV more electronegative than the NADH/NAD+ couple. St-Pierre et al. [8] found that both Complex I and Complex III of intact mitochondria from rat skeletal muscle, heart, and liver produced $O_2^{\cdot-}$, although its levels are very low at physiological conditions.

Kudin et al. [9] also suggested that Complex I is a major site of $O_2^{\cdot-}$ production by rat and human brain mitochondria. However, in contrast to the findings by Kushnareva et al. [7], these authors determined the reduction potential of the $O_2^{\cdot-}$ production site as equal to -295 mV. In accordance with this determination, the site of $O_2^{\cdot-}$ generation at Complex I should be the flavin mononucleotide moiety. Chen et al. [4] investigated $O_2^{\cdot-}$ production in mitochondria by the use of the electron spin resonance (ESR) spin trapping method. These authors concluded that $O_2^{\cdot-}$ was produced by the FMN group and [4Fe–4S] components of NADH dehydrogenase of Complex I. Kussmaul and Hirst [10] have studied the mechanism of O_2·⁻ generation by isolated Complex I from bovine heart mitochondria. They suggested that O_2·⁻ is formed by the transfer of one electron from fully reduced flavin to dioxygen. The formed unstable flavin radical supposedly transfers an electron to the Fe–N centers. They also concluded that Complex I produced mostly O_2·⁻ and not H_2O_2. Galkin and Brandt [11] showed that, in Complex I (NADH:ubiquinone oxidoreductase) from aerobic yeast *Yarrowia lipolytica*, O_2·⁻ was produced by the oxidation of FMNH2 hydroquinone or FMN semiquinone.

In a recent work, Kudin et al. [12] have shown that at least 50% of total ROS generation in succinate-oxidizing homogenates of brain tissue can be attributed to Complex I of the mitochondrial respiratory chain. They found a linear relationship between the rate of oxygen consumption and ROS generation, with succinate as mitochondrial substrate. They also proposed that, under the conditions of dioxygen saturation, about 1% consumed dioxygen by mitochondria converted into O_2·⁻. The FMN moiety of Complex I is again supposed to be the major site of O_2·⁻ generation. Lambert et al. [13] studied the relationship between the rate of O_2·⁻ production by Complex I and NADPH redox state in rat skeletal muscle mitochondria. They found that the highest rate of O_2·⁻ production was observed during succinate oxidation, while the O_2·⁻ production during pyruvate oxidation was over fourfold lower. It was concluded that less than 10% of O_2·⁻ originated from the flavin site during reverse electron transport from succinate.

In 2008, Lambert et al. [14] investigated the effects of diphenyleneiodonium (DPI) on O_2·⁻ production by Complex I in mitochondria isolated from rat skeletal muscle. DPI strongly inhibited O_2·⁻ production by Complex I driven by reverse electron transport from succinate, but it had no effect on O_2·⁻ production by the forward electron transport from NAD-linked substrates in the presence of rotenone or antimycin. Thus, DPI is apparently unable to inhibit O_2·⁻ production by the flavin site of Complex I.

The second major site of mitochondrial O_2·⁻ production is believed to be ubiquinone-Complex III. There are numerous studies that show the importance of this site as a producer of O_2·⁻. For example, in 1976, Boveris et al. [15] found that removal of ubiquinone by acetone treatment decreased the ability of mitochondrial preparations to generate O_2·⁻ and hydrogen peroxide (H_2O_2). On these grounds, these authors suggested that ubisemiquinone or hydroubiquinone might be responsible for succinate-dependent ROS production by the mitochondrial inner membrane.

Further examinations of the mechanism of ROS generation at Complex III were connected with the widely discussed and accepted Q-cycle hypothesis. In 1976, Mitchell [16] (a future Nobel laureate) suggested the existence in mitochondria of the so-called Q cycle, which occurred in the mitochondrial respiratory chain and involved Complex III (Figure 3.3). Shortly, operation of the now modified Q cycle in Complex III leads to the reduction of cytochrome *c*, the oxidation of ubihydroquinone to ubiquinone, and the transfer of four protons into the intermembrane space according to the two-cycle process.

In a recent review, Crofts et al. [17] concluded that the present data strongly supported a mechanism in which the rate-determining step for the overall process was

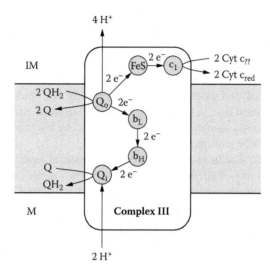

FIGURE 3.3   Schematic representation of Complex III of the electron transport chain.

the transfer of the first electron from ubihydroquinone to the oxidized Fe–S protein at the Q(o)-site. This process involves a proton-coupled electron transfer down a hydrogen bond between the ubihydroquinone and a histidine ligand of the [2Fe–2S] cluster. When oxidation of the semiquinone is prevented, it participates in bypass reactions, including $O_2\cdot^-$ generation if dioxygen is available. The Q-cycle hypothesis was and is widely discussed in literature, but it is beyond the scope of this chapter, which is dedicated to the examination of ROS and RNS formation and their signaling functions in mitochondria.

The generation of $O_2\cdot^-$ by one-electron oxidation of ubisemiquinone formed in the center *o* of the Q cycle in the cytochrome bc1 site of the mitochondrial respiratory chain has also been proposed by many authors [18–22]. Raha et al. [23] have shown that $O_2\cdot^-$ produced in the Q-cycle was effectively eliminated by manganese superoxide dismutase (MnSOD) in intact mitochondria. Muller et al. [24] suggested that $O_2\cdot^-$ was formed in the reaction with unbound ubisemiquinone that escaped from Complex III. Lambert and Brand [25] demonstrated that the site of $O_2\cdot^-$ generation by Complex I is in the region of the ubisemiquinone-binding sites and not upstream at the flavin or low-potential FeS centers. Ohnishi et al. [26] suggested that a major site of $O_2\cdot^-$ generation is not flavin but the protein-associated ubisemiquinone, which is spin-coupled with the Fe–S cluster N2. Forquer et al. [27] concluded that the similar transition states for the Q-cycle and $O_2\cdot^-$ production by the cytochrome bc1 complex correspond to a model with electron transfer from ubisemiquinone to dioxygen.

Cape et al. [28] identified the ubisemiquinone radical generated by the Q(o) site using continuous-wave and pulsed electron paramagnetic resonance (EPR) spectroscopy. It was found that ubisemiquinone was buried in the protein, probably in or near the Q(o) site. Drose and Brandt [29] have shown that $O_2\cdot^-$ formation at the ubihydroquinone oxidation center of the membrane-bound or purified cytochrome bc1 complex was stimulated by ubiquinone, indicating that, in a reverse reaction, the electron may be transferred to dioxygen from reduced cytochrome bc1 via ubiqui-none rather than during the forward Q-cycle reaction.

As can be seen from Figure 3.1, Complex II is also able to initiate O_2·$^-$ formation by the succinate-FAD/FeS-Q pathway. In 1998, Zhang et al. [30] suggested that the reduced FAD of succinate dehydrogenase is an electron donor capable of produc-ing O_2·$^-$ by reacting with dioxygen in the absence of other electron acceptors and in the presence of cytochrome c. Liu et al. [31] proposed that the physiologically relevant O_2·$^-$ generation supported by the Complex II substrate succinate occurred at the FMN of Complex I through reversed electron transfer and not at the ubiqui-none of Complex III. In 2006, Chen et al. [32] confirmed that the involvement of succinate dehydrogenase moiety in O_2·$^-$ production should not be ruled out. Using 3-nitropropionic acid (3-NPA), an inhibitor of succinate dehydrogenase (SDH) at Complex II, Bacsi et al. [33] found that mitochondria generated O_2·$^-$ from a site between the Q pool and the 3-NPA block in the respiratory Complex II.

The important distinction between Complex I and Complex III is a difference in the distribution of O_2·$^-$ production in mitochondria. Han et al. [34] have studied O_2·$^-$ formation in the mitochondrial matrix and intermembrane space. Earlier, it was generally accepted that O_2·$^-$ generated by mitochondria was released into the mito-chondrial matrix, where it is converted to H_2O_2 by MnSOD. The release of O_2·$^-$ into the intermembrane space was considered an unlikely event due to scavenging by cytochrome c presenting in this compartment at a high concentration. However, these authors found that ubisemiquinone at the outer site of the Complex III ubiquinone pool was able to transfer an electron to dioxygen, forming O_2·$^-$ with its subsequent release into the intermembrane space [35].

At the same time, Muller and Remmen [36] showed that Complex-I-dependent O_2·$^-$ is exclusively released into the matrix and unable to escape from intact mitochondria. In their opinion, the site of O_2·$^-$ production at Complex I is the Fe–S clusters of the matrix-protruding hydrophilic arm. In contrast, Complex III can supposedly release O_2·$^-$ to both sides of the inner mitochondrial membrane. It was proposed that the site of O_2·$^-$ production in Complex III is ubihydroquinone, which is situated immediately next to the intermembrane space. Kudin et al. [37] also showed that the O_2·$^-$ genera-tion site in Complex I, probably the FMN moiety, released O_2·$^-$ predominantly to the mitochondrial matrix space, while the bc(1)-complex-dependent O_2·$^-$ generation site (ubisemiquinone at center o) released O_2·$^-$ to the cytosolic space.

3.1.2   COMPARISON OF COMPLEX I AND COMPLEX III AS PRODUCERS OF SUPEROXIDE IN MITOCHONDRIA

The following data suggest three major routes of O_2·$^-$ formation in mitochondria by one electron reduction of dioxygen:

1. Complex I $\rightarrow$ O_2·$^-$
2. Complex I $\rightarrow$ Complex III $\rightarrow$ O_2·$^-$
3. Complex II $\rightarrow$ Complex III $\rightarrow$ O_2·$^-$
   $\rightarrow$ Complex I $\rightarrow$ O_2·$^-$ (Reversed electron transfer)

The one-electron transfer from the FMN and Fe–S complexes of Complex I to dioxygen is an exothermal process ($\Delta E^\circ < 0$), and therefore, this reaction thermodynamically and kinetically is quite favorable and must proceed with a very high rate (the rate constants for exothermal one-electron transfer reactions are usually close to the diffusion limit of $10^9 \, M^{-1}s^{-1}$).

$$FMN \text{ or } Fe2+S + O_2 \Leftrightarrow FMN+ \text{ or } Fe3+S + O_2 \cdot^- \qquad (3.1)$$

This situation is more complicated for Complex III, where ubiquinone is considered to be the $O_2 \cdot^-$-producing center. In this case, three reduced forms of ubiquinone— ubisemiquinone radical anion ($Q \cdot^-$), neutral ubisemiquinone radical ($HQ \cdot$), and ubihydroquinone (QH_2)—could be electron donors in reactions with dioxygen:

$$Q^- + O_2 \Leftrightarrow Q + O_2 \cdot^- \qquad (3.2)$$

$$HQ \cdot + O_2 \Rightarrow Q + H ++ O_2 \cdot^- \qquad (3.3)$$

$$QH_2 + O_2 \Rightarrow HQ \cdot + H + O_2 \cdot^- \qquad (3.4)$$

It is impossible to estimate the difference of one-electron reduction potentials for Reactions 3.3 and 3.4, but in any case, these reactions are irreversible because their back reactions could not be simple elementary reactions. However, from the kinetic point of view, Reactions 3.3 and 3.4 cannot compete with Reactions 3.1 and 3.2 because their rates must be very small ones. (For example, Sugioka et al. [38] showed that the rate of coenzyme Q1H2 [an analog of Q_{10}] autoxidation is extremely small, being equal to $1.5 \, M^{-1}s^{-1}$ at pH 7.5 and 25°C.)

Reaction 3.2 is a reversible process, and its rate might be high enough. However, ubiquinone Q is a lipid (membrane)-soluble compound, and its reduction potential cannot be determined in water. Earlier, we developed the spectrophotometrical method of determination of the equilibrium constant of Reaction 3.2 by direct measurement of the concentrations of $O_2 \cdot^-$ and ubisemiquinone in aprotic medium (dimethylformamide), a model of lipid membrane [39,40]. It was found that K_{eq} for Reaction 3.2 is equal to 0.35. As $[Q] \gg [O_2 \cdot^-]$ in mitochondria, the equilibrium of Reaction 3.2 should be sharply shifted to the left and, therefore, free ubisemiquinone radicals cannot be efficient producers of $O_2 \cdot^-$. However, experimental findings point to the participation of ubisemiquinone in $O_2 \cdot^-$ production by Complex III; therefore, its redox properties apparently changed inside Complex III. The participation of ubihydroquinone and neutral ubisemiquinone in the reduction of dioxygen into $O_2 \cdot^-$ seems to be unlikely processes.

3.1.3 STIMULATION AND INHIBITION OF ROS PRODUCTION IN MITOCHONDRIA

ROS ($O_2 \cdot^-$ and H_2O_2) production by mitochondria can be inhibited by mitochondrial antioxidants glutathione, MnSOD, copper zinc superoxide dismutase (CuZnSOD), and other antioxidant molecules. Han et al. [35] studied the effect of glutathione depletion on ROS formation in mitochondria. $O_2 \cdot^-$ release into the intermembrane space by Complex III was not influenced by GSH status but H_2O_2 release, which was formed by dismutation of matrix-directed $O_2 \cdot^-$ from Complex I or III, was increased at the 50% depletion of matrix GSH. Jo et al. [41] showed that mitochondrial $NADP^+$-dependent isocitrate dehydrogenase (IDPm) was an important component of the regulation of mitochondria glutathione levels.

Mitochondria contain two superoxide dismutases (SODs): MnSOD in the matrix and CuZnSOD in intermembrane space [42]. Both enzymes regulate $O_2 \cdot^-$ levels in these compartments. Inarrea et al. [42] have found that CuZnSOD is inactive in intact mitochondria and activated only upon the oxidation of its thiol groups, suggesting the redox regulation of this enzyme in intermembrane space. In a subsequent work [43], these authors demonstrated that $O_2 \cdot^-$ and probably H_2O_2 activated CuZnSOD in intermembrane space by oxidation of its SH groups, while thioredoxin reductase deactivated CuZnSOD by reducing back the dithiol groups.

Because the one-electron transfer reaction of ubiquinone with $O_2 \cdot^-$ is a completely reversible process (see earlier, Reaction 3.2), ubiquinone in mitochondria is able to exhibit both antioxidant and prooxidant effects. For example, Yamamura et al. [44] suggested that ubiquinone plays dual roles in mitochondrial generation of intracellular signaling. Ubiquinone acted as a prooxidant participating in redox signaling and as an antioxidant that inhibited permeability transition and cytochrome c release, thereby attenuating death signaling toward apoptosis and necrosis. Koopman et al. [45] investigated the inhibitory effect of a mitochondria-targeted derivative of coenzyme Q10 (MitroQ) on $O_2 \cdot^-$ production by Complex I in the presence of rotenone in human skin fibroblasts. It was found that MitoQ did not affect $O_2 \cdot^-$ production but inhibited mitochondrial outgrowth.

Another factor affecting ROS generation in mitochondria is calcium uptake. This problem is far from being completely understood; it has been argued that there is still no convincing mechanism to explain how calcium signals can enhance $O_2 \cdot^-$ generation because a rise in mitochondrial calcium concentration depolarizes mitochondria and consequently should decrease $O_2 \cdot^-$ formation [46]. Thus, Sadek et al. [47] and Matsuzaki and Szweda [48] have shown that the addition of calcium to solubilized mitochondria or submitochondrial particles resulted in loss in complex I activity. SOD prevented inhibition, indicating the involvement of $O_2 \cdot^-$ in the inactivation process. On the other hand, Heinen et al. [49] have shown that opening of mitochondrial calcium-induced K^+ influx increased respiration and enhanced ROS production by mitochondria. Later on, these authors concluded [50] that, under conditions allowing reverse electron flow, matrix K(+) influx through mitochondrial calcium-sensitive K(+) channel (mtBK(Ca) channels) reduced mitochondrial ROS production by accelerating forward electron flow. Kanno et al. [51] found that, in the presence of inorganic phosphate, calcium increased oxygen consumption and ROS production by isolated mitochondria. It was also shown that calcium-induced ROS generation resulted in the oxidation of free thiol groups in adenine nucleotide translocase (ANT) in mitochondrial membrane.

Hennet et al. [52] have shown that the addition of tumor necrosis factor-alpha (TNF-α) to L929 cells induced mitochondrial $O_2 \cdot^-$ production. Ko et al. [53] confirmed their findings. 12(S)-Hydroxyeicosatetraenoic acid (12-HETE) is another promoter of mitochondrial NOS.

3.1.4 Uncoupled Proteins in Mitochondrial Electron Transport

Uncoupling proteins (UCPs) 1, 2, 3, 4, and 5 (UCP1, UCP2, UCP3, UCP4, and UCP5) are members of the anion carrier protein family located in the inner mitochondrial membrane. The role of USP1 is well established: it is responsible for rapid respiration not coupled with ATP synthesis and the stimulation of the thermogenic process [54]. On the other hand, it has been shown that USP2 and other USP homologues can participate in electron transfer and be activated by $O_2\cdot^-$. In 2002, Echtay et al. [55] demonstrated that $O_2\cdot^-$ increased mitochondrial proton conductance by the activation of UCP1, UCP2, and UCP3. Furthermore, it has been shown that exogenous $O_2\cdot^-$ can enter mitochondria and activate UCPs in the matrix [56].

Krauss et al. [57] showed that UCP2-mediated proton leak in thymocytes and pancreatic b cells was activated by endogenous produced $O_2\cdot^-$. Removal of endogenous $O_2\cdot^-$, or UCP2, caused decreased proton leak and increased mitochondrial membrane potential. However, it should be noted that the UCP2 function can reduce mitochondrial $O_2\cdot^-$ by decreasing mitochondrial membrane potential. Therefore, it is possible that a feedback loop exists in cells: The expression of UCPs abnormally increases $O_2\cdot^-$ levels and induces UCP-dependent uncoupling that, in turn, reduces mitochondrial membrane potential and $O_2\cdot^-$ formation.

An important suggestion has been made by Brownlee [58] who proposed that UCP2 activation follows the hyperglycemia-induced $O_2\cdot^-$ formation by the mitochondrial electron transfer chain. On the other hand, Echtay et al. [59] pointed out that uncoupling proteins 2 and 3 are highly active H^+ transporters when activated by coenzyme Q (ubiquinone).

In our opinion, the participation of ubiquinone in $O_2\cdot^-$ stimulation of the UCP-mediated proton leak makes it possible to represent a mechanism of this process. We proposed [60,61] that the activation of UCP2 by $O_2\cdot^-$ leading to proton leak and the inhibition of oxidative phosphorylation proceeded through the one-electron transfer reaction of $O_2\cdot^-$ with the ubiquinone–UCP2 complex:

$$O_2\cdot^- + Q(UCP2) \Rightarrow O_2 + Q\cdot - (UCP2) \tag{3.5}$$

Then, the bound negatively charged semiquinone radical anion extracts a proton and initiates proton leak:

$$Q\cdot^- (UCP2) + H + \Rightarrow HQ(\cdot)(UCP2) \tag{3.6}$$

The neutral HQ(·)(UCP2) complex may dissociate to UCP2 and neutral ubisemiquinone; the latter can be converted into ubiquinone and ubihydroquinone.

It is uncertain whether the expression of UCPs really influences $O_2\cdot^-$ levels. Thus, Fink et al. [62] showed that overexpression of both UCP1 and UCP2 induced uncoupling properties in mitochondria but did not affect $O_2\cdot^-$ levels in bovine aortic endothelial cells. Silva et al. [63] found that endogenously generated matrix $O_2\cdot^-$ did not regulate UCP activity. In contrast, Giardina et al. [64] recently showed that the levels of UCP2 in macrophages increased rapidly with increased mitochondrial $O_2\cdot^-$ production but not in response to $O_2\cdot^-$ produced outside mitochondria or in response to H_2O_2. These findings support UCP2 function in regulating mitochondrial reactive oxygen production through a feedback loop.

3.1.5   MITOCHONDRIAL ROS SIGNALING IN PATHOLOGICAL STATES

The formation of a relatively harmless physiological radical $O_2^{\cdot-}$ may initiate significant disturbances in mitochondria through the generation of its descendants—reactive free radicals (hydroxyl and peroxy radicals) as well as the reactive species formed by the reaction of $O_2^{\cdot-}$ with NO. Therefore, $O_2^{\cdot-}$ and NO signaling can be responsible for toxic effects in mitochondria. Correspondingly, $O_2^{\cdot-}$ is able to mediate both normal physiological and damaging signaling processes.

One of the first examples of the damaging effects of free-radical formation in mitochondria was reported in 1985 by Dean and Pollak [65]. These authors suggested that endogenous free-radical generation may influence proteolysis in mitochondria both by direct reaction with proteins or by making proteins more susceptible to proteinases. In 1993, Ambrosio et al. [66] proposed that electron flow through the respiratory chain may be an important source of oxygen radicals and that resumption of mitochondrial respiration upon reoxygenation might contribute to reperfusion injury. Vasquez-Vivar et al. [67] examined the interaction of $O_2^{\cdot-}$ with aconitase and suggested that this reaction may lead to the reduction of $[4Fe-4S]^{2+}$ aconitase to $[3Fe-4S]^+$ aconitase accompanied by the formation of high-reactive damaging hydroxyl radicals in mitochondria.

Pain et al. [68] have shown that opening of mitochondrial K_{ATP} channels leads to the preconditioned state by generating ROS. Samavati et al. [69] studied the role of mitochondria-derived ROS in the activation of extracellular signal-regulated kinases (ERKs), key regulatory enzymes mediating cell survival, proliferation, and differentiation. They found that the opening of mitochondrial K_{ATP} channels, which increased mitochondrial $O_2^{\cdot-}$ production, enhanced phosphorylation and activity of ERK kinase.

In an important 2006 work, Andrukhiv et al. [70] showed that the opening of the mitochondrial ATP-sensitive K(+) channel (mitoK$_{ATP}$) increased $O_2^{\cdot-}$ generation from Complex I of the electron transport chain through matrix alkalinization. The effect of matrix alkalinization on ROS formation is understandable because a lifetime of $O_2^{\cdot-}$ should sharply increase with increase in pH. Votyakova and Reynolds [71] investigated the effects of the mitochondrial membrane potential $\Delta\Psi_m$ on ROS production by rat brain mitochondria. The highest rates of mitochondrial ROS generation were observed while mitochondria were respiring on the Complex II substrate succinate when the majority of ROS was derived from reverse electron transport to Complex I.

Brownlee and coworkers [72,73] demonstrated that hyperglycemia induced a decrease in glyceraldehyde-3-phosphate dehydrogenase activity in bovine aortic endothelial cells through increased mitochondrial $O_2^{\cdot-}$ production. Later on, Nishikawa et al. [74] showed that hyperglycemia-induced production of ROS was abrogated by the inhibitors of mitochondrial metabolism, overexpression of UCP-1, or MnSOD. Normalization of mitochondrial ROS production by each of these agents prevented glucose-induced activation of protein kinase C (PKC), formation of

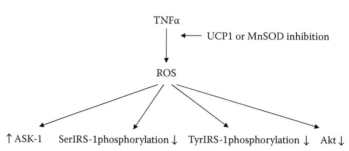

FIGURE 3.4 TNFα stimulation of ROS-initiated signaling cascade contributed to impaired insulin signaling.

advanced glycation endproduct (AGE), and accumulation of sorbitol in bovine vascular endothelial cells. Brodsky et al. [75] demonstrated that mitochondrial generation of $O_2\cdot^-$ was augmented in endothelial cells in the presence of a high glucose concentration that led to the suppression of NO production by mtNOS. On the other hand, Ahmad et al. [76] proposed that mitochondrial $O_2\cdot^-$ and H_2O_2 stimulated glucose-deprivation-induced cytotoxicity and metabolic oxidative stress in human cancer cells.

Imoto et al. [77] showed that TNF-α increased mitochondrial ROS production, which was suppressed by overexpression of either UCP1 or MnSOD in cultured human hepatoma cells. TNF-α significantly activated apoptosis signal-regulating kinase 1 (ASK1), increased serine phosphorylation of insulin receptor substrate (IRS-1), and decreased insulin-stimulated tyrosine phosphorylation of IRS-1 and serine phosphorylation of Akt kinase (Figure 3.4). All these effects were inhibited by the overexpression of either UCP1 or MnSOD. It was supposed that TNF-α stimulation of ROS-initiated signaling cascade contributed, at least in part, to impaired insulin signaling.

In 1998 Duranteau et al. [78] demonstrated that hypoxia activated the generation of ROS in cardiomyocytes and that ROS originated from the mitochondrial electron transport chain. Kulisz et al. [79] have shown that an increase in ROS formation during hypoxia induced p38-MAPK phosphorylation. It was proposed that $O_2\cdot^-$ produced in mitochondria during hypoxia was released into the cytosol, where it subsequently converted to H_2O_2, which then acted as the downstream signal leading to p38 phosphorylation.

Nojiri et al. [80] reported that $O_2\cdot^-$ overproduction is responsible for heart failure with impaired mitochondrial respiration. These authors suggested that hypoxia induced an increase in mitochondrial ROS, which activated hypoxic pulmonary vasoconstriction (HPV) through an oxidant-signaling pathway. Waypa et al. [81] demonstrated that hypoxia increased calcium in pulmonary artery smooth-muscle cells by augmenting ROS signaling from the mitochondria. Bell et al. [82] proposed that the Q_o site of Complex III increased cytosolic ROS under hypoxic conditions, which inhibited the prolyl hydroxylase (PHD) ability to degrade the hypoxia-inducible factor (HIF).

Gao and Wolin [83] recently investigated the effect of hypoxia on the relationship between cytosolic and mitochondrial NADPH-stimulated $O_2\cdot^-$ generation in endothelium-denuded bovine coronary arteries (BCA). Hypoxia apparently increased mitochondrial and decreased cytosolic-nuclear $O_2 \cdot^-$ formation under conditions associated with increased cytosolic NADH, mitochondrial NADPH, and hyperpolarization of mitochondria. Rotenone enhanced mitochondrial $O_2 \cdot^-$, suggesting that hypoxia probably increased $O_2 \cdot^-$ generation by Complex I.

Chen et al. [84] have studied the role of mitochondrial Complex II in postischemic heart. They found that protein S-glutathionylation of special thiols of Complex II is an important factor of regulation of ROS overproduction under the conditions of oxidative stress. Indo et al. [85] demonstrated that the enhanced amount of ROS was generated from mitochondria in cells with impaired electron transport chain and mitochondrial DNA damage.

The deficiency of NADH:ubiquinone oxidoreductase or Complex I is the common cause of disorders of the oxidative phosphorylation system in humans. Verkaart et al. [86] have shown that $O_2 \cdot^-$ production increased in inherited Complex I deficiency and that this increase was primarily a consequence of the reduction in cellular Complex I activity and not due to the leakage of electrons from mutationally malformed complexes.

Kohli et al. [87] studied the involvement of mitochondrial ROS in the development of hepatic steatosis (the accumulation of excess neutral fat within hepatocytes) in obese persons, which is thought to be a consequence of abnormal dietary intake of macronutrients leading to the development of obesity and fatty liver disease. They suggested that H_2O_2 formed by dismutation of $O_2 \cdot^-$ generated by mitochondrial Complex III is an initiator of activation of the PI3-kinase pathway. Activation of PI3 kinase depended on Akt kinase and its natural inhibitor phosphatase and tensin homolog (PTEN) acting by the degradation of PI3 kinase. It was found that PTEN was inactivated through phosphorylation during nutrient perturbation, which amplified the PI3-kinase signal transduction (Figure 3.5). It was also found that the inhibition of PTEN expression in hepatocytes resulted in an increase in Akt expression. Thus, PTEN turns out to be a brake for PI3-kinase signaling during normal cellular respiration. These findings also suggest that acute regulation of PTEN function during nutrient perturbation occurs through its phosphorylation.

Redout et al. [88] reported that the development of pulmonary arterial hypertension (PAH) that induced right-ventricular (RV) heart failure was associated with the enhanced ROS production by small Nox and mitochondria. An increase in the expression and activity of mitochondrial Complex II may be particularly important for ventricular ROS production in heart failure.

Caldiz et al. [89] investigated the role of mitochondrial reactive oxygen species in the activation of the slow force response (SFR) to stretch in feline myocardium. Their findings suggest a pivotal role of mitochondrial ROS in the genesis of the SFR to stretch. Kulich et al. [90] have shown that early intracellular ROS stimulated by 6-hydroxydopamine, followed by a delayed phase of mitochondrial ROS production, was associated with phosphorylation of a mitochondrial ERK. It was suggested that the inhibition of the ERK pathway protected from 6-hydroxydopamine toxicity and that phosphorylated ERK accumulation in mitochondria of degenerating human Parkinson's disease neurons may be an important factor of Parkinsonian oxidative neuronal injury.

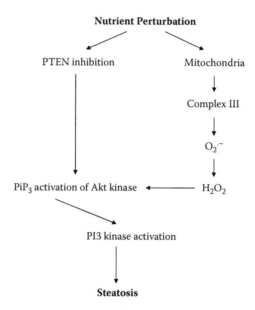

FIGURE 3.5 Mitochondrial ROS signaling in stimulation of steatosis: Nutrient perturbation initiation of overproduction of superoxide Complex III and the inhibition of PTEN, a natural inhibitor of Akt kinase results in PI3 kinase activation leading to steatosis.

Although mechanisms of troglitazone-associated idiosyncratic liver injury are not known, there are some evidences that oxidant stress and mitochondrial injury might be a potential hazard. Lim et al. [91] proposed that troglitazone-induced enhanced mitochondrial generation of O_2·$^-$ might activate the thioredoxin-2 (Trx2)/Ask1 signaling pathway, leading to cell death. Javadov et al. [92] proposed that mitochondrial O_2·$^-$ production is responsible for the inhibition of Na^+/H^+ exchanger isoform 1 (NHE-1) and the attenuation of cardiomyocyte hypertrophy by suppression of MAPK activation.

Cassina et al. [93] found that mitochondrial O_2·$^-$ formation increased in astrocytes expressing the mutations of superoxide dismutase-1 (SOD1). Taken together, their results indicate that mitochondrial dysfunction in astrocytes critically influences motor neuron survival and supports the potential pharmacological utility of mitochondrial-targeted antioxidants in amyotrophic lateral sclerosis (ALS) treatment.

Viel et al. [94] suggested that endothelin-1 (ET-1)-induced generation of ROS in the aorta and resistance arteries of DOCA-salt rats originated from XO and mitochondria. Their findings demonstrate the involvement of ETA-receptor-modulated O_2·$^-$ production derived from both XO and mitochondria in arteries from DOCA-salt rats.

The role of O_2·$^-$ overproduction in pulmonary hypertension and cancer has been recently reviewed by Archer et al. [95].

3.1.6 MITOCHONDRIAL ROS SIGNALING UNDER PHYSIOLOGICAL CONDITIONS

O_2·$^-$ mediates numerous enzymatic catalytic processes in mitochondria. In 1997, Madesh et al. [96] showed that liver mitochondrial phospholipase A2 can be activated by $O_2\cdot^-$-generating systems. Later on, Yamagishi et al. [97] demonstrated that the circulating hormone leptin induced mitochondrial $O_2\cdot^-$ production in aortic endothelial cells by increasing fatty acid oxidation through protein kinase A (PKA). There was a relationship between $O_2\cdot^-$ formation and the activation of PKC in mitochondria. Thus, cephaloridine (CER), an inductor of nephrotoxicity, caused the enhancement of mitochondrial $O_2\cdot^-$ and PKC activation [98]. Kimura et al. [99] investigated the involvement of mitochondria-derived ROS in intracellular signal transduction and vasoconstriction by Ang II. They found that Ang II stimulated mitochondrial ROS generation through the opening of mitoKATP channels in the vasculature, resulting in the reduction of $\Delta\Psi_m$ and redox-sensitive activation of MAPK kinases, especially p38 and JNK.

Hongpaisan et al. [100] have studied the effect of calcium on $O_2\cdot^-$ signaling in mitochondria. It was found that $O_2\cdot^-$ was selectively produced by mitochondria near the plasmalemmal sites of calcium entry after large calcium increase. It was followed by the upregulation of the two kinases CaMKII and PKA, whose activities were directly or indirectly phosphorylation dependent. The whole mechanism involved the inactivation of protein phosphatases by $O_2\cdot^-$. Enhanced $O_2\cdot^-$ production also promoted PKC activity by a phosphatase-independent pathway.

Manipulating MnSOD activity, these authors were also able to estimate which ROS ($O_2\cdot^-$ or H_2O_2) are dominant signal molecules in these enzymatic processes. It is frequently proposed that, due to a quick dismutation of $O_2\cdot^-$ into H_2O_2, the last is a functionally more important signaling species. However, it was found in this study that enhancing SOD activity by overexpression or by application of a mimetic that should accelerate the conversion of $O_2\cdot^-$ into H_2O_2 did not enhance the rates of these processes. Thus, it appears that, at least in hippocampal cells, $O_2\cdot^-$ is the principal ROS modulator of calcium-dependent signaling cascades. This conclusion is very important for the consideration of mechanisms of $O_2\cdot^-$ and H_2O_2 signaling mechanisms (see Chapter 7).

Storz et al. [101] demonstrated that the release of mitochondrial ROS activated the signal pathway, in which the serine/threonine protein kinase D (PKD) activated the NFκB transcription factor, leading to the induction of the *SOD2* gene, which encoded mitochondrial MnSOD. This PKD-mediated mitochondrion-to-nucleus signaling pathway caused the detoxification of mitochondrial ROS (Figure 3.6).

Kaminski et al. [102] found that PKC-theta-dependent ROS generation by mitochondrial Complex I was essential for activation-induced T-cell death (AICD). Saitoh et al. [103] have shown that $O_2\cdot^-$, and H_2O_2 formed by its dismutation, are produced proportionally to cardiac metabolism. Their findings suggest that H_2O_2 is produced in proportion to metabolism that couples coronary blood flow to myocardial oxygen consumption.

In a recent review, Jones [104] proposed a new hypothesis concerning the role of $O_2\cdot^-$ in mitochondria. He considered two major redox circuits in mitochondria: high-flux pathways that are included mechanisms for ATP production and low-flux pathways that utilize sulfur switches of proteins for metabolic regulation and cell signaling. $O_2\cdot^-$, participating in the high-flux mitochondrial electron transfer chain, links the high-flux and low-flux pathways. Increase in $O_2\cdot^-$ production in the high-flux pathway creates aberrant "short-circuit" pathways between otherwise noninteracting

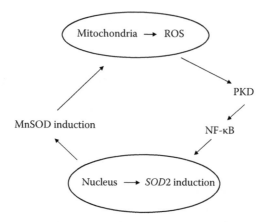

FIGURE 3.6   Mitochondria-to-nucleus signaling and SOD2 gene expression.

components. In accordance with this new hypothesis, $O_2 \cdot^-$ is not a by-product of electron transfer but rather is a positive signal to coordinate energy metabolism. Thus, this hypothesis highlights an important role of $O_2 \cdot^-$ formation in both physiological and pathophysiological mitochondrial processes.

3.2   SIGNALING FUNCTIONS OF NITRIC OXIDE AND PEROXYNITRITE IN MITOCHONDRIA

As shown in Chapter 2, mitochondria contain NOS (mtNOS); therefore, the origin of NO there is well established. In earlier works, the effects of NO in mitochondria were studied in the experiments with donors of NO and peroxynitrite. In 1994, several groups of authors [105–108] reported that there are two different inhibitory effects of NO in mitochondria: the direct inhibition of mitochondrial electron transport chain by the interaction of NO with electron carriers, and the interaction with $O_2 \cdot^-$ to form damaging peroxynitrite (Figure 3.7).

Lizasoain et al. [109] found that both NO and peroxynitrite inhibited respiration by brain submitochondrial particles, the former reversibly at cytochrome c oxidase

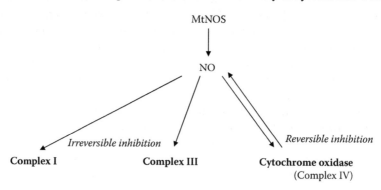

FIGURE 3.7   Signaling functions of nitric oxide in mitochondria.

(COX), and the latter irreversibly at Complexes I–III. Glutathione and glucose prevented the inhibition by peroxynitrite but not by NO. Therefore, unless peroxynitrite is formed within mitochondria, it is unlikely to inhibit respiration in cells directly because of reactions with cellular thiols and carbohydrates. Poderoso et al. [110] have shown that the inhibitory effect of NO in rat heart mitochondria was accompanied by increased H_2O_2 production. Cassina and Radi [111] proposed that peroxynitrite is a major reactive species responsible for NO-dependent inactivation of mitochondrial electron transport components. Clementi et al. [112] suggested that, although NO may regulate cell respiration physiologically by the inhibition of COX, long-term exposure to NO led to persistent inhibition of Complex I through its S-nitrosylation. Parihar et al. [113] have recently shown that NO irreversibly inhibited purified cytochrome oxidase in a reverse-oxygen-concentration-dependent manner by the formed peroxynitrite, which was a genuine species involved in irreversible cytochrome oxidase inactivation.

Huang et al. [114] have studied the intracellular mechanisms by which NO and peroxynitrite inhibited smooth-muscle cell growth, and the potential role of peroxynitrite formation in the antiproliferative effects of NO. Both exogenous NO and peroxynitrite decreased cell growth and DNA synthesis of cultured rat aortic smooth-muscle cells. However, peroxynitrite-induced growth arrest was irreversible and induced apoptosis and cytotoxicity, while exogenous NO induced reversible cytostasis by a redox-sensitive mechanism independent of peroxynitrite formation. Thus, the antiproliferative effects of NO and peroxynitrite on smooth-muscle cells use divergent intracellular pathways with distinct redox sensitivities.

Real estimation of NO inhibitory effects in mitochondria became possible after the study of mtNOS under in vitro and in vivo conditions [115–117]. Ghafourifar and Richter [115] showed that mitochondrial NOS activity was associated with the inner mitochondrial membrane and not with the matrix fraction. MtNOS was constitutively active in succinate-energized mitochondria and exerted substantial control over mitochondrial respiration and membrane potential. Giulivi et al. [116] showed that the rate of NO production by rat liver mitochondria is close to that of $O_2\cdot^-$. Therefore, at saturating concentrations of L-arginine, the NO steady-state concentration might be in the range of 0.1–0.5 µM, and the reaction of NO with cytochrome oxidase could be a main catabolic pathway. Lopez-Figueroa et al. [117] demonstrated the in vivo production of NO within mitochondria.

Riobo et al. [118] have studied the effect of long-term incubation of NO with mitochondrial membranes. It was found that NO resulted in a persistent inhibition of NADH:ubiquinone reductase (Complex I) activity, whereas succinate:COX reductase activity, including Complex II and Complex III, remained unaffected. SOD and uric acid inhibited this selective effect of NO.

Schild et al. [119] demonstrated that NO concentration produced by mitochondrial NOS increased under hypoxic conditions in isolated rat liver mitochondria. NO mediated the impairment of active (State 3) respiration after reoxygenation. These findings suggest that mitochondria are exposed to high amounts of NO generated by mtNOS upon hypoxia/reoxygenation. Increased NO levels, in turn, inhibit mitochondrial respiration and may cause oxidative stress that leads to irreversible impairment of mitochondria. Han et al. [120] showed that shear stress induced mitochondrial RNS formation that inhibited electron transfer chain at multiple sites. This could be a mechanism by which shear stress modulates endothelial cell signaling and function.

In 1999, Poderoso et al. [121] made a brave attempt to estimate the contents of NO and $O_2 \cdot^-$ in mitochondria and the rates of their reactions with COX and ubihydroquinone. Their calculated concentrations of $[O_2 \cdot^-] = 10^{-10}$ M and $[NO] = 5 \times 10^{-8}$ M seem to be a reasonable approximation. Unfortunately, the calculations were based on the proposal that the oxidation reactions of ubihydroquinone and COX with NO are exothermic processes and their rates are high enough. However, the new determination of the one-electron reduction potential of NO equal to -0.8 V [122] denies the NO ability to oxidize both compounds.

NO performs various functions in mitochondria. Sasaki et al. [123] showed that NO is able to activate directly mitoK$_{ATP}$ channels. Correspondingly, there is a relationship between NO-induced cardioprotection and mitoK$_{ATP}$ channels. It was suggested [124] that NO participates in S-nitrosation of the protein thiol groups. However, NO is a very poor nitrosating agent, and therefore, the other more reactive nitrogen compounds, for example peroxynitrite, might react with the thiol groups. Recently, Dahm et al. [125] found that S-nitrosothiols but not NO or peroxynitrite are responsible for persistent S-nitrosation of Complex I and the other mitochondrial membrane proteins.

3.2.1 INHIBITION OF COX AND DIOXYGEN CONSUMPTION BY NITRIC OXIDE IN MITOCHONDRIA

It has been noted that NO reacts with different components of the mitochondrial transport chain and inhibits their activities through interaction with the SH groups and iron centers of cytochromes. However, the most important mode of NO activity in mitochondria is the reversible inhibition of COX and regulation of dioxygen consumption. In 1994, Cleeter et al. [105] and Brown and Cooper [108] showed that NO reversibly inhibited COX, the terminal enzyme of the mitochondrial respiratory chain. In subsequent years, the inhibition of COX by NO has been discussed by numerous works due to its importance in many physiological and pathophysiological processes, including aging (see Chapter 6). For example, Sarkela et al. [126] showed that NO affected the rate of $O_2 \cdot^-$ production by changing the rate of dioxygen consumption at the cytochrome oxidase level.

In vivo sources of NO in the reaction with COX may not only be mtNOS but also endothelial NO synthase because the relatively long-living NO is able to diffuse from the endothelium to mitochondria. Thus, Hinze and coworkers [127,128] proposed that endogenous endothelial NOS–derived NO was a physiological regulator of myocardial oxygen consumption and that eNOS controlled myocardial dioxygen consumption and modulated mitochondrial respiration. Palacios-Callender et al. [129] also showed that endogenous NO regulated $O_2 \cdot^-$ production in the whole cells at low oxygen concentrations by modifying the redox state of COX. Li et al. [130] suggested that intramitochondrial SOD2 and intracellular myoglobin may affect NO bioavailability and regulate cardiac dioxygen consumption.

Despite numerous studies, the detailed molecular mechanism for the reversible inhibition of mitochondrial respiration by NO is still uncertain. It is known that the rate constants for the binding of NO and O_2 to the reduced binuclear center Cu_B/a_3 of COX are similar, and a more tight bonding of dioxygen to this center must favor the complex of COX with dioxygen over the complex of COX with NO. However, the inhibition of COX by NO was observed even at the high $[O_2/NO]$ ratios. Antunes et al. [131] have studied this paradox and concluded that all known features of the inhibition of COX by NO probably originated from a competition between NO and O_2 for the reduced binuclear center Cu_B/a_3 of COX.

In contrast to the reversible inhibition of COX by NO, which is an important regulatory process of mitochondrial respiration, peroxynitrite inhibits COX irreversibly together with the other destructive oxidation processes in mitochondria. Barone et al. [132] showed that, when isolated COX respired with ascorbate as a reducing substrate, the conversion of peroxynitrite to NO was observed. They demonstrated that this process was a consequence of the direct interaction of peroxynitrite with NO.

Guidarelli and Cantoni [133] have shown that peroxynitrite stimulated the release of arachidonic acid (AA) in cells in an indirect way through the enhancement of $O_2^{·-}$ formation at the level of Complex III of the respiratory chain. It was therefore proposed that the formation of NO, $O_2^{·-}$, and peroxynitrite in mitochondria resulted in the existence of a cycle in which peroxynitrite inhibited Complex III and promoted the further formation of $O_2^{·-}$. In addition, an increase in peroxynitrite generation caused the activation of mitochondrial phospholipase A2 (PLA_2).

REFERENCES

1. HJ Forman and A Azzi. On the virtual existence of superoxide anions in mitochondria: Thoughts regarding its role in pathophysiology. *FASEB J* 11: 374–375, 1997.
2. HJ Forman and JA Kennedy. Role of superoxide radical in mitochondrial dehydrogenase reactions. *Biochem Biophys Res Commun* 60: 1044–1050, 1974.
3. JF Turrens and A Boveris. Generation of superoxide anion by the NADH dehydrogenase of bovine heart mitochondria. *Biochem J* 191: 421–427, 1980.
4. Y-R Chen, C-L Chen, L Zhang, KB Green-Church, and JL Zweier. Superoxide generation from mitochondrial NADH dehydrogenase induces self-inactivation with specific protein radical formation. *J Biol Chem* 280: 37339–37348, 2005.
5. ML Genova, B Ventura, G Giuliano, C Bovina, G Formiggini, G Parenti Castelli, and G Lenaz. The site of production of superoxide radical in mitochondrial Complex I is not a bound ubisemiquinone but presumably iron-sulfur cluster N2. *FEBS Lett* 505: 364–368, 2002.
6. A Herrero and G Barja. Localization of the site of oxygen radical generation inside the Complex I of heart and nonsynaptic brain mammalian mitochondria. *J Bioenerg Biomembr* 32: 609–615, 2000.
7. Y Kushnareva, AN Murphy, and A Andreyev. Complex I-mediated reactive oxygen species generation: Modulation by COX and NAD(P)+ oxidation-reduction state. *Biochem J* 368: 545–553, 2002.
8. J St-Pierre, JA Buckingham, SJ Roebuck, and MD Brand. Topology of superoxide production from different sites in the mitochondrial electron transport chain. *J Biol Chem* 277: 44784–44790, 2002.

9. AP Kudin, NY-B Bimpong-Buta, S Vielhaber, CE Elger, and WS Kunz. Characterization of superoxide-producing sites in isolated brain mitochondria. *J Biol Chem* 279: 4127–4135, 2004.

10. L Kussmaul and J Hirst. The mechanism of superoxide production by NADH:ubiquinone oxidoreductase (complex I) from bovine heart mitochondria. *Proc Natl Acad Sci USA* 103: 7607–7612, 2006.

11. A Galkin and U Brandt. Superoxide radical formation by pure complex I (NADH: Ubiquinone oxidoreductase) from *Yarrowia lipolytica. J Biol Chem* 280: 30129–30135, 2005.

12. AP Kudin, D Malinska, and WS Kunz. Sites of generation of reactive oxygen species in homogenates of brain tissue determined with the use of respiratory substrates and inhibitors. *Biochim Biophys Acta* 1777: 689–695, 2008.

13. AJ Lambert, JA Buckingham, and MD Brand. Dissociation of superoxide production by mitochondrial complex I from NAD(P)H redox state. *FEBS Lett* 582: 1711–1714, 2008.

14. AJ Lambert, JA Buckingham, HM Boysen, and MD Brand. Diphenyleneiodonium acutely inhibits reactive oxygen species production by mitochondrial complex I during reverse, but not forward electron transport. *Biochim Biophys Acta* 1777: 397–403, 2008.

15. A Boveris, E Cadenas, and AO Stoppani. Role of ubiquinone in the mitochondrial generation of hydrogen peroxide. *Biochem J* 156: 435–444, 1976.

16. P Mitchell. Possible molecular mechanisms of the protonmotive function of copper zinc superoxide dismutase systems. *J Theor Biol* 62: 327–367, 1976.

17. AR Crofts, JT Holland, D Victoria, DR Kolling, SA Dikanov, R Gilbreth, S Lhee, R Kuras, and MG Kuras. The Q-cycle reviewed: How well does a monomeric mechanism of the bc(1) complex account for the function of a dimeric complex? *Biochim Biophys Acta* 1777: 1001–1019, 2008.

18. M Ksenzenko, AA Konstantinov, GB Khomutov, AN Tikhonov, and EK Ruuge. Effect of electron transfer inhibitors on superoxide generation in the cytochrome bc1 site of the mitochondrial respiratory chain. *FEBS Lett* 155: 19–24, 1983.

19. JF Turrens, A Alexandre, and AL Lehninger. Ubisemiquinone is the electron donor for superoxide formation by complex III of heart mitochondria. *Arch Biochem Biophys* 237: 408–414, 1985.

20. AA Konstantinov, AV Peskin, EY Popova, GB Khomutov, and EK Ruuge. Superoxide generation by the respiratory chain of tumor mitochondria. *Biochim Biophys Acta* 894: 1–10, 1987.

21. G Lenaz, M Battino, C Castelluccio, R Fato, M Cavazzoni, H Rauchova, C Bovina, G Formiggini, and G Parenti Castelli. Studies on the role of ubiquinone in the control of the mitochondrial respiratory chain. *Free Radic Res Commun* 8: 317–327, 1990.

22. Q Chen, EJ Vazquez, S Moghaddas, CL Hoppel, and EJ Lesnefsky. Production of reactive oxygen species by mitochondria: Central role of complex III. *J Biol Chem* 278: 36027–36031, 2003.

23. S Raha, GE McEachern, AT Myint, and BH Robinson. Superoxides from mitochondrial complex III: The role of manganese superoxide dismutase. *Free Radic Biol Med* 29:170–180, 2000.

24. F Muller, AR Crofts, and DM Kramer. Multiple Q-cycle bypass reactions at the Q(o) site of the cytochrome bc(1) complex. *Biochemistry* 41: 7866–7874, 2002.

25. AJ Lambert and MD Brand. Inhibitors of the quinone-binding site allow rapid superoxide production from mitochondrial NADH:ubiquinone oxidoreductase (complex I). *J Biol Chem* 279: 39414–39420, 2004.

26. ST Ohnishi, T Ohnishi, S Muranaka, H Fujita, H Kimura, K Uemura, K Yoshida, and K Utsumi. A possible site of superoxide generation in the complex I segment of rat heart mitochondria. *J Bioenerg Biomembr* 37:1–15, 2005.

27. I Forquer, R Covian, MK Bowman, B Trumpower, and DM Kramer. Similar transition states mediate the Q-cycle and superoxide production by the cytochrome bc1 complex. *J Biol Chem* 281: 38459–38465, 2006.
28. JL Cape, MK Bowman, and DM Kramer. A semiquinone intermediate generated at the Qo site of the cytochrome bc1 complex: Importance for the Q-cycle and superoxide production. *Proc Natl Acad Sci USA* 104: 7887–7892, 2007.
29. S Drose and U Brandt. The mechanism of mitochondrial superoxide production by the cytochrome bc1 complex. *J Biol Chem* 2008 Jun 3. [Epub ahead of print].
30. L Zhang, L Yu, and CA Yu. Generation of superoxide anion by succinate-cytochrome c reductase from bovine heart mitochondria. *J Biol Chem* 273: 33972–33976, 1998.
31. Y Liu, G Fiskum, and D Schubert. Generation of reactive oxygen species by the mitochondrial electron transport chain. *J Neurochem* 80: 780–787, 2002.
32. YR Chen, CL Chen, A Yeh, X Liu, and JL Zweier. Direct and indirect roles of cytochrome b in the mediation of superoxide generation and no catabolism by mitochondrial succinate–cytochrome c reductase. *J Biol Chem* 281: 13159–13168, 2006.
33. A Bacsi, M Woodberry, W Widger, J Papaconstantinou, S Mitra, JW Peterso, and I Boldogh. Localization of superoxide anion production to mitochondrial electron transport chain in 3-NPA-treated cells. *Mitochondrion* 6: 235–244, 2006.
34. D Han, E Williams, and E Cadenas. Mitochondrial respiratory chain-dependent generation of superoxide anion and its release into the intermembrane space. *Biochem J* 353: 411–416, 2001.
35. D Han, R Canali, D Rettori, and N Kaplowitz. Effect of glutathione depletion on sites and topology of superoxide and hydrogen peroxide production in mitochondria. *Mol Pharmacol* 64: 1136–1144, 2003.
36. FL Muller and LY Remmen. Complex III releases superoxide to both sides of the inner mitochondrial membrane. *J Biol Chem* 279: 49064–49073, 2004.
37. AP Kudin, G Debska-Vielhaber, and WS Kunz. Characterization of superoxide production sites in isolated rat brain and skeletal muscle mitochondria. *Biomed Pharmacother* 59:163–168, 2005.
38. K Sugioka, M Nakano, H Totsune-Nakano, H Minakami, S Tero-Kubota, and Y Ikegami. Mechanism of O_2- generation in reduction and oxidation cycle of ubiquinones in a model of mitochondrial electron transport systems. *Biochim Biophys Acta* 936: 377–385, 1988.
39. IB Afanas'ev and NI Polozova. Equilibrium constants for the reactions of superoxide with natural quinones and their analogues. *Zh Org Khim* 15: 1802–1806, 1979.
40. IB Afanas'ev. *Superoxide Ion: Chemistry and Biological Implications*, vol. 1. Boca Raton, FL: CRC Press, 1989, 161.
41. S-H Jo, M-K Son, H-J Koh, S-M Lee, I-H Song, Y-O Kim, Y-S Lee, K-S Jeong, WB Kim, J-W Park, BJ Song, and T-L Huhe. Control of mitochondrial redox balance and cellular defense against oxidative damage by mitochondrial NADP+-dependent isocitrate dehydrogenase. *J Biol Chem* 276:16168–16176, 2001.
42. P Inarrea, H Moini, D Rettori, D Han, J Martinez, I Carcia, E Fernandez-Vizarra, M Iturralde, and E Cadenas. Redox activation of mitochondrial intermembrane space Cu,Zn-superoxide dismutase. *Biochem J* 387:203–209, 2005.
43. P Inarrea, H Moini, D Han, D Rettori, I Aguilo, MA Alava, M Iturralde, and E Cadenas. Mitochondrial respiratory chain and thioredoxin reductase regulate intermembrane Cu,Zn–superoxide dismutase activity: Implications for mitochondrial energy metabolism and apoptosis. *Biochem J* 405: 173–179, 2007.
44. T Yamamura, H Otani, Y Nakao, R Hattori, M Osako, H Imamura, and DK Das. Dual involvement of coenzyme Q10 in redox signaling and inhibition of death signaling in rat heart mitochondria. *Antioxid Redox Signal* 3: 103–112, 2001.

45. WJ Koopman, S Verkaart, HJ Visch, FH van der Westhuizen, MP Murphy, LW van den Heuvel, JA Smeitink, and PH Willems. Inhibition of complex I of the electron transport chain causes O^{2-}-mediated mitochondrial outgrowth. *Am J Physiol Cell Physiol* 288: C1440–C1450, 2005.

46. C Camello-Almaraz, PJ Gomez-Pinilla, MJ Pozo, and P Camello. Mitochondrial reactive oxygen species and Ca^{2+} signalling. *Am J Physiol Cell Physiol* 291: 1082–1088, 2006.

47. HA Sadek, PA Szweda, and LI Szweda. Modulation of mitochondrial complex I activity by reversible Ca(2+) and NADH mediated superoxide anion dependent inhibition. *Biochemistry* 43: 8494–8502, 2004.

48. S Matsuzaki and LI Szweda. Inhibition of complex I by Ca^{2+} reduces electron transport activity and the rate of superoxide anion production in cardiac submitochondrial particles. *Biochemistry* 46: 1350–1357, 2007.

49. A Heinen, AK Camara, M Aldakkak, SS Rhodes, ML Riess, and DF Stowe. Mitochondrial Ca^{2+}-induced K^+ influx increases respiration and enhances ROS production while maintaining membrane potential. *Am J Physiol Cell Physiol* 292: C148–C156, 2007.

50. A Heinen, M Aldakkak, DF Stowe, SS Rhodes, ML Riess, SG Varadarajan, and AK Camara. Reverse electron flow-induced ROS production is attenuated by activation of mitochondrial Ca^{2+} sensitive K^+ channels. *Am J Physiol Cell Physiol* 293: H1400–H1407, 2007.

51. T Kanno, EE Sato, S Muranaka, H Fujita, T Fujiwara, T Utsumi, M Inoue, and K Utsumi. Oxidative stress underlies the mechanism for Ca(2+)-induced permeability transition of mitochondria. *Free Radic Res* 38: 27–35, 2004.

52. T Hennet, C Richter, and E Peterhaus. Tumor necrosis factor-α induces superoxide anion generation in mitochondria of L929 cells. *Biochem J* 289: 587–592, 1993.

53. S Ko, TT Kwok, KP Fung, YM Choy, CY Lee, and SK Kong. Tumour necrosis factor induced an early release of superoxide and a late mitochondrial membrane depolarization in L929 cells. Increase in the production of superoxide is not sufficient to mimic the action of TNF. *Antioxid Redox Signal* 3: 461–472, 2001.

54. KS Echtay. Mitochondrial uncoupling proteins: What is their physiological role? *Free Radic Biol Med* 43: 1351–1371, 2007.

55. KS Echtay, D Roussel, J St-Pierre, MB Jekabsons, S Cadenas, LA Stuart, JA Harper, SJ Roebuck, A Morrison, S Pickering, JC Clapman, and MD Brand. Superoxide activates mitochondrial uncoupling proteins. *Nature* 415(6867): 96–99, 2002.

56. KS Echtay, MP Murphy, RA Smith, DA Talbot, and MD Brand. Superoxide activates mitochondrial uncoupling protein 2 from the matrix side: Studies using targeted antioxidants. Regulation of the expression of cyclooxygenase-2 by nitric oxide in rat peritoneal macrophages. *J Biol Chem* 277: 47129–47135, 2002.

57. S Krauss, C-Y Zhang, L Scorrano, LT Dalgaard, J St-Pierre, ST Grey, and BB Lowell. Superoxide-mediated activation of uncoupling protein 2 causes pancreatic b cell dysfunction. *J Clin Invest* 112: 1831–1842, 2003.

58. M Brownlee. Radical explanation for glucose-induced b cell dysfunction. *J Clin Invest* 112: 1788–1790, 2003.

59. KS Echtay, E Winkler, K Frischmuth, and M Klingenberg. Uncoupling proteins 2 and 3 are highly active H^+ transporters and highly nucleotide sensitive when activated by coenzyme Q (ubiquinone). *Proc Nat Acad Sci USA* 98: 1416–1421, 2001.

60. IB Afanas'ev. Interplay between superoxide and nitric oxide in aging and diseases. *Biogerontology* 5: 267–270, 2004.

61. I.B. Afanas'ev. Superoxide and nitric oxide in pathological conditions associated with iron overload: The effects of antioxidants and chelators. *Curr Med Chem* 12: 2731–2739, 2005.

62. BD Fink, KJ Reszka, JA Herlein, MM Mathahs, and WI Sivitz. Respiratory uncoupling by UCP1 and UCP2 and superoxide generation in endothelial cell mitochondria. *Am J Physiol Endocrinol Metab* 288: 71–79, 2005.
63. JP Silva, IG Shabalina, E Dufour, N Petrovic, EC Backlund, K Hultenby, R Wibom, J Nedergaard, B Cannon, and NG Larsson. SOD2 overexpression: Enhanced mitochondrial tolerance but absence of effect on UCP activity. *EMBO J* 24: 4061–4070, 2005.
64. TM Giardina, JH Steer, SZ Lo, and DA Joice. Uncoupling protein-2 accumulates rapidly in the inner mitochondrial membrane during mitochondrial reactive oxygen stress in macrophages. *Biochim Biophys Acta* 1777: 118–129, 2008.
65. RT Dean and JK Pollak. Endogenous free radical generation may influence proteolysis in mitochondria. *Biochem Biophys Res Commun* 126: 1082–1089, 1985.
66. G Ambrosio, JL Zweier, C Duilio, P Kuppusamy, G Santoro, PP Elia, I Tritto, P Coirillo, M Condorelli, M Chiariello, and JT Flaherty. Evidence that mitochondrial respiration is a source of potentially toxic oxygen free radicals in intact rabbit hearts subjected to ischemia and reflow. *J Biol Chem* 268: 18532–18541, 1993.
67. J Vasquez-Vivar, B Kalyanaraman, and MC Kennedy. Mitochondrial aconitase is a source of hydroxyl radical: An electron spin resonance investigation. *J Biol Chem* 275: 14064–14069, 2000.
68. T Pain, X-M Yang, SD Critz, Y Yue, A Nakano, GS Liu, G Heusch, MV Cohen, and JM Downey. Opening of mitochondrial K_{ATP} channels triggers the preconditioned state by generating free radicals. *Circ Res* 87: 460–466, 2000.
69. L Samavati, MM Monick, S Sanlioglu, GR Buettner, LW Oberley, and GW Hunninghake. Mitochondrial K(ATP) channel openers activate the ERK kinase by an oxidant-dependent mechanism. *Am J Physiol Cell Physiol* 283: C273–C281, 2002.
70. A Andrukhiv, AD Costa, IC West, and KD Garlid. Opening mitoKATP increases superoxide generation from Complex I of the electron transport chain. *Am J Physiol Heart Circ Physiol* 291: 2067–2074, 2006.
71. TV Votyakova and IJ Reynolds. DeltaPsi(m)-dependent and -independent production of reactive oxygen species by rat brain mitochondria. *J Neurochem* 79: 266–277, 2001.
72. T Nishikawa, D Edelstein, XL Du, S Yamagishi, T Matsumura, Y Kaneda, MA Yorek, D Beebe, PJ Oates, HP Hammes, I Giardino, and M Brownlee. Normalizing mitochondrial superoxide production blocks three pathways of hyperglycaemic damage. *Nature* 404: 787–790, 2000.
73. XL Du, D Edelstein, L Rossetti, IG Fantus, H Goldberg, F Ziyadeh, J Wu, and M Brownlee. Hyperglycemia-induced mitochondrial superoxide overproduction activates the hexosamine pathway and induces plasminogen activator inhibitor-1 expression by increasing Sp1 glycosylation. *Proc Natl Acad Sci USA* 97: 12222–12226, 2000.
74. T Nishikawa, D Kukidome, K Sonoda, K Fujisawa, T Matsuhisa, H Motoshima, T Matsumura, and E Araki. Impact of mitochondrial ROS production on diabetic vascular complications. *Diabetes Res Clin Pract* 77, Suppl 1: S41–S45, 2007.
75. SV Brodsky, S Gao, H Li, and MS Goligorsky. Hyperglycemic switch from mitochondrial nitric oxide to superoxide production in endothelial cells. *Am J Physiol Heart Circ Physiol* 283: H2130–H2139, 2002.
76. IM Ahmad, N Aykin-Burns, JE Sim, SA Walsh, R. Higashikubo, GR Buettner, S Venkataraman, MA Mackey, SW Flanagan, LW Oberley, and DR Spitz. Mitochondrial O_2. and H_2O_2 mediate glucose deprivation-induced cytotoxicity and oxidative stress in human cancer cells. *J Biol Chem* 280: 4254–4263, 2005.
77. K Imoto, D Kukidome, T Nishikawa, T Matsuhisa, K Sonoda, K Fujisawa, M Yano, H Motoshima, T Taguchi, K Tsuruzoe, T Matsumura, H Ichijo, and E Araki. Impact of mitochondrial reactive oxygen species and apoptosis signal-regulating kinase 1 on insulin signaling. *Diabetes* 55: 1197–1204, 2006.

78. J Duranteau, NS Chandel, A Kulisz, Z Shao, and PT Schumacker. Intracellular signaling by reactive oxygen species during hypoxia in cardiomyocytes. *J Biol Chem* 273: 11619–11624, 1998.
79. A Kulisz, N Chen, NS Chandel, Z Shao, and PT Schumacker. Mitochondrial ROS initiate phosphorylation of p38 MAP kinase during hypoxia in cardiomyocytes. *Am J Physiol Lung Cell Mol Physiol* 282: L1324–L1329, 2002.
80. H Nojiri, T Shimizu, M Funakoshi, O Yamaguchi, H Zhou, S Kawakami, Y Ohta, M Sami, T Tachibana, H Ishikawa, H Kurosawa, RC Kahn, K Otsu, and T Shirasawa. Oxidative stress causes heart failure with impaired mitochondrial respiration. *J Biol Chem* 281: 33789–33801, 2006.
81. GB Waypa, R Guzy, PT Mungai, MM Mack, JD Marks, MW Roe, and PT Schumacker. Increases in mitochondrial reactive oxygen species trigger hypoxia-induced calcium responses in pulmonary artery smooth muscle cells. *Circ Res* 99: 970–978, 2006.
82. EL Bell, TA Klimova, J Eisenbart, CT Moraes, MP Murphy, GRS Budinger, and NS Chandel. The Qo site of the mitochondrial complex III is required for the transduction of hypoxic signaling via reactive oxygen species production. *J Cell Biol* 177: 1029–1036, 2007.
83. Q Gao and MS Wolin. Effects of hypoxia on relationships between cytosolic and mitochondrial NAD(P)H redox and superoxide generation in coronary arterial smooth muscle. *Am J Physiol Heart Circ Physiol* Jun 2008; 10.1152/ajpheart.00316.2008.
84. Y-R Chen, C-L Chen, DR Pfeiffer, and JL Zweier. Mitochondrial complex II in post-ischemic heart: Oxidative injury and the role of protein S-glutathionylation. *J Biol Chem* 282: 32640–32654, 2007.
85. HP Indo, M Davidson, HC Yen, S Suenaga, K Tomita, T Nishii, M Higuchi, Y Koga, T Ozawa, and HJ Majima. Evidence of ROS generation by mitochondria in cells with impaired electron transport chain and mitochondrial DNA damage. *Mitochondrion* 7: 106–718, 2007.
86. S Verkaart, WJ Koopman, SE van Emst-de Vries, LG Nijtmans, LW van den Heuvel, JA Smeitink, and PH Willems. Superoxide production is inversely related to complex I activity in inherited complex I deficiency. *Biochim Biophys Acta* 1772: 373–381, 2007.
87. R Kohli, X Pan, P Malladi, MS Wainwright, and PF Whitington. Mitochondrial reactive oxygen species signal hepatocyte steatosis by regulating the phosphatidylinositol 3-kinase cell survival pathway. *J Biol Chem* 282: 21327–21336, 2007.
88. EM Redout, MJ Wagner, MJ Zuidwijk, C Boer, RJ Musters, C van Hardeveld, WJ Paulus, and WS Simonides. Right-ventricular failure is associated with increased mitochondrial complex II activity and production of reactive oxygen species. *Cardiovasc Res* 75: 770–781, 2007.
89. CI Caldiz, CD Garciarena, RA Dulce, LP Novaretto, AM Yeves, IL Ennis, HE Cingolani, GC de Cingolani, and NG Pérez. Mitochondrial reactive oxygen species activate the slow force response to stretch in feline myocardium. *J Physiol* 584: 895–905, 2007.
90. SM Kulich, C Horbinski, M Patel, and CT Chu. 6-Hydroxydopamine induces mitochondrial ERK activation. *Free Radic Biol. Med* 43: 372–383, 2007.
91. PL Lim, J Liu, ML Go, and UA Boelsterli. The mitochondrial superoxide/thioredoxin-2/Ask1 signaling pathway is critically involved in troglitazone-induced cell injury to human hepatocytes. *Toxicol Sci* 101: 341–349, 2008.
92. S Javadov, D Baetz, V Rajapurohitam, A Zeidan, LA. Kirshenbaum, and M Karmazyn. Antihypertrophic effect of Na+/H+ exchanger isoform-1 inhibition is mediated by reduced mitogen-activated protein kinase activation secondary to improved mitochondrial integrity and decreased generation of mitochondrial-derived reactive oxygen species. *J Pharmacol Exp Ther* 317: 1036–1043, 2007.

93. P Cassina, A Cassina, M Pehar, R Castellanos, M Gandelman, A de León, KM Robinson, RP Mason, JS Beckman, L Barbeito, and R Radi. Mitochondrial dysfunction in SOD1G93A-bearing astrocytes promotes motor neuron degeneration: Prevention by mitochondrial-targeted antioxidants. *J Neurosci* 28: 4115–4122, 2008.

94. EC Viel, K Benkirane, D Javeshghani, RM Touyz, and EL Schiffrin. Xanthine oxidase and mitochondria contribute to vascular superoxide anion generation in DOCA-salt hypertensive rats. *Am J Physiol Heart Circ Physiol* May 2008; 10.1152/ajpheart.00304.2008.

95. SL Archer, M Gomberg-Maitland, ML Maitland, S Rich, JGN Garcia, and EK Weir. Mitochondrial metabolism, redox signaling, and fusion: A mitochondria-ROS-HIF-1α-Kv1.5 O2-sensing pathway at the intersection of pulmonary hypertension and cancer. *Am J Physiol Heart Circ Physiol* 294: H570–H578, 2008.

96. M Madesh and KA Balasubramanian. Activation of liver mitochondrial phospholipase A2 by superoxide. *Arch Biochem Biophys* 346: 187–192, 1997.

97. SI Yamagishi, D Edelstein, XL Du, Y Kaneda, M Guzman, and M Brownlee. Leptin induces mitochondrial superoxide production and monocyte chemoattractant protein-1 expression in aortic endothelial cells by increasing fatty acid oxidation via protein kinase A. *J Biol Chem* 276: 25096–25100, 2001.

98. Y Kohda and M Gemba. Enhancement of protein kinase C activity and chemiluminescence intensity in mitochondria isolated from the kidney cortex of rats treated with cephaloridine. *Biochem Pharmacol* 64: 543–549, 2002.

99. S Kimura, GX Zhang, A Nishiyama, T Shokoji, L Yao, YY Fan, M Rahman, and Y Abe. Mitochondria-derived reactive oxygen species and vascular map kinases: Comparison of angiotensin II and diazoxide. *Hypertension* 45: 710–716, 2005.

100. J Hongpaisan, CA Winters, and SB Andrews. Strong calcium entry activates mitochondrial superoxide generation, upregulating kinase signaling in hippocampal neurons. *J Neurosci* 24: 10878–10887, 2004.

101. P Storz, H Doppler, and A Toker. Protein kinase D mediates mitochondrion-to-nucleus signaling and detoxification from mitochondrial reactive oxygen species. *Mol Cell Biol* 25: 8520–8530, 2005.

102. M Kaminski, M Kiessling, D Suss, PH Krammer, and K Gulow. Novel role for mitochondria: Protein kinase Ctheta-dependent oxidative signaling organelles in activation-induced T-cell death. *Mol Cell Biol* 27: 3625–3639, 2007.

103. S-ichi Saitoh, C Zhang, JD Tune, B Potter, T Kiyooka, PA Rogers, JD Knudson, GM Dick, A Swafford, and WM Chilian. Hydrogen peroxide: A feed-forward dilator that couples myocardial metabolism to coronary blood flow. *Arterioscler Thromb Vasc Biol* 26: 2614–2621, 2006.

104. DP Jones. Disruption of mitochondrial redox circuitry in oxidative stress. *Chem Biol Interact* 163: 38–53, 2006.

105. MWJ Cleeter, JM Cooper, VM Darley-Usmar, S Moncada, and AHV Schapira. Reversible inhibition of cytochrome c oxidase, the terminal enzyme of the mitochondrial respiratory chain, by nitric oxide: Implications for neurodegenerative diseases. *FEBS Lett* 345: 50–54, 1994.

106. R Radi, M Rodriguez, L Castro, and R Telleri. Inhibition of mitochondrial electron transport by peroxynitrite. *Arch Biochem Biophys* 308: 89–95, 1994.

107. M Schweizer and C Richter. Nitric oxide potently and reversibly deenergizes mitochondria at low oxygen tension. *Biochem Biophys Res Commun* 204: 169–175, 1994.

108. GC Brown and CE Cooper. Nanomolar concentrations of nitric oxide reversibly inhibit synaptosomal respiration by competing with oxygen at cytochrome oxidase. *FEBS Lett* 356: 295–298, 1994.

109. I Lizasoain, MA Moro, RG Knowles, V Darley-Usmar, and S Moncada. Nitric oxide and peroxynitrite exert distinct effects on mitochondrial respiration which are differentially blocked by glutathione or glucose. *Biochem J* 314: 877–880, 1996.

110. JJ Poderoso, MC Carreras, C Lisdero, N Riobo, F Schopfer, and A Boveris. Nitric oxide inhibits electron transfer and increases superoxide radical production in rat heart mitochondria and submitochondrial particles. *Arch Biochem Biophys* 328: 85–92, 1996.

111. A Cassina and R Radi. Differential inhibitory action of nitric oxide and peroxynitrite on mitochondrial electron transport. *Arch Biochem Biophys* 328: 309–316, 1996.

112. E Clementi, GC Brown, M Feelisch, and S Moncada. Persistent inhibition of cell respiration by nitric oxide: Crucial role of S-nitrosylation of mitochondrial complex I and protective role of glutathione. *Proc Natl Acad Sci USA* 95: 7631–7636, 1998.

113. A Parihar, P Vaccaro, and P Ghafourifar. Nitric oxide irreversibly inhibits cytochrome oxidase at low oxygen concentrations: Evidence for inverse oxygen concentration-dependent peroxynitrite formation. *IUBMB Life* 60: 64–67, 2008.

114. J Huang, SC Lin, A Nadershahi, SW Watts, and R Sarkar. Role of redox signaling and poly (adenosine diphosphate-ribose) polymerase activation in vascular smooth muscle cell growth inhibition by nitric oxide and peroxynitrite. *J Vasc Surg* 47: 599–607, 2008.

115. P Ghafourifar and C Richter. Nitric oxide synthase activity in mitochondria. *FEBS Lett* 418: 291–296, 1997.

116. C Giulivi, JJ Poderoso, and A Boveris. Production of nitric oxide by mitochondria. *J Biol Chem* 273: 11038–11043, 1998.

117. MO Lopez-Figueroa, C Caamano, MI Morano, LC Ronn, H Akil, and SJ Watson. Direct evidence of nitric oxide presence with mitochondria. *Biochem Biophys Res Commun* 272: 129–133, 2000.

118. NA Riobo, E Clementi, M Melani, A Boveris, E Cadenas, S Moncada, and JJ Poderoso. Nitric oxide inhibits mitochondrial NADH:ubiquinone reductase activity through peroxynitrite formation. *Biochem J* 359: 139–145, 2001.

119. L Schild, T Reinheckel, M Reiser, TFW Horn, G Wolf, and W Augustin. Nitric oxide produced in rat liver mitochondria causes oxidative stress and impairment of respiration after transient hypoxia. *FASEB J* 17: 2194–2201, 2003.

120. Z Han, YR Chen, CI Jones, G Meenakshisundaram, JL Zweier, and BR Alevriadou. Shear-induced reactive nitrogen species inhibit mitochondrial respiratory complex activities in cultured vascular endothelial cells. *Am J Physiol Cell Physiol* 292: C1103–C1112, 2007.

121. JJ Poderoso, C Lisdero, F Schopfer, N Riobo, MC Carreras, E Cadenas, and A Boveris. The regulation of mitochondrial oxygen uptake by redox reactions involving nitric oxide and ubiquinol. *J Biol Chem* 274: 37709–37716, 1999.

122. MD Bartberger, W Liu, E Ford, KM Miranda, C Switzer, JM Fukuto, PJ Farmer, DA Wink, and KN Houk. The reduction potential of nitric oxide (NO) and its importance to NO biochemistry. *Proc Natl Acad Sci USA* 99: 10958–10963, 2002.

123. M Sasaki, T Sato, A Ohler, B O'Rourke, and E Marban. Activation of mitochondrial ATP-dependent potassium channels by nitric oxide. *Circulation* 101: 439–445, 2000.

124. M Steffen, TM Sarkela, AA Gybina, TW Steele, NJ Trasseth, D Kuehl, and C Giulivi. Metabolism of S-nitrosoglutathione in intact mitochondria. *Biochem J.* 56: 395–402, 2001.

125. CC Dahm, K Moore, and MP Murphy. Persistent S-nitrosation of complex I and other mitochondrial membrane proteins by S-nitrosothiols but not nitric oxide or peroxynitrite: Implications for the interaction of nitric oxide with mitochondria. *J Biol Chem* 281: 10056–10065, 2006.

126. TM Sarkela, J Berthiaume, S Elfering, AA Gybina, and C Giulivi. The modulation of oxygen radical production by nitic oxide in mitochondria. *J Biol Chem* 276: 6945–6949, 2001.

127. KE Loke, PI McConnell, JM Tuzman, EG Shesely, CJ Smith, CJ Stackpole, CI Thompson, G Kaley, MS Wolin, and TH Hintze. Endogenous endothelial nitric oxide synthase–derived nitric oxide is a physiological regulator of myocardial oxygen consumption. *Circ Res* 84: 840–845, 1999.

128. J-N Trochu, J-B Bouhour, G Kaley, and TH Hintze. Role of endothelium-derived nitric oxide in the regulation of cardiac oxygen metabolism: Implication in health and disease. *Circ Res* 87: 1108–1117, 2000.

129. M Palacios-Callender, M Quintero, VS Hollis, RJ Springett, and S Moncada. Endogenous NO regulates superoxide production at low oxygen concentrations by modifying the redox state of cytochrome c oxidase. *Proc Natl Acad Sci USA* 101: 7630–7635, 2004.

130. W Li, T Jue, J Edwards, X Wang, and TH Hintze. Changes in NO bioavailability regulate cardiac O_2 consumption: Control by intramitochondrial SOD2 and intracellular myoglobin. *Am J Physiol Heart Circ Physiol* 286: H47–H54, 2004.

131. F Antunes, A Boveris, and E Cadenas. On the mechanism and biology of cytochrome oxidase inhibition by nitric oxide. *Proc Natl Acad Sci USA* 101: 16774–16779, 2004.

132. MC Barone, VM Darley-Usmar, and PS Brookes. Reversible inhibition of cytochrome c oxidase by peroxynitrite proceeds through ascorbate-dependent generation of nitric oxide. *J Biol Chem* 278: 27520–27524, 2003.

133. A Guidarelli and O. Cantoni. Pivotal role of superoxides generated in the mitochondrial respiratory chain in peroxynitrite-dependent activation of phospholipase A2. *Biochem J* 366: 307–314, 2002.

4 ROS and RNS Signaling in Catalysis of Heterolytic Reactions by Kinases, Phosphatases, and Other Enzymes

ROS and RNS signaling in the catalysis of heterolytic reactions by protein kinases and other enzymes is one of the most unexpected biological functions of superoxide ($O_2^{\cdot-}$), nitric oxide (NO), and reactive diamagnetic molecules originated from these physiological free radicals (first of all, hydrogen peroxide and peroxynitrite [$ONOO^-$]). Mechanisms of these catalytic processes will be discussed later (Chapter 7); however, it should be noted here that it was difficult to imagine earlier that free radicals, well-known damaging species, could participate in heterolytic reactions of hydrolysis, etherification, and others. Obviously, the most probable mechanisms of such processes should be electron transfer reactions mediated by ROS and RNS, which will be considered later. In this chapter, we will discuss the experimental findings of the role of ROS and RNS signaling in reactions catalyzed by protein kinases, protein phosphatases, and phospholipases.

4.1 PROTEIN KINASES

Protein kinases catalyze lipid hydrolysis and phosphorylation of various substrates and enzymes. Activation of these enzymes, for example protein kinase C (PKC), depends on the phosphorylation of the threonine and serine residues. Excellent reviews [1,2] were published some time ago that tapped into a huge literature on the role of reactive oxygen species (ROS) in enzymatic processes. At present, a number of experimental studies continue to be performed, but many questions on ROS and reactive nitrogen species (RNS) signaling remain unanswered. They will be discussed in the following text.

4.1.1 Protein Kinase C

The participation of ROS in the activation of PKC has been known for a long time. In 1989, Kaas et al. [3] reported that hepatocyte cytosolic PKC was activated by quinone-generated ROS, and this effect was due to the redox modification of the

Protein kinase C

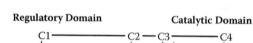
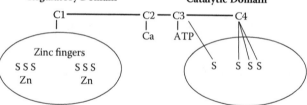

FIGURE 4.1   Scheme of protein kinase C.

thiol residues of the enzyme. Larsson and Cerutti [4] also showed that treatment of cells with low $O_2\cdot^-$ concentration, but not with hydrogen peroxide (H_2O_2), resulted in phosphorylation of cytoplasmic extracts containing PKC that probably involved the oxidation of sulfhydryl groups in the regulatory lipid-binding domain of PKC.

However, both $O_2\cdot^-$ and H_2O_2 are apparently able to activate different isoforms of PKC. Thus, Konishi et al. [5,6] have shown that H_2O_2 activated PKC through tyrosine phosphorylation. Klann et al. [7] demonstrated the activation of protein kinase by $O_2\cdot^-$. Knapp and Klann [8] suggested that $O_2\cdot^-$ is able to stimulate autonomous PKC activity by thiol oxidation and release of zinc from the cysteine-rich region of the enzyme. Korichneva et al. [9] proposed that PKC activation depends on the nature, the time of exposure, and "stringency" of oxidant molecules. Thus, a long duration of treatment with H_2O_2 led to the inactivation of PKC, while a short mild treatment induced PKC activation.

To explain different effects of various oxidants on the activation and inactivation of PKC, Gopalakrishna and Gundimeda [10] proposed that the effects of oxidants and antioxidants on PKC depended on a PKC domain under attack. Protein kinase has regulatory and catalytic domains containing thiol residues (Figure 4.1). It was suggested that oxidants activated PKC by the oxidation of thiol groups of the regulatory domain after releasing zinc, while "oxidized antioxidants" inhibited PKC by the interaction with the thiols of the catalytic domain.

It should be noted that the concept of "stringency" of oxidants developed by Korichneva et al. [9] or the role of "oxidized antioxidants" in PKC inhibition [10] are tentative hypotheses. It seems more proper to consider the chemical structures of reagents capable of oxidizing the thiol groups and releasing zinc. Zinc is not a transition metal and therefore cannot be oxidized. Only $O_2\cdot^-$ as an anion and a "supernucleophile" [11] is able to remove the Zn^{2+} cation with subsequent oxidation of SH groups. We venture to propose that the activation of PKC by prooxidants, for example H_2O_2 or tamoxifen [12], might be mediated by the $O_2\cdot^-$ formed during catalysis, whereas the other prooxidants or "oxidized antioxidants" unable to produce $O_2\cdot^-$ can only inhibit PKC.

In recent work, Gopalakrishna et al. [13] demonstrated that $O_2\cdot^-$ produced by the xanthine/xanthine oxidase system and $CoCl_2$, a transition metal redox catalyst, activated PKC. $CoCl_2$ competed with the redox-inert zinc in the zinc-thiolates of the PKC regulatory domain and induced the oxidation of thiol cysteine residues. Most of the aforementioned work was performed in cell-free systems. However, it is known that there is another pathway of PKC activation by $O_2^{\cdot-}$ in cells: through the inhibition of phosphatases and corresponding suppression of dephosphorylation of protein kinases. However, Hongpaisan et al. [14] demonstrated that direct PKC activation by $O_2^{\cdot-}$ also takes place in cells. It was found that although mitochondrial $O_2^{\cdot-}$ generation activated some protein kinases by the inhibition of phosphatase in hippocampal neurons, PKC activation was independent of phosphatase activity.

There are many examples of ROS signaling during PKC activation. Van Marwijk Kooy et al. [15] have shown that exposure of platelets to UVB radiation activated PKC by oxygen radicals causing exposure of fibrinogen-binding sites and subsequent platelet aggregation. PKC was found to be involved in the activation of NFκB by $O_2^{\cdot-}$ in human endothelial cells [16]. Kabir et al. [17] demonstrated that mitochondrial ROS generator-antimycin A can precondition the myocardium via PKCε activation. It was suggested that $O_2^{\cdot-}$ activated PKCε kinase through direct tyrosine phosphorylation. These findings support the conjecture that $O_2^{\cdot-}$ is able to activate PKC, although PKC can in its turn catalyze $O_2^{\cdot-}$ generation (see the following text). Wu et al. [18] suggested that ROS mediated the sustained activation of PKCα and extracellular signal-regulated kinase (ERK) for migration of human hepatoma cells Hepg2. Kuribayashi et al. [19] have shown that in human squamous carcinoma cells, $O_2^{\cdot-}$ activated PKCζ isoenzyme, which phosphorylated RhoGDI-1, the member of Rho family GTPases, and induced the liberation of RhoGTPases from RhoGDI-1, leading to their activation.

O'Brian and coworkers [20,21] have shown that PKC-regulatory oxidative modifications produced by physiological disulfides exhibited opposing effects on the activity of PKCδ and PKCε isoforms. For example, glutathione inactivated the oncogenic isozyme PKCε via S-glutathiolation, while PKCδ, a proapoptotic isozyme, exhibited its resistance to inactivation.

ROS signaling is an important factor of PKC/PKD cascade, a protective mechanism against cell death. Protein kinase D (PKD) is a serine protein kinase with structural properties distinct from those of all PKC isoforms. This enzyme is activated within cells by ROS via a PKC-dependent signal transduction pathway. In 2000 Waldron and Rozengurt [22] showed that the treatment of intact cells with H_2O_2 activated PKD via a PKC-dependent signal transduction pathway. They suggested that H_2O_2 penetrated the plasma membrane and induced the activation of tyrosine kinases Src and PLC. Then H_2O_2-induced signals mediated by both Src and PLC led to PKC-dependent PKD activation. Later on, Waldron et al. [23] demonstrated the critical role of activation loop Ser744 and Ser748 phosphorylation in ROS-induced PKD activation (Figure 4.2A). These results also supported the role of tyrosine kinase Src in oxidative stress-induced activation of the PKC/PKD kinase cascade.

Wang et al. [24] and Storz et al. [25] studied the effects of the $O_2^{\cdot-}$ generator menadione on AP-1-induced hepatocyte death. They proposed an alternative mechanism of PKC/PKD cascade through the ROS activation of PKD by two coordinated signaling events, the first of which is Tyr463 phosphorylation, which is mediated by the Src-Abl signaling pathway; the second step is phosphorylation of the activation loop Ser738/Ser742, which is mediated by the Src-PKCδ pathway (Figure 4.2B). Resistance to death from $O_2^{\cdot-}$ required both PKC/PKD and ERK1/2 activation, in

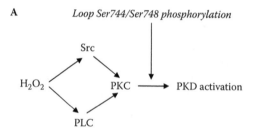

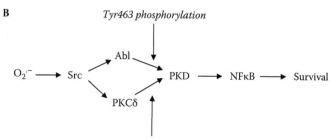

FIGURE 4.2 PKC–PKD cascade. (A) Mechanism of ROS signaling in PKC/PKD cascade according to RT Waldron and E Rozengurt. *J Biol Chem.* 275: 17114–17121, 2000 and RT Waldron, O Rey, E Zhukova, E Rozengurt. *J Biol Chem.* 279: 27482–27493, 2004. (B) Mechanism of ROS signaling in PKC/PKD cascade according to RT Waldron, O Rey, E Zhukova, E Rozengurt. *J Biol Chem.* 279: 27482–27493, 2004 and P Storz, H Doppler, A Toker. *Mol Cell Biol* 24: 2614–2626, 2004. Src: a member of the family of proto-oncogenic tyrosine kinases; Abl: tyrosine kinase.

order to downregulate proapoptotic JNK/c-Jun signaling. In recent work, Doppler and Storz [26] identified Tyr95, a previously undescribed phosphorylation site in PKD1 that is regulated by ROS and results in the PKD1 activation loop phosphorylation and activation.

At the same time, numerous findings demonstrate the formation of $O_2{\cdot}^-$ in processes catalyzed by PKC [27-41]. For example, Ohara et al. [30] have shown that lysophosphatidylcholine increased $O_2{\cdot}^-$ production through PKC activation in rabbit aortas. Li et al. [31] found that PKCα, and not PKCβI or PKCβII, is required for $O_2{\cdot}^-$ production and maximal oxidation of LDL by activated human monocytes. Armstead and Mayhan [32] have shown that PKC activation increased $O_2{\cdot}^-$ production observed after fluid percussion brain injury (FPI). These data also suggest that $O_2{\cdot}^-$ formation links PKC activation to impaired KATP channel function after FPI. Ungvari et al. [33] demonstrated that high pressure itself can enhance arterial $O_2{\cdot}^-$ production by activating directly a PKC-dependent NADPH oxidase (Nox). Siow et al. [34] demonstrated that homocysteine stimulated $O_2{\cdot}^-$ production in monocytes through the activation of Nox, where PKCβ played an important role in phosphorylation of p47[phox] and p67[phox] units of Nox. Zhou et al. [35] found that PKCβ was responsible for the enhanced $O_2{\cdot}^-$ production in diabetic vessels from streptozotocin-induced diabetic rats. $O_2{\cdot}^-$ overproduction resulted in the impairment of arachidonic-acid-mediated dilation in small coronary arteries. Li et al. [36] suggested that PKCα, but not PKCβ, is required for the phosphorylation of cPLA(2) protein upon activation of human monocytes. Both PKCα and cPLA(2) participated in $O_2{\cdot}^-$ production through the generation of arachidonic acid (AA).

Kim et al. [37] studied the mechanism of cell death in response to glucose depletion (GD), a common characteristic of the tumor microenvironment. It was found that ROS stimulated GD-induced necrosis in lung carcinoma cells. Inhibition of ROS production prevented necrosis and switched the cell death mode to apoptosis. PKC-dependent extracellular regulated kinase 1/2 (ERK1/2) activation also switched GD-induced necrosis to apoptosis through the inhibition of ROS production, supposedly by induction of MnSOD. Korchak et al. [38] demonstrated that both PKCα and PKCβ activated $O_2{\cdot}^-$ production in phagocytes but played different roles in calcium signaling for phagocytic responses. Miller et al. [39] found that Nox-produced $O_2{\cdot}^-$ mediated paraquat cytotoxicity in BV-2 microglial cells through PKCδ- and ERK-dependent pathways. It is of interest that there is a gender dependence in the impairment of endothelium-dependent vasodilation after acute exposure to high glucose in rat aorta, which stimulated $O_2{\cdot}^-$ overproduction, owing to differences in PKCβ2 expression [40]. Xia et al. [41] demonstrated that ROS generated by Nox together with PKCβ1 and PKC-zeta play important roles in high glucose-stimulated VEGF expression and secretion by mesangial cells.

4.1.2 PROTEIN KINASE AKT/PROTEIN KINASE B

Another protein kinase, which is activated by ROS, is serine/threonine kinase Akt/protein kinase B. This enzyme plays a key role in many cellular processes, including cell survival and protein synthesis. It has been shown [42] that H_2O_2 is able to activate Akt/PBK. Ushio-Fukai et al. [43] have shown that both angiotensin II (Ang II) and exogenous H_2O_2 induced rapid phosphorylation of Akt/PKB, suggesting that Ang II-induced Akt/PKB activation was mediated by ROS. Akt/PKB stimulation by Ang II and H_2O_2 stimulation were abrogated by the phosphatidylinositol 3-kinase (PI3-K) inhibitors, indicating that PI3-K kinase is another upstream mediator of Akt/PKB activation in vascular smooth muscle cells (VSMCs). Shi et al. [44] found that ROS activated Akt phosphorylation in hepatoma cells through the PIK/Akt pathway. Akt/PKB activation by H_2O_2 also depended on the phosphorylation of epidermal growth factor receptor (EGFR) [45].

However, ROS can apparently activate Akt/PKB independently of PI3-K kinase. Gorin et al. [46] have shown that Ang II caused rapid activation of Akt/PKB but delayed activation of phosphoinositide 3-kinase (PI3-K) in glomerular mesangial cells. Thus, the activation of Akt/PKB by Ang II indicates that in this case PI3-K is not an upstream mediator of Akt/PKB activation. Furthermore, AA mimicked the effect of Ang II, while the antioxidants N-acetylcysteine and diphenylene iodonium inhibited both AA- and Ang II-induced Akt/PKB activation. These findings demonstrate that Ang II induces protein synthesis and hypertrophy in glomerular mesangial cells through the AA/redox-dependent pathway and Akt/PKB activation independent of PI3-K. In subsequent work [47], these authors studied the mechanism of $O_2{\cdot}^-$-mediated Akt/PKB activation by Ang II. They demonstrated that Ang II

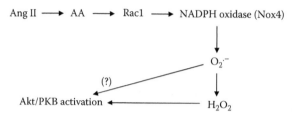

Note: Gorin et al. consider Akt/PKB activation only by hydrogen peroxide, but superoxide itself can be another activation agent.

FIGURE 4.3   Mechanism of Akt/PKB activation by Ang II and arachidonic acid [46].

rapidly activated Rac1 and that the activation was suppressed by phospholipase A2 and mimicked by AA. The mechanism of AA activation of a Rac1-regulated Nox and subsequent ROS-mediated Akt/PKB activation is presented in Figure 4.3.

Hingtgen et al. [48] also suggested that the $O_2{}^{\cdot-}$ generated by an Nox2-containing Nox was a major mediator of Ang II-induced Akt activation and cardiomyocyte hypertrophy, and that dysregulation of this signaling cascade may play an important role in cardiac hypertrophy. Li and Malik [49] have shown that Ang II-induced Akt/PKB activation in VSMC was mediated by AA metabolites formed by lipoxygenase and cytochrome P-450 and was independent of H_2O_2 production. Lim and Clement [50] demonstrated an important role of PTEN (phosphatase and tensin homolog) phosphatase in Akt activation. It was found that exposure of serum-derived mouse embryonic fibroblasts to fetal bovine serum led to a rapid and strong phosphorylation of Akt that depended on the ascorbate-reversible $O_2{}^{\cdot-}$-mediated oxidation of PTEN. These findings support the role of $O_2{}^{\cdot-}$ as a physiologically relevant second messenger for Akt activation through S-nitrosylation of PTEN.

There is a mutual dependence of Akt and Nox in phagocyte and nonphagocyte cells. Akt is able to activate Nox-dependent generation of $O_2{}^{\cdot-}$ in leukocytes by phosphorylation of Ser304 and Ser328 of p47^{phox} [51]. On the other hand, Nox, which is a major source of $O_2{}^{\cdot-}$ in endothelial cells and cardiomyocytes, stimulates myocardial ischemia-reperfusion (I/R) and angiogenesis via mitogen-activated protein kinases (MAPKs) and Akt/protein kinase B pathways [52].

Wang et al. [53] have shown that macrophage colony-stimulating factor (M-CSF)-stimulated ROS production induced Erk, Akt, and p38 activation during macrophage/monocyte survival. Interestingly, NAC and DPI decreased cell survival, and Akt1 and p38 MAPK phosphorylation. These findings demonstrate that ROS generated by the Nox complex contribute to monocyte/macrophage survival induced by M-CSF via regulation of Akt and p38 MAPK. Li et al. [54] reported that $ONOO^-$ activated Nrf2 (a nuclear erythroid 2 p45-related redox-sensitive leucine zipper transition factor) through PI3K/Akt signaling, which led to upregulation of heme oxygenase-1 (HO-1) expression.

These findings demonstrated the activation of Akt/PKB by H_2O_2 through PI3K-dependent and independent pathways. However, such a mechanism is realized only with small amounts of H_2O_2. In contrast, elevated toxic concentrations of H_2O_2 caused the downregulation of Akt by dephosphorylation and caspase-3-independent proteolysis [55]. In recent work, Kou et al. [56] found that exogenous xanthine oxidase increased cellular ROS production and caused $O_2\cdot^-$-dependent inhibition of Akt phosphorylation and enhancement of p38 MAPK phosphorylation in human umbilical vein cells.

Akt/PKB is a survival kinase, but its survival effect can be the origin of some deadly pathologies such as cancer. Thus, Govindarajan et al. [57] recently demonstrated that Akt was associated with malignant transformation of melanoma through the stabilization of cells with extensive mitochondrial DNA mutation, which can generate ROS or the expression of $O_2\cdot^-$ generating Nox Nox4. Thus, Akt kinase can serve as a molecular switch that increases angiogenesis and the generation of $O_2\cdot^-$, fostering more aggressive tumor behavior. Ichihara et al. [58] have shown that doxorubicin-induced cardiotoxicity in mice depended on Akt activation, which was inhibited by PEG-SOD.

4.1.3   MAP Kinases

Mitogen-activated protein (MAP) kinases are serine/threonine protein kinases that respond to mitogen and regulate various cellular activities, including mitosis, differentiation, apoptosis, etc. There are several distinct groups of MAPKs that have been characterized in mammals, including extracellular signal-regulated kinases ERK 1/2 activated by growth factors and tumor promoters; c-Jun N-terminal kinases (JNKs), also known as stress-activated protein kinases (SAPKs); and p38 kinases, which are responsive to stress stimuli such as cytokines, UV irradiation, heat shock, etc. However, the most interesting properties of these enzymes (at least, for this chapter) are signaling by ROS and RNS, which are the important mediators of MAPK activity. As was already noted earlier, the discovery of ROS and RNS signaling in the heterolytic processes (hydrolysis and etherification) catalyzed by kinases and phosphatases was previously completely unexpected and is an important peculiarity of these enzymatic reactions. Now, we consider ROS and RNS signaling in processes catalyzed by major MAPKs.

4.1.3.1   Extracellular Signal-Regulated Kinases ERK1/2

It was shown several years ago that ERK1/2 kinases are surviving mediators, which are activated by ROS. As usual, two ROS, namely, $O_2\cdot^-$ and H_2O_2, are major signaling compounds in the activation of ERK and other MAP kinases. Although several works suggested that H_2O_2 could be a major signaling ROS in MAPK-catalyzed processes, an important work by Baas and Berk [59] demonstrated that, at least in VSMCs, $O_2\cdot^-$ and not H_2O_2 mediated ERK1/2 activation. Importantly, $O_2\cdot^-$ stimulated the ERK activation in the early stage of VSMC growth when phosphatase expression was absent. Therefore, the activation of ERK1/2 cannot be explained by phosphatase inhibition.

Bhunia et al. [60] also demonstrated the importance of $O_2\cdot^-$ signaling in the activation of ERK MAPK. They found that lactosylceramide (LacCer) stimulated endogenous $O_2\cdot^-$ production in human aortic smooth muscle cells through the activation of Nox. This process was inhibited by diphenylene iodonium (DPI), and by antioxidants, N-acetyl-cysteine (NAC) or pyrrolidine dithiocarbamate. It was also suggested that $O_2\cdot^-$ stimulated ERK (p44) activation through $p21^{rasz}$GTP (Figure 4.4). Janssen-Heininger

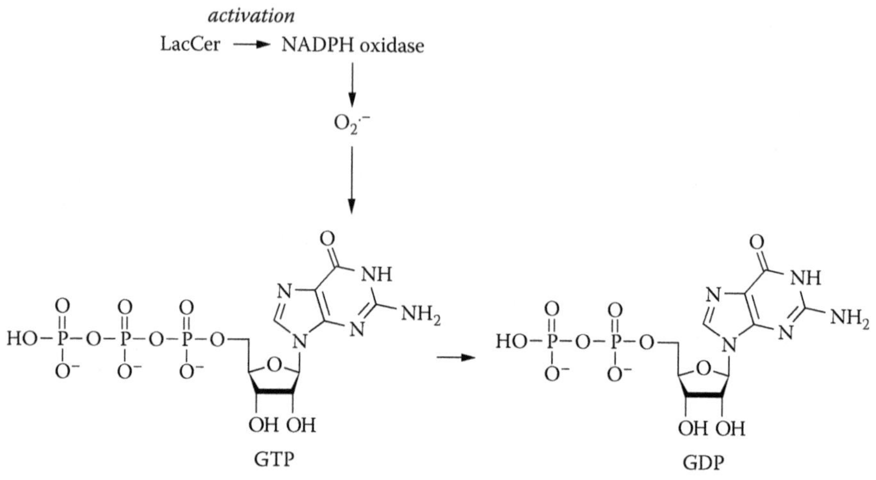

$$O_2^{\cdot-} + G(phos)(phos)(phosOH) \longrightarrow G(phos)(phosOH) + {}^-PO_3$$

$$G(phos)(phosOH) \longrightarrow ERK1/2 \text{ activation} \qquad\qquad [60]$$

FIGURE 4.4 Superoxide signaling in ERK1/2 activation through p21^{rasz}GTP.

et al. [61] also showed that the ROS and RNS activation of ERK and JNK kinases in rat lung epithelial cells depended on Ras (a regulatory GTPhydrolase). Svegliati et al. [62] found that platelet-derived growth factor (PDGF) and ROS regulated Ras protein levels in primary human fibroblasts through ERK1/2 activation.

Small GTPase Rac is able to stimulate ERK activation in keratinocytes. Rygiel et al. [63] found that Tiam1, an activator of the small GTPase Rac, influenced Nox-mediated ROS production, which regulated ERK phosphorylation and the susceptibility of keratinocytes to apoptotic signaling.

Several studies demonstrated H_2O_2 signaling in ERK activation. In 1996, Guyton et al. [64] showed that H_2O_2 activated ERK in various cell types although its activation rapidly changed to ERK inactivation. These authors suggested that the hydrogen-peroxide-stimulated activation of ERK might play a critical role in cell survival following oxidant injury. Goldstone and Hunt [65] suggested that redox regulation was important for the ERK pathway during lymphocyte activation. Gurjar et al. [66] have shown that interleukin-1β (IL-1β) stimulated biphasic ERK activation in VSMC cells: The transient activation that reached a maximum at 15 min and declined to baseline levels within 1 h, and a second phase of sustained ERK activation lasting up to 8 h. Treatment of VSMC with the $O_2^{\cdot-}$ scavenger N-acetyl-L-cysteine (NAC) selectively inhibited the sustained phase of ERK activation, suggesting an important role for $O_2^{\cdot-}$ mediation. Madamanchi et al. [67] found that ROS activated ERK1/2 and p38 kinases in SOD-deficient mouse aortic smooth muscle cells. They suggested that the enhancement of ERK and p38 activation was apparently due to an increase in $O_2^{\cdot-}$ formation.

Yu et al. [68] demonstrated that the inhibition of high-glucose-level-induced and Ang II-induced activation of ERK1/2 MAP kinase by 3-hydroxy-3-methylglutaryl CoA reductase inhibitor (statin) in cultured human mesangial cells originated from inhibition of Nox activity. Exposure of the cells to high glucose levels or Ang II significantly increased ERK activity that was completely blocked by treatment with pitavastatin as well as with Nox inhibitor diphenylene iodonium. It was concluded that pitavastatin attenuated high-glucose-induced and Ang II-induced ERK activity in mesangial cells through inhibition of $O_2\cdot^-$ generation. Ortego et al. [69] also showed that $O_2\cdot^-$ increased ERK1/2 and p38 activation and phosphorylation in monocytes, which were reduced by statins. Xie et al. [70] demonstrated that ERK (p42/44 MAPK) is a critical component of the Ang-stimulated $O_2\cdot^-$-sensitive signaling pathway in cardiac microvascular endothelial cells, which plays a key role in the regulation of osteopontin gene expression.

Greene et al. [71] have shown that the phosphorylation of cytosolic and nuclear ERK1/2 in VSMCs in response to bradykinin was mediated by Nox-produced ROS, which were considered important mediators in the signal transduction pathway through which bradykinin promoted VSMC proliferation during vascular injury. Nishiyama et al. [72] suggested that ROS, generated by Nox, activated ERK1/2 kinase and contributed to the progression of renal injury in Dahl salt-sensitive hypertensive rats. ROS-induced activation of ERK1/2 and JNKs, but not p38 kinase, was involved in the induction of osteopontin gene expression by Ang II and interleukin-1β in adult rat cardiac fibroblasts [73]. Gong et al. [74] suggested that mitochondrial ATP-sensitive potassium channel (mKATP) may mediate ERK1/2 activation during anoxia preconditioning (APC) by generating ROS, which then triggers the delayed protection of APC in rat cardiomyocytes.

Mukhin et al. [75] studied the effect of serotonin and 5-hydroxytryptamine (5-HT)(1A) receptor) on ERK1/2 activation through a G(i)βγ-mediated pathway. It was concluded that ERK activation was mediated by ROS because it was inhibited by antioxidants and stimulated by cysteine-reactive oxidants. Furthermore, 5-HT(1A) receptor generated both $O_2\cdot^-$ and H_2O_2. It was also suggested that Nox produced $O_2\cdot^-$ after stimulation with the transfected 5-HT(1A) receptor in Chinese hamster ovary cells. Later on, Simon et al. [76] reported that 5-serotonin stimulated c-fos and cyclin D1 expression through a ROS-dependent mechanism that requires Ras, Rac1, and ERK MAPK.

Gao et al. [77] studied the release of vasoactive factors by perivascular adipose tissue (PVAT) in response to perivascular nerve activation by electrical field stimulation (EFS). In Wistar-Kyoto rats, rings of superior mesenteric artery (MA) with intact PVAT (PVAT+) showed a greater contractile response to EFS than rings with PVAT removed (PVAT−). SOD reduced the contractile response to EFS more in PVAT+ than in PVAT−. Exogenous $O_2\cdot^-$ enhanced the contractile response to EFS and to phenylephrine in PVAT−, and this augmentation was suppressed by inhibition of tyrosine kinase and ERK MAPK. These findings suggest that PVAT enhances the arterial contractile response to perivascular nerve stimulation through the production of $O_2\cdot^-$ mediated by Nox and the activation of tyrosine kinase and ERK1/2 pathway.

Wang et al. [78] studied the role of vascular endothelial growth factor (VEGF) released by osteoblasts during bone formation. It was found that shock wave (SW) elevation of

VEGF-A expression in human osteoblasts was mediated by Ras-induced $O_2^{\cdot-}$ and ERK activation. It was concluded that $O_2^{\cdot-}$ mediated the SW-induced ERK activation.

Important data concerning different signaling pathways of $O_2^{\cdot-}$ and H_2O_2 in ERK activation have been obtained by Milligan et al. [79]. They found that the ability of H_2O_2 to activate ERK1/2 in mesothelial cells was similar to that found with tumor necrosis factor (TNF) stimulation. The activation was inhibited by various reactive oxygen scavengers. Superoxide also stimulated ERK1/2 activity, but in contrast to H_2O_2, it did not stimulate IκB-α proteolysis and just slightly induced NFκB nuclear translocation. These results suggest that $O_2^{\cdot-}$ and H_2O_2 stimulate ERK activation by different signaling pathways.

Superoxide and H_2O_2 signaling pathways also differ in normal hepatocytes [80]. Menadione-produced $O_2^{\cdot-}$ induced JNK phosphorylation, caspase-9, -6, -3 activation, PARP cleavage, and apoptosis. $O_2^{\cdot-}$-induced apoptosis was dependent on JNK activity. Menadione also induced the phosphorylation of ERK1/2 and thereby attenuated cell death. In contrast, H_2O_2 increased necrotic cell death and did not activate MAPKs signaling. Thus, only $O_2^{\cdot-}$, and not H_2O_2, was able to mediate ERK and JNK activation. Yamakawa et al. [81] found that lysophosphatidylcholine (lysoPC)-induced activation of ERK1/2 was mediated by H_2O_2 and $O_2^{\cdot-}$ generated by NADH/NADPH oxidase in VSMCs. Kefaloyianni et al. [82] also showed that H_2O_2 induced strong activation of ERK, JNKs, and p38-MAPK in skeletal myoblasts. ERK and JNK activation by H_2O_2, but not p38-MAPK activation, was mediated by Src kinase and, at least in part, by EGFR.

Lee et al. [83] also demonstrated that $O_2^{\cdot-}$ was an important signaling species in ERK activation by serotonin. They found that serotonin rapidly elevated phosphorylation and activation of extracellular signal-regulated kinases (ERK1 and ERK2) from smooth muscle cells and that the enhanced phosphorylation was blocked by the antioxidants Tiron, N-acetyl-L-cysteine (NAC), and Ginkgo biloba extract. Therefore, it was concluded that $O_2^{\cdot-}$ signaling was important in ERK1 and ERK2 phosphorylation and activation. In a subsequent work [84], these authors suggested that 5-hydroxytryptamine (another $O_2^{\cdot-}$ producer) activated ERK in bovine pulmonary artery smooth muscle cells, preferably through H_2O_2 formed by $O_2^{\cdot-}$ dismutation.

Devadas et al. [85] have shown that T cell receptor (TCR) induced rapid generation of both H_2O_2 and $O_2^{\cdot-}$. Furthermore, their findings suggested that $O_2^{\cdot-}$ and H_2O_2 were produced separately by distinct TCR-stimulated pathways: TCR-stimulated activation of the Fas ligand (FasL) promoter and subsequent cell death depended on $O_2^{\cdot-}$ production, while nuclear factors of activated T cells (NFAT)-induced activation or interleukin 2 transcription were independent of all ROS. Anti-CD3-induced phosphorylation of ERK1/2 required H_2O_2 generation but was unaffected by $O_2^{\cdot-}$. Thus, $O_2^{\cdot-}$ and H_2O_2 regulate two distinct redox pathways, a proapoptotic (FasL) and a proliferative pathway (ERK).

Rhyu et al. [86] have shown that the transforming growth factor TGF-β1–induced ROS production activated ERK directly or through the Smad pathway in proximal tubular epithelial cells. Lin et al. [87] demonstrated that TGF-β1, fibronectin expression, Ras, ERK, p38, and c-Jun activation were enhanced in glomerular mesangial cells cultured in high-glucose or advanced glycation end products (AGE). Superoxide, and not NO or H_2O_2, mediated the effects of AGE and high glucose.

Locher et al. [88] investigated the mechanism of native low-density lipoprotein (LDL)-stimulated proliferation of human VSMC. It was found that native LDL induced rapid formation of $O_2 \cdot^-$ (inhibited by the antioxidants N-acetylcysteine, Tiron, nordihydroguaiaretic acid, diphenylene iodonium, or superoxide dismutase) and subsequent activation of ERK 1/2 signaling pathways. Wang et al. [89] have shown that extracorporeal shock wave (ESW) (an alternative noninvasive method for the promotion of bone growth and tendon repair) enhanced early $O_2 \cdot^-$ production, which induced tyrosine kinase-mediated ERK activation, resulting in osteoprogenitor cell growth and maturation into bone nodules. Other promoters of ROS-induced ERK activation were peroxisome-proliferator-activated receptor-gamma (PPARγ) agonists. Huang et al. [90] proposed a novel mechanism, independent of Ras activation, by which $O_2 \cdot^-$ production initiated PPARγ-dependent activation of the Raf-MEK-ERK1/2 signaling pathway. Susa and Wakabayashi [91] have shown that extracellular alkalosis induced phosphorylation of ERK and enhanced serum-induced ERK phosphorylation in cultured rat aortic smooth muscle cells. ERK activation was inhibited by superoxide dismutase; 4,5-dihydroxy-1,3-benzene-disulfonic acid, a cell-permeable antioxidant; and diphenyliodonium, a Nox inhibitor. These findings suggested that extracellular alkalosis-ERK activation was initiated by ROS produced by Nox.

As was noted earlier, ERK is generally considered to be a survival mediator participating in protection by growth factors against cell death. Nonetheless, cell death can apparently also be mediated by ERK. Thus, Zhang et al. [92] have shown that hyperoxia stimulated Nox-production of reactive oxygen species, which mediated cell death of lung epithelium through ERK activation and the release of cytochrome c with subsequent activation of caspases 9 and 3.

Similarly, Kulich and Chu [93] suggested that sustained ERK activation contributed to toxicity of 6-hydroxydopamine (6-OHDA) in neuronal cells. They found that antioxidative compounds catalase and metalloporphyrin, but not SOD, protected against 6-OHDA and inhibited the development of sustained ERK phosphorylation in these cells. Therefore, ROS-stimulated ERK signaling cascades may contribute to neuronal toxicity. Badrian et al. [94] found that exposure of myocytes to the $O_2 \cdot^-$ generator menadione resulted in significantly higher death of MEK-ERK–expressing myocytes. These findings show that MEK-ERK signaling could increase or decrease cell survival, depending on the stress stimulus applied.

Ang II is another stimulator of ERK activation. Sano et al. [95] found that Ang II stimulated ROS production in cardiac fibroblasts via the AT1 receptor and Nox, and that ROS activated ERK, p38 MAPK, and JNK, which were significantly inhibited by antioxidants. Similarly, Frank et al. [96] showed the involvement of ROS in the activation of tyrosine kinase and ERK by Ang II in VSMCs. In contrast, Kyaw et al. [97] found that although Ang II stimulated rapid and significant activation of ERK1/2, JNK, and p38 MAPK in cultured rat aortic smooth muscle cells, ERK1/2 activation was not affected by antioxidants studied and was probably not mediated by ROS. However, later on, Laplante et al. [98] showed that the effect of Ang II on blood pressure was mediated by ERK-MAPK activation and enhanced $O_2 \cdot^-$ production through the activation of AT1 receptors. These findings suggest an interactive mechanistic relationship between $O_2{\cdot}^-$ production and ERK activation in the development of Ang II-indiced hypertension.

Papparella et al. [99] investigated the role of $O_2{\cdot}^-$ and Nox in Ang II-stimulated ERK1/2 activation in fibroblasts from hypertensive patients. Ang II increased intracellular $O_2{\cdot}^-$ production and ERK1/2 phosphorylation in hypertensive patients (HT) in comparison to normotensive controls (NT). These findings suggested that Ang II-induced ERK1/2 activation in hypertension included an exaggerated response of p47phox, Nox, and $O_2{\cdot}^-$. Recently, Ding et al. [100] confirmed an association between ERK1/2-signaling pathway, the generation of $O_2{\cdot}^-$, and spontaneous tone in isolated aorta from Ang II-infused hypertensive rats. It was showed that Ang II infusion induced the production of $O_2{\cdot}^-$ and spontaneous tone and that both effects depended on ERK-MAPK activation. Li et al. [101] studied the effect of Ang II in aortae from p47$^{phox-/-}$ mice. They showed that p47^{phox} was an important factor of Ang II-induced Nox activation and $O_2{\cdot}^-$ production. There was significant pre-activation of ERK1/2, p38, and JNK in p47^{phox} vessels, while the enhanced ERK1/2 phosphorylation depended on increased $O_2{\cdot}^-$ production, which was reduced by $O_2{\cdot}^-$ scavenger tiron.

Ross and Armstead [102] suggested that activation of PTK and ERK MAPK by $O_2{\cdot}^-$ produced the impairment of KATP and KCa channel-mediated vasodilation. Later on, Philip and Armstead [103] studied the effect of XO-generated $O_2{\cdot}^-$ on N-methyl-D-aspartate (NMDA)-induced pial artery dilation in piglets equipped with a closed window. Exposure of the cerebral cortex to xanthine oxidase reversed NMDA dilation to vasoconstriction, and this impairment was partially prevented by the inhibition of ERK and p38 kinases.

Oeckler et al. [104] examined the effects of passive stretch in endothelium-removed bovine coronary artery on ROS generation and the activation of MAPKs. Passive stretch increased $O_2{\cdot}^-$ production and ERK phosphorylation, which was attenuated by inhibitors of tyrosine kinases src, the epidermal growth factor receptor (EGFR), and Nox. H_2O_2 also stimulated contraction through EGFR phosphorylation and ERK. Thus, stretch apparently enhanced force generation via ERK signaling through an EGFR/src-dependent mechanism activated by ROS derived from a stretch-mediated activation of Nox (Figure 4.5). Cyclic strain also stimulated ROS formation and rapid phosphorylation and activation of ERK1/2, JNK1/2, and p38 MAPKs in smooth muscle cells [105]. However, while p38 inhibitors blocked cyclic strain-induced cell alignment, the inhibitors of ERK1/2 and JNK did not. Therefore, only p38 MAPK is apparently a critical component of the ROS-sensitive signaling pathway that is playing a crucial role in vascular alignment induced by cyclic stain.

Guest et al. [106] demonstrated that rat aortic smooth muscle (RASM) cells subjected to cyclic strain rapidly activated both ERK1/2(MAPK) and p38(MAPK). Cyclic strain also increased production of $O_2{\cdot}^-$ via an Nox-dependent mechanism. Superoxide apparently mediated ERK activation, which was inhibited by the antioxidant N-acetylcysteine, while this antioxidant had no effect on p38 activation.

Czaja et al. [107] have shown that the treatment of hepatocytes with nontoxic menadione concentrations resulted in brief ERK and JNK MAPK activation, while the treatment with toxic menadione concentrations induced a prolonged activation

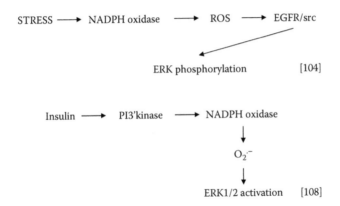

FIGURE 4.5   ERK activation by stress and insulin-stimulated superoxide formation.

of both ERK and JNK. Inhibition of ERK induced hepatocyte death from previously nontoxic menadione concentrations in association with sustained JNK activation. ERK activation was initiated by both $O_2\cdot^-$ producer menadione and H_2O_2, but by different pathways.

Ceolotto et al. [108] demonstrated that insulin is a promoter of Nox-dependent $O_2\cdot^-$ generation in human fibroblasts that involved activation of PI3'-kinase and stimulated ERK1/2 activation (Figure 4.5). This effect of insulin may contribute to the pathogenesis and progression of cardiovascular disease.

Gelain et al. [109] have shown that in Sertoli cells, retinol induced the Src-dependent activation of ERK1/2 and the ERK1/2-mediated phosphorylation of the transcription factor CREB. The effect of retinol was mediated by ROS because it was completely suppressed by the antioxidants Trolox, mannitol, and SOD. Moon et al. [110] demonstrated that overexpression of the disialoganglioside synthase (CD3) gene suppressed cell proliferation, cell cycle progression, and MMP-9 expression in VSMCs through the generation of ROS. Superoxide and H_2O_2 were generated at increased levels in GD3 synthase gene transfectants in comparison with empty vector (EV)-transfected VSMC. This phenomenon was blocked by antioxidants such as N-acetyl-L-cysteine (NAC) and pyrrolidine dithiocarbamate (PDTC). It was also found that a blockade of ROS with antioxidants reversed ERK phosphorylation.

Hasan and Schafer [111] investigated the role of ROS in the regulation of Egr-1, the regulator of cell proliferation and apoptosis, by free heme released from hemoglobin in the vasculature during cardiovascular pathologies. They found that hemin (ferriprotoporphyrin IX, the oxidized form of heme) increased Egr-1 expression and ERK1/2 phosphorylation, and the antioxidant N-acetyl cysteine, the Nox inhibitors apocynin and diphenyleneiodonium chloride, and the $O_2\cdot^-$ scavenger tiron inhibited both ERK1/2 activation and Egr-1 expression in VSMCs. Wakade et al. [112] found that the neuroprotector and anticancer drug tamoxifen significantly suppressed $O_2\cdot^-$ production, reduced oxidative protein/DNA damage and caspase-3 activation, and decreased the activation of extracellular signal regulated kinases (ERKs) in the ipsilateral cortex of tamoxifen-treated rats following permanent middle cerebral artery occlusion. Furthermore, it was found that $O_2 \cdot^-$ was responsible, at least in part, for early ERK activation following cerebral ischemia.

Navarro et al. [113] have shown that doxorubicin (DOX) increased phosphorylation of enzymes comprising the MAP kinase cascades in primary hepatocyte cultures. Thus, ERK was phosphorylated by the DOX treatment, while the cell-permeable $O_2 \cdot^-$ dismutase mimetic MnTBAP and the flavin-containing enzyme inhibitor diphenyleneiodonium reverted DOX-induced effects. These findings suggest that $O_2 \cdot^-$, probably generated by DOX, is responsible for the MAP kinase cascade activation.

Finlay et al. [114] have studied $O_2 \cdot^-$ signaling in rat smooth muscle cells expressing tumor suppressor tuberin or in tuberin-deficient cells. Tuberin-expressing cells responded to the platelet-derived growth factor (PDGF) stimulation by activating the classic mitogen-activated extracellular signal-regulated kinase kinase (MEK)-1-dependent phosphorylation of ERK MAP kinase. In contrast, in tuberin-deficient cells, PDGF stimulation resulted in MEK-1–independent p42/44 MAPK (ERK1/2) phosphorylation via $O_2 \cdot^-$ mediation. It has previously been shown that ERK1/2 promotes TPA-induced myeloid cell differentiation. Now, Traore et al. [115] suggested that TPA-stimulated ROS generation activated ERK1/2 via a redox-mediated inhibition of ERK1/2-directed phosphatase in cells.

Tephly and Carter [116] found that $O_2 \cdot^-$ production by alveolar macrophages depended on both Nox activity and the activity of the mitochondrial respiratory chain. They also suggested that in alveolar macrophages, $O_2 \cdot^-$ mediated chemokine expression after TNF-α stimulation in an ERK-dependent manner. Datla et al. [117] demonstrated that the overexpression of Nox4 NADPH oxidase enhanced receptor tyrosine kinase phosphorylation and ERK activation in endothelial cells. Inhibition of the ERK pathway also reduced the endothelial angiogenic responses. Thus, ROS production by Nox4-type NADPH oxidase stimulated endothelial angiogenic responses, at least partly, through the enhanced activation of receptor tyrosine kinases and the downstream ERK pathway.

Inaba et al. [118] have shown that neutrophils from rheumatoid arthritis patients exhibited increased spontaneous $O_2 \cdot^-$ release accompanied by increased basal phosphorylation of ERK and p38 kinases, accelerated spontaneous apoptosis, and enhanced $O_2 \cdot^-$ release in response to FMLP as compared to healthy normal neutrophils.

The small GTPase p21 Ras plays a central role in the control of cell survival and apoptosis. Sabbatini et al. [119] studied the effects of Ras/ERK1/2 signaling inhibition on oxidative damage in cultured renal and endothelial cells and on renal ischemia-reperfusion injury in the rat. Primary human renal tubular and human endothelial cells underwent cell death when subjected to oxidative stress, which induced ERK1/2 and phosphoinositide 3-kinase (PI3-kinase) activation. Ras/ERK1/2 signaling significantly reduced acute postischemic renal injury.

It has been shown that extracellular superoxide dismutase (EC-SOD) expression decreased in myocardial infarction (MI)-induced failing heart. Van Deel et al. [120] demonstrated that mice with EC-SOD gene deficiency (EC-SOD KO) exhibited an increase in the phosphorylation of p38, ERK, and c-Jun kinases. These findings suggested that EC-SOD KO increased the activation of MAPK signaling pathways, supposedly owing to the enhancement of $O_2 \cdot^-$ production.

Wojcicka et al. [121] investigated the role of ERK and oxidative stress in the pathogenesis of arterial hypertension induced by chronic leptin administration to rats. It was found that after leptin administration, ERK phosphorylation level increased in renal and aortic tissues. The SOD mimetic tempol normalized blood pressure, whereas the ERK inhibitor PD98059 exerted a hypotensive effect. The Nox inhibitor, apocynin, suppressed leptin-induced ERK activation. These findings indicated that Nox-produced $O_2 \cdot^-$ activated ERK after leptin administration to rats. Huddleston et al. [122] found that $O_2 \cdot^-$ enhanced the activation of ryanodine receptors (RyRs) in the mouse hippocampus. Superoxide also enhanced ERK phosphorylation. Thus, $O_2 \cdot^-$-induced potentiation required the redox targeting of RyR3 and the subsequent activation of ERK.

Lyng et al. [123] has studied MAPK signaling pathways in bystander cells exposed to irradiated cell-conditioned medium (ICCM). Human keratinocytes were irradiated (0.005–5 Gy) using a cobalt-60 teletherapy unit. It was found that exposure of cells to ICCM induced ERK and JNK phosphorylation.

All the foregoing data demonstrate the signaling function of $O_2 \cdot^-$ and H_2O_2 in ERK phosphorylation and activation. There are not so many studies on the ERK activation mediated by NO or the other RNS and, in some cases, ERK activation by RNS also depends on the formation of $O_2 \cdot^-$ or H_2O_2. The regulatory effects of NO on ERK activation/inactivation processes have been studied in cancer cells. Pervin et al. [124] have shown that the expression of MKP-1 phosphatase by NO led to dephosphorylation of ERK1/2, which was an initial step in the development of the apoptotic pathway in breast cancer cells. However, NO can apparently initiate both positive or negative cell growth in tumors. Thus, Thomas et al. [125] demonstrated that at low steady-state concentrations of NO (<50 nM), the phosphorylation of ERK was induced via a guanylate cyclase-dependent mechanism in human breast cells. However, ERK phosphorylation was transient during NO exposure. In subsequent work [126], it has been suggested that the activation of ERK (and some other enzymes) by NO depends on concomitant $O_2 \cdot^-$ formation; the latter reacted with NO to form $ONOO^-$ and decreased NO-dependent ERK activation.

Scorziello et al. [127] have studied the effect of preconditioning (IPC) on NO-dependent processes in neurons. It was found that IPC reduced cytochrome c release into the cytosol, improved mitochondrial function, and decreased ROS formation. Moreover, it stimulated nNOS expression, NO production, and ERK1/2 activation. Downstream ERK1/2 cascade was stimulated by NO through Ras activation.

Wang et al. [128] identified p42/44 MAPK kinase (ERK) as a downstream target of eNOS-derived $O_2 \cdot^-$ production. They concluded that this pathway appears to be distinct from the effects of either H_2O_2 or NO. It follows that $O_2 \cdot^-$ and NO produced by eNOS activate ERK by different ways. The ability of eNOS to activate ERK1/2 through the release of $O_2 \cdot^-$ suggests that under pathological conditions eNOS may contribute directly to endothelial dysfunction.

Mizuno et al. [129] investigated the effect of exogenous NO on cell proliferation and the expression of phosphorylated p42/44 mitogen-activated protein kinase (ERK) in human pulmonary arterial smooth muscle cells (HPASMC). NO donors transiently increased the phosphorylation of ERK and then suppressed it. It was proposed that exogenous NO is able to transiently activate p42/44 MAPK via the induction of p53, and then suppress it through inactivation of the Ras and Raf cascades.

It is known that Ang II is a powerful activator of MAPK cascades in cardiovascular tissues through a redox-sensitive mechanism. Although NO is able to suppress the vasoconstrictive and proarteriosclerotic actions of ANG II, its role in Ang II-induced redox-sensitive signal transduction is not yet clear. Zhang et al. [130] have shown that acute intravenous administration of the inhibitor N^G-nitro-L-arginine methyl ester (L-NAME) to catheterized, conscious rats enhanced phosphorylation of aortic ERK1/2 and p38 kinases, which was suppressed only partially by a superoxide dismutase mimetic Tempol. Furthermore, three different inhibitors of NO synthase (NOS) suppressed Ang II-induced MAP kinase phosphorylation in rat VSMCs, which was closely linked to $O_2 \cdot^-$ generation in these cells. These results show the involvement of endogenous NOS in ANG II-induced signaling pathways, leading to activation of ERK and p38 MAP kinases, taking into account that NO may have dual effects on the redox activation vascular MAP kinases.

Peroxynitrite is usually considered a toxic species, mainly because of decomposition to reactive hydroxyl radicals; therefore, its signaling function in enzymatic processes is uncertain. For example, it has been shown that $ONOO^-$ inactivates tryptophan hydroxylase via sulfhydryl oxidation [131]. However, Zouki et al. [132] found that $ONOO^-$ activated ERK in neutrophils supposedly through the Ras/Raf-1/MEK signal transduction pathway. Later on, Upmacis et al. [133] also showed that $ONOO^-$ stimulated ERK, p38, and phospholipase A2 (cPLA2) phosphorylation in rat arterial smooth muscle cells.

4.1.3.2 p38 Kinase

It has already been mentioned that p38 MAPKs are a class of MAPKs that are responsive to stress stimuli, such as cytokines, UV irradiation, heat shock, and osmotic shock, and are involved in cell differentiation and apoptosis. p38 kinase usually participates in signaling processes together with the other MAPKs such as ERK and Jun1/2, and, therefore, the role of ROS, in individual members of these cascades, might be uncertain.

In 1997–1998 it was shown that hypoxia/reoxygenation and oxidative stress activate p38 [134–137]. Ushio-Fukai et al. [137] have shown that p38MAPK is a critical component of H_2O_2-sensitive signaling pathways activated by Ang II in VSMCs and plays a crucial role in vascular hypertrophy. It is of interest that although Ang II increased the intracellular H_2O_2 formation and rapid phosphorylation of both p42/44 (ERK) and p38 kinases, exogenous H_2O_2 activated only p38MAPK, and the NADPH inhibitor diphenylene iodonium attenuated only Ang II-stimulated phosphorylation of p38MAPK, but not that of ERK. Similarly, catalase almost completely suppressed intracellular Ang II-induced H_2O_2 formation, resulting in the inhibition of phosphorylation of p38MAPK, but not ERK MAPK. Therefore, it is possible that p38 and ERK phosphorylation was mediated by different ROS. Clerk et al. [136] also showed that p38 kinase was activated by H_2O_2. Chiu et al. [138] suggested that TGF-β–induced p38 activation was mediated by the Rac1-regulated generation of ROS in cultured human keratinocytes. ROS apparently participated in intracellular redox signaling regulating LPS-induced activation of the MAPK (p38) pathway and

MAPK(p38)-mediated regulation of LPS-dependent inflammatory cytokine production in the alveolar epithelium [139].

Herrera et al. [140] showed that TGF-β induced ROS production in fetal hepatocytes, which might be responsible for p38 activation. Activation of p38MAPK occurred later and coincided with the maximal production of ROS. Wang and Doerschuk [141] found that the activation of p38 MAPK mediated by xanthine oxidase-produced ROS was required for cytoskeletal remodeling in endothelial cells during ICAM-1 cross-linking or neutrophil adherence.

Babilonia et al. [142] demonstrated that low K intake significantly increased the phosphorylation of p38 MAPK and ERK, but had no effect on phosphorylation of c-Jun N-terminus kinase in renal cortex and outer medulla. The stimulatory effect of low K intake on p38 and ERK was abolished by the treatment of rats with tempol. The mediation of p38 and ERK phosphorylation during low K intake by $O_2 \cdot^-$ and the other ROS was supported by the finding that H_2O_2 also increased the phosphorylation of ERK and p38 in the cultured mouse collecting duct cells. It was concluded that low K intake-induced increase in $O_2 \cdot^-$ levels was responsible for stimulation of p38 and ERK. In subsequent work [143], these authors confirmed that the addition of H_2O_2 inhibited small-conductance K (SK) channels in the cortical collecting duct by the activation of PTK, P38, and ERK activities.

Becuwe et al. [144] have shown that exogenous AA induced the MnSOD gene and activated the p38 MAPK pathway by ROS formed by AA oxidation in human hepatoma cells. Gaitanaki et al. [145] found that both H_2O_2 and $O_2 \cdot^-$ produced by xanthine oxidase activated p38 kinase in isolated perfused amphibian heart.

There are various potential ROS producers in cells, for example, mitochondria and Nox, which can be responsible for p38 activation during oxidative stress. Hsieh et al. [146] have shown that high glucose induced ROS generation and p38 MAPK phosphorylation in rats. These effects of high glucose were blocked by antioxidants (taurine and tiron), the inhibitors of mitochondrial electron transport chain complex I and II, and MnSOD mimetic. Therefore, in this case, mitochondria was a producer of $O_2 \cdot^-$, which initiated p38 activation. In subsequent work [147], these authors demonstrated that prolonged exposure of rat immortalized renal proximal tubular cells and nondiabetic rat renal proximal tubular cells to high concentrations of glucose or Ang II resulted in $O_2 \cdot^-$ production and ERK (p44/42 MAPK) phosphorylation. Kulisz et al. [148] proposed that hypoxia induced p38 MAPK phosphorylation by augmenting mitochondrial ROS generation in cardiomyocytes. An increase in p38 phosphorylation was observed in a PO2-dependent manner and was inhibited by mitochondrial inhibitors. Exogenous H_2O_2 also induced p38 phosphorylation. It has been suggested that physiological hypoxia stimulated p38 phosphorylation through a mechanism that required electron flux in the proximal region of the mitochondrial electron transport chain, which was mediated by either H_2O_2 or $O_2 \cdot^-$. Emerling et al. [149] also showed that the activation of p38MAPK and the induction of the transcription factor hypoxia-inducible factor 1 HIF-1 are dependent of mitochondrial ROS.

Pawate et al. [150] demonstrated that Nox-derived ROS stimulated lipopolysaccharide (LPS)- and interferon-γ (IFN)-induced signaling cascades, leading to gene expression in glial cells, which were mediated by p38, ERK, and c-Jun MAPKs. Harfouche et al. [151] studied ROS involvement in signaling by the angiopoietin-1

(Ang-1)/tie-2 receptor pathway. It was found that exposure of human umbilical vein endothelial cells to Ang-1 induced Nox-dependent rapid and transient $O_2{\cdot}^-$ production. Overexpression of SOD and catalase, as well as preincubation with selective inhibitors of Nox, enhanced basal p38 phosphorylation. It was concluded that the activation of tie-2 receptors by Ang-1 triggers production of ROS through activation of Nox and that ROS generation by Ang-1 promotes endothelial cell migration while negatively regulating ERK1/2 phosphorylation.

Xu et al. [152] found that hypoxia activated two upstream signaling pathways, Akt kinase and p38 MAPK. Both hypoxia-induced Akt and p38 MAPK functional activity as well as IL-8 mRNA and protein expression were reduced by the inhibition of PI3K and p38MAPK. These findings suggest that hypoxia activates the PI3K/Akt and p38 MAPK pathways in human ovarian carcinoma cells. Another important factor of p38 activation in hypoxia is carbon monoxide (CO), which is produced endogenously in the breakdown of heme [153]. Zuckerbraun et al. [154] have studied the mechanism of CO activation. They found that CO-induced p38 activation depended on the inhibition of cytochrome c oxidase (COX) and the generation of low levels of ROS in mitochondria.

Choi et al. [155] studied the contribution of p38 MAPK and the events upstream/downstream of p38 leading to dopaminergic neuronal death. They used primary cultures of mesencephalic neurons treated with 6-hydroxydopamine. It was found that phosphorylation of p38 preceded apoptosis. SOD mimetic and the NO chelator blocked 6-hydroxydopamine-induced phosphorylation of p38, suggesting the participation of $O_2{\cdot}^-$ and NO in p38 activation.

It has been shown that Ang II activates p38 MAPK and increases ROS formation. To study the mechanism of Ang II activity, Bao et al. [156] infused Sprague–Dawley rats and MAPKAP kinase-2 knockout mice with Ang II. Ang II infusion increased the levels of phosphorylated p38 kinase, production of $O_2{\cdot}^-$, and expression of Nox subunit gp91 in the aorta. These findings suggest that Ang II-induced hypertension, organ damage, and ROS production are possibly mediated by p38 kinase. Brezniceanu et al. [157] have shown that in rat immortalized renal proximal tubular cells and freshly isolated mouse renal proximal tubules, the transforming growth factor-β1 (TGF-β1) stimulated ANG gene expression and that its action was mediated by ROS generation and p38 MAPK activation.

It has been shown that MAPKs, especially p38 kinase, are activated in chronic heart failure (CHF). Accordingly, Widder et al. [158] have shown that the aortic levels of phosphorylated p38 kinase protein and $O_2{\cdot}^-$ generation were significantly enhanced in rats with CHF. Chronic p38 kinase inhibition prevented endothelial vasomotor dysfunction through the reduction of $O_2{\cdot}^-$ production. Alvarez-Maqueda et al. [159] found that homocysteine, a risk factor of cardiovascular disease, increased intracellular H_2O_2 production by neutrophils and enhanced the activation and phosphorylation of p38-MAPK and ERK1/2 kinases. These findings suggest that homocysteine enhances the oxidative stress of neutrophils and highlights the potential role of phagocytic cells in vascular wall injury through $O_2{\cdot}^-$ release in hyperhomocysteinemia conditions.

Peng et al. [160] investigated the role of gp91phox-containing NADH oxidase signaling in cardiomyocyte TNF-α expression and myocardial dysfunction induced by lipopolysaccharide (LPS). It was found that gp91^{phox}-containing NADH oxidase

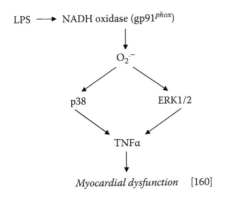

LPS ⟶ NADH oxidase (gp91^{phox})

$O_2^{\cdot-}$

p38   ERK1/2

TNFα

Myocardial dysfunction   [160]

FIGURE 4.6 p38 and ERK1/2 activation by LPS-stimulated NADPH oxidase-mediated superoxide production.

is responsible for LPS-induced TNF-α expression in the heart. This signaling pathway was mediated by $O_2^{\cdot-}$ generation and the activation of ERK1/2 and p38 MAPK kinases (Figure 4.6).

Nediani et al. [161] studied increased Nox activity in the right ventricle (RV) and the left ventricle (LV) of end-stage failing human hearts. When compared to control ventricles, diseased RV and LV showed a significant increase in Nox $O_2^{\cdot-}$ production that induced the activation of ERK and p38, but not JNK MAPKs. Rocic et al. [162] demonstrated that Nox-stimulated ROS production mediated p38 activation in human coronary artery endothelial cell (HCAEC) tube formation.

Mitra and Abraham [163] found that the addition of the $O_2^{\cdot-}$ producer paraquat to resting or LPS-exposed neutrophils enhanced activation of p38 MAPK, but not that of Akt or ERK1/2. These results demonstrated that the increased intracellular $O_2^{\cdot-}$ concentrations enhanced the nuclear accumulation of NFκB and the expression of NFκB-dependent proinflammatory cytokines through a p38 MAPK-dependent mechanism.

It has been shown that p38 MAPK is related to phosphorylation and activation of cytosolic phospholipase A(2) (cPLA(2)) and release of AA. Nito et al. [164] demonstrated that p38 MAPK and cPLA(2) activation markedly increased during transient focal cerebral ischemia in rats. Their findings suggest that the p38 MAPK/cPLA(2) pathway may promote blood–brain barrier (BBB) disruption with secondary vasogenic edema and that $O_2^{\cdot-}$ can stimulate this pathway after ischemia-reperfusion injury.

It has been suggested that the small GTPase p21ras is able to mediate the activation of MAPKs by ROS. Thus, Adachi et al. [165] demonstrated that H_2O_2 generated by Nox catalyzed glutathiolation of p21ras Cys118 to form the reactive intermediate GSS-p21ras, which activated p38 and Akt phosphorylation in VSMCs. (However, it should be noted that in subsequent work [166], these authors concluded that the formation of GSS-p21ras by the reaction of glutathione with p21ras initiated by ONOO$^-$ resulted in the activation of ERK and Akt but not p38 kinase in endothelial cells.)

Similar to the other kinases, p38 is able to catalyze ROS production. Azhar et al. [167] investigated the effect of oxidative stress on antisteroidogenic action through modulation of oxidant-sensitive MAPK signaling pathways. Treatment of mouse adrenocortical cells with $O_2{\cdot}^-$ or H_2O_2 significantly inhibited steroid production and resulted in increased phosphorylation and activation of p38 MAPK but not the phosphorylation of either ERK or JNK kinases. Pretreatment of cells with MnTMPyP (SOD mimetic) completely prevented the $O_2{\cdot}^-$ and H_2O_2-mediated inhibition of steroid production. Therefore, it was concluded that p38 MAPK activation mediated ROS-induced inhibition of adrenal steroidogenesis. Suzuki et al. [168] found that IL-1β induced phosphorylation and activation of p38 MAPK and phosphorylation of MAPK kinase-3/6 (MKK3/6) in human neutrophils. At the same time, ERK was faintly phosphorylated and JNK was not phosphorylated at all. IL-1β also induced phosphorylation of ERK, JNK, and p38 MAPK in human endothelial cells. Present findings suggest that the MKK3/6-p38 MAPK cascade is selectively activated by IL-1β and that activation of this cascade mediated $O_2{\cdot}^-$ release in human neutrophils. Saito et al. [169] have shown that PMA induced neutrophil death through p38 kinase, which was activated by Nox-generated $O_2{\cdot}^-$.

It has been shown that p38 kinase can be activated by NO. Huwiler and Pfeilschifter [170] demonstrated that the high concentrations of exogenously applied NO activated p38 MAPK in mesangial cells. In addition, there was a parallel activation of ERK. Browning et al. [171] also showed that NO activated p38 kinase in human neutrophils and 293T fibroblasts. The mechanism of p38 activation by NO is uncertain. It is possible that NO directly affects p38 kinase or inactivates the opposing phosphatase by nitrosylation, with subsequent increased phosphorylation and activation of this enzyme [170]. It was also shown that exogenous NO caused phosphorylation of p38 in neutrophils [172].

It is of interest that the activated p38 MAPK regulated $ONOO^-$ generation from LPS+IFNγ-stimulated astrocytes [173]. Peroxynitrite generation was blocked by Nox inhibitor, diphenyleneiodonium chloride, and NOS inhibitor, N-ω-nitro-l-arginine methyl ester, indicating the participation of both these enzymes in $ONOO^-$ generation. Thus, the stimulation of $ONOO^-$ generation by activated p38 MAPK apparently takes place through dual regulatory mechanisms, involving iNOS induction and Nox activation.

4.1.3.3   c-Jun N-Terminal Kinases (JNKs), Also Called Stress-Activated Protein Kinases (SAPKs)

Similar to the other MAPKs, c-Jun kinases (JNK) are activated by ROS of different origin such as ischemia/reperfusion, mitochondria, or exogenous hydroperoxide. In 1996, Lo et al. [174] investigated JNK activation by cytokines and ROS in chondrocytes. Treatment of bovine chondrocytes with both IL-1 and TNF-α led to rapid induction of JNK activity, which was suppressed by antioxidants. Moreover, JNK activation was also observed during direct addition of H_2O_2 and NO to the chondrocyte cultures. These findings suggest that both ROS and RNS are able to mediate JNK activation in chondrocytes.

Using the cell culture model of ischemia and reperfusion in cardiac myocytes, Laderoute and Webster [175] have shown that reoxygenation, but not hypoxia alone, caused c-Jun activation. c-Jun activation was reduced by preincubating myocytes during the hypoxia phase with the spin-trap agent α-phenyl N-*tert*-butyl nitrone or with N-acetylcysteine. Their inhibitory effects implicate ROS as the initiators of a kinase pathway involving the stress-activated protein kinases (JNKs/SAPKs) in reoxygenated cardiac myocytes. Lee and Koretzky [176] showed that CD40 stimulation of resting splenic B lymphocytes and murine B lymphoma cells produced ROS and activated JNK. Preincubation with the antioxidant N-acetyl-L-cysteine and cell transfection with MnSOD inhibited JNK activation, providing further support for ROS involvement in this pathway.

Cui and Douglas [177] have shown that AA induced phosphorylation and activation of JNK. It is interesting that free radical scavenger N-acetylcysteine blocked AA-induced JNK activation, while H_2O_2 activated it. Furthermore, AA activated Nox in parallel to $O_2 \cdot^-$ generation and JNK activation. It was concluded that in kidney epithelial cells JNK activation by AA was mediated by Nox-dependent $O_2 \cdot^-$ generation.

It has been shown that in the rat passive Heymann nephritis model of membranous nephropathy, complement C5b-9 induced sublethal glomerular epithelial cell (GEC) injury and proteinuria. Peng et al. [178] demonstrated that C5b-9 activated JNK kinase in cultured rat GECs. Incubation of GECs with the C5b-9 complement also stimulated production of $O_2 \cdot^-$, which was inhibited by antioxidants N-acetylcysteine, glutathione, and α-tocopherol. At the same time, JNK was inhibited by H_2O_2. Thus, C5b-9 activation of JNK kinase in cultured rat GECs is apparently dependent on ROS signaling. Pani et al. [179] reported that the activation of JNK-1 in ConA-stimulated thymocytes apparently depended on the formation of oxygen radicals. Joneson and Bar-Sagi [180] demonstrated that Racl-mediated $O_2 \cdot^-$ production induced JNK activation in quiescent fibroblasts.

Benhar et al. [181] demonstrated that the sensitization of transformed cells to stress stimuli is due to the potentiation of JNK and p38 pathways. It was shown that the enhanced ROS production was involved in the JNK/p38 activation mechanism in transformed cells. JNK/p38 activation was inhibited by antioxidants and, in particular, by inhibitors of the mitochondrial respiratory chain and Nox. Taken together, these findings suggested that ROS-dependent potentiation of stress kinase pathways accounted for the sensitization of transformed cells to stress and anticancer drugs.

It is known that oxidized low-density lipoprotein (oxLDL) activates many signal transduction pathways in endothelial cells, including the stress-activated protein kinase JNK and the ERK. These MAP kinases determine cell survival in response to environmental stress. Go et al. [182] have shown that JNK signaling is activated by both ROS and RNK derived from Nox, NOSs, peroxides, and oxLDL (Figure 4.7). It was found that upon exposure of endothelial cells to oxLDL, both ERK and JNK are activated through independent signal transduction pathways.

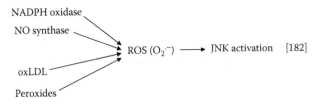

FIGURE 4.7   ROS signaling in JNK activation.

Matesanz et al. [183] have shown that xanthine oxidase-derived extracellular $O_2^{\cdot-}$ activated JNK and p38, but not ERK and Akt kinases, in human VSMCs. Viedt et al. [184] proposed that ROS generation through the activation of Nox by Ang II is a link between hormone receptor interaction and the stimulation of JNK and p38 MAPKs. Ding et al. [185] found that, in Ang II-stimulated cultured human mesangial cells, the upstream mediators of the JNK activation were ROS and the epidermal growth factor receptor (EGFR). ROS production was suppressed by NADPH inhibitors diphenyleneiodonium sulfate and apocynin, but not the inhibitors of the other oxidant-producing enzymes, including the mitochondrial complex I, xanthine oxidase, cyclooxygenase, lipoxygenase, and NOS. These data suggest that the ROS/EGFR/JNK pathway is important in transducing the proliferative effect of Ang II in cultured HMCs.

TGF-β-induced p38 and JNK phosphorylation in murine embryonic fibroblasts was mediated by ROS [186]. The activation of JNK and p38 enzymes was inhibited by diphenyleneiodonium and the SOD and catalase mimetic MnTBaP. Mantena and Katiyar [187] have shown that the treatment of human epidermal keratinocytes (NHEK) with grape seed proanthocyanidins (GSPs) inhibited UVB-induced free radical-mediated damage and the phosphorylation of ERK1/2, JNK, and p38 kinases. H_2O_2 also induced phosphorylation of ERK1/2, JNK, and p38, which was inhibited by antioxidants. It was concluded that ROS mediated UVB-stimulated activation of MAPKs in human epidermal keratinocytes.

Lin et al. [188] investigated the effects of CuZnSOD and/or catalase on oxLDL-induced cell proliferation and MAPK signaling in human aortic smooth muscle cells (HASMCs). It was found that CuZnSOD and/or catalase overexpression suppressed ERK1/2 and JNK phosphorylation. These results suggest that CuZnSOD and catalase suppress cell proliferation caused by oxLDL stimulation through the inhibition of ROS-dependent ERK1/2 and JNK phosphorylation. Pechtelidou et al. [189] investigated the activation of JNK1/2 signaling pathway in cardiac myoblasts during treatment with exogenous H_2O_2 both under transient and sustained stimulation. Both sustained and transient H_2O_2 treatment resulted in JNK activation, whereas catalase and superoxide dismutase prevented oxidative stress-induced cell death. It was concluded that in H9c2 myoblasts, the sustained activation of JNK1/2 signaling pathway during oxidative stimulation was followed by an apoptotic phenotype, while transient JNK1/2 activation correlated well with cell survival, suggesting a dual role for this signaling pathway in the cells.

Several works demonstrate the signaling role of RNS in the activation of the stress-activated protein kinases (SAPK), including JNK MAPK. In 1996 Pfeilschifter and Huwiler [190] showed that exposure of rat glomerular mesangial cells and primary cultures of bovine glomerular endothelial cells to the NO producers, including MAHMA-NONOate, S-nitrosoglutathione, and spermine-NO, resulted in a time- and concentration-dependent activation of stress-activated protein kinases (SAPK) as measured by the phosphorylation of c-Jun in a solid phase kinase assay. Pretreatment of the cells with the tyrosine kinase inhibitor genistein strongly suppressed NO-induced c-Jun phosphorylation. Furthermore, N-acetylcysteine markedly reduced the activation of SAPK in response to NO. These studies identify SAPK as targets for NO, which may be critical for NO-induced apoptosis of glomerular mesangial and endothelial cells.

However, subsequent studies showed that NO can negatively regulate JNK. For example, Park et al. [191] have shown that endogenously produced NO suppressed JNK activation in intact cells. Treatment of murine microglial cells, murine macrophage cells, and rat alveolar macrophages with IFN-γ induced endogenous NO production, suppressing, at the same time, JNK activation. The IFNγ-induced suppression of JNK1 activation was completely prevented by N^G-nitro-L-arginine, an NOS inhibitor. It was suggested that endogenous NO inhibited JNK activation in macrophage cells through the reaction with the thiol residues of the enzyme.

Go et al. [192] found that both NO and $O_2{\cdot}^-$ are required for activation of JNK by shear stress in endothelial cells. Peroxynitrite alone also activated JNK. These findings suggest that $ONOO^-$ or the other RNS are signaling molecules capable of JNK activation. In subsequent work [193], these authors showed that shear stress stimulated JNK by activating the cascade of PI3K, Akt, and NO production from eNOS.

Numerous data cited earlier show the direct and indirect activation of JNK kinase by ROS. However, similar to other protein kinases, JNK is apparently able to catalyze ROS production under specific conditions. It has been shown that JNK participates in both cell death and survival responses to different stimuli. Ventura et al. [194] investigated the function of JNK in tumor necrosis factor (TNF-α)-stimulated cell death using mouse fibroblasts. They found that JNK can suppress TNF-α-stimulated apoptosis but, at the same time, it can also potentiate TNF-α-stimulated necrosis by increasing the production of reactive oxygen species. Thus, JNK-dependent ROS production can shift the balance of TNFα-stimulated cell death from apoptosis to necrosis. There are other works supporting a catalytic role for JNK in ROS production. Antosiewicz et al. [195] found that diallyl trisulfide (DATS), which is a highly promising anticancer constituent of processed garlic, induced ROS generation mediated by the JNK signaling axis in human prostate cancer cells. Nakajima et al. [196] showed that ROS-dependent JNK activation was followed by further enhancement of ROS.

4.1.4   cAMP-Dependent Protein Kinases, I and II (PKAI and PKAII)

Dimon-Gadal et al. [197] studied the in vitro effects of radiolytically generated hydroxyl radicals and $O_2{\cdot}^-$ on the activation of cAMP-dependent protein kinases, I and II (PKAI and -II, respectively). As expected, hydroxyl radicals exhibited damaging effects and decreased the kinase phosphotransferase activities stimulated by either cAMP or its site-specific analogs. Superoxide affected only PKAI, modifying both cAMP-binding sites A and B of the regulatory subunit, but had a smaller effect on the catalytic subunit. These results suggest that ROS can alter the structure of PKA enzymes.

4.1.5   AMP-Activated Protein Kinase (AMPK)

AMP-activated protein kinase (AMPK) is a highly conserved eukaryotic protein that acts as a cellular energy sensor. Toyoda et al. [198] proposed that acute oxidative stress can increase the rate of glucose transport via an AMPK-mediated mechanism. They showed that H_2O_2 activated AMPKα1 in skeletal muscle through an AMP-independent mechanism, resulting in an increase in the rate of glucose transport, at least in part, via an AMPKα1-mediated mechanism.

Sandström et al. [199] demonstrated that endogenously produced ROS (H_2O_2) stimulated the contraction-mediated activation of AMPK in fast-twitch muscle. An et al. [200] suggested that nicotine-induced AMPK phosphorylation appeared to be mediated by ROS because nicotine significantly increased $O_2{\cdot}^-$ generation and $ONOO^-$ mimicked the effects of nicotine on AMPK, while N-acetylcysteine suppressed nicotine-enhanced AMPK phosphorylation. Taken together, these findings suggested that the nicotine activation of AMPK is mediated by $ONOO^-$, resulting in enhanced threonine phosphorylation. Zhang et al. [201] found that the thromboxane A2 receptor (TPr) stimulation activated AMPK in VSMCs. Exposure of VSMCs to H_2O_2 caused time- and dose-dependent AMPK activation and the phosphorylation of AMPK-Thr172. It was concluded that TPr stimulation triggered ROS-mediated LKB1-dependent AMPK activation.

4.1.6 Protein Tyrosine Kinase (PTK)

Protein tyrosine kinases (PTKs) are a subgroup of protein kinases that transfer phosphate groups from ATP to the tyrosine residues of proteins. Similar to the other protein kinases, ROS signaling is an important factor of PTK catalysis. Earlier, it was shown that tyrosine kinases presented in human lymphocytes (p56lck and p59fyn) [202], human neutrophils (p56/59hck, p72syk, and p77btk) [203], or human T-cell leukemia line [204] were activated by ROS production. Yan and Berton [205] also found that ROS (including, most probably, $O_2{\cdot}^-$) produced by adherent neutrophils activated Src family kinases. It is possible that the inability of ROS to activate tyrosine kinases in vitro indicates its indirect activation by the inactivation of phosphatases. However, Wu and de Champlain [206] found that $O_2{\cdot}^-$ activated tyrosine kinase in smooth muscle cells from rat mesenteric artery.

Tang et al. [207] have shown that H_2O_2 at low-to-moderate concentrations temporarily inactivated Src tyrosine kinases under in vivo but not in vitro conditions. These authors found that Src kinases localized to focal adhesions and the plasma membrane were rapidly and permanently inactivated by H_2O_2 through the reduction of phosphorylation of the conserved tyrosine residue at the activation loop. At the same time, the cytoplasmic Src kinases were gradually activated by H_2O_2. Babilonia et al. [208] also showed that $O_2{\cdot}^-$, and possibly H_2O_2, enhanced the expression of PTK in mouse collecting duct principal cell line. It has been suggested that $O_2{\cdot}^-$ mediates the effect of low K intake on c-Src expression in the cortical collecting duct. All these findings indicate that ROS are able to regulate PTKs directly without the participation of phosphatases.

Abid et al. [209] demonstrated that the activation of PI3K-Akt-forkhead and p38 MAPK, but not ERK1/2 or JNK, by VEGF was mediated by $O_2{\cdot}^-$-produced Nox in human umbilical vein endothelial cells or human coronary artery endothelial cells. It was also proposed that the starting point of $O_2{\cdot}^-$ signaling was the activation of PTK Src. Bouaouina et al. [210] showed that both p38 MAPK and Src kinases are involved in β_2 integrin activation by TNF-α and ROS in human neutrophils. Their results demonstrated that constitutively activated Src tyrosine kinases and a redox-regulated activation of p38 MAPK are involved in TNF inside-out signaling, leading to β_2 integrin activation.

PTKs are apparently also able to catalyze $O_2\cdot^-$ formation. For example, Kim et al. [211] found that the inhibitors of PTKs diminished the stimulated $O_2\cdot^-$, H_2O_2, and NO production in macrophages stimulated with IL-1, LPS, or FMLP.

4.1.7 JANUS KINASE (JAK)

Janus kinases (JAKs) are a family of intracellular nonreceptor tyrosine kinases that transduce cytokine-mediated signals via the JAK-signal transducers and activators of transcription (STATs) pathway. (JAK is now considered an abbreviation for "Janus kinase." The name is taken from the two-faced Roman god of doorways, Janus, because JAKs possess two near-identical phosphate-transferring domains. One domain exhibits kinase activity, while the other negatively regulates the kinase activity of the first.)

ROS signaling is an important characteristic of JAK-STATs catalytic pathway. Schieffer et al. [212] investigated the role of $O_2\cdot^-$ generated by the Nox system in the Ang II activation of the JAK/STAT cascade. Ang II stimulation of rat aortic smooth muscle cells enhanced $O_2\cdot^-$ production and induced phosphorylation of JAK2, STAT1a/b, and STAT3. These findings suggest that stimulation of the JAK/STAT cascade by Ang II required $O_2\cdot^-$ produced by the Nox system. Moriwaki et al. [213] determined the effects of IFN-γ on $O_2\cdot^-$ production and phosphorylation of signal transducer and activator of transcription STAT-1α in human mesangial cells (HMCs). Increase in $O_2\cdot^-$ formation was completely abolished by diphenyleneiodonium and the Janus-activated kinase (JAK)2 inhibitor AG490. Therefore, it was concluded that IFN-γ stimulated $O_2\cdot^-$ production through the JAK-STAT pathway and Nox.

4.1.8 APOPTOSIS SIGNAL-REGULATING KINASE 1 (ASK1)

ASK1 is an MAPK, which is associated with the initiation of apoptosis. ASK1 is a member of the mitogen-activated protein kinase kinase kinase (MAPKKK) super-family that activates during TNF-α-induced apoptosis. Apoptosis signal-regulating kinase-1 (ASK1) lies upstream of a major redox-sensitive pathway leading to the activation of Jun NH_2-terminal kinase (JNK) and the induction of apoptosis. In 1998, Gotoh and Cooper [214] showed that H_2O_2 induced ASK1 activation in 293 cells. TNF-α-induced activation of ASK1 was inhibited by antioxidants. H_2O_2-induced apoptosis was markedly enhanced by the expression of ASK1. Thus, these findings suggest that TNF-α-induced activation of ASK1 is mediated by ROS. It was also found that TNF-α or H_2O_2 treatment increased the dimeric form of ASK1, while the pretreatment with antioxidative *N*-acetylcysteine decreased it.

Du et al. [215] demonstrated that methylglyoxal (MG)-induced apoptosis was associated with rapid production of $O_2\cdot^-$ followed by an increase in temporal activation of ASK1. NAC and PDTC blocked the activation of ASK1 and MG-induced apoptosis completely. Moreover, SOD-mimic MnTBAP and catalase together inhibited MG-induced ASK1 activation and apoptosis induction. Correspondingly, MG-mediated ASK1 activation was enhanced by diethyldithiocarbamate (DDC). These results suggest that at the early stage ASK1 activation is mediated by $O_2 \cdot^-$ with subsequent progression of apoptosis in MG-treated Jurkat cells.

Goldman et al. [216] proposed a mechanism by which H_2O_2 induced ASK1 activation through its phosphorylation at Ser967. They found that treatment of COS7 cells with H_2O_2 initiated dephosphorylation of Ser967, resulting in the dissociation of the ASK1·14-3-3 complex with concomitant increase of ASK1 catalytic activity and ASK1-mediated activation of JNK and p38 pathways. Zhou et al. [217] have shown that in Ang II-treated cardiac myocytes Nox produced $O_2 \cdot^-$, which stimulated the apoptosis signal-regulating kinase 1 (ASK1)-p38 kinase pathway. Nadeau et al. [218] found that cell exposure to H_2O_2 caused the rapid oxidation of ASK1, leading to its multimerization through the formation of interchain disulfide bonds. During the subsequent reduction, the thiol-disulfide oxidoreductase thioredoxin-1 (Trx1) became covalently bound to ASK1. These findings indicate that ASK1 oxidation is required as a step subsequent to activation for signaling downstream of ASK1 after H_2O_2 treatment.

4.2 PHOSPHOLIPASES

Activated phospholipases (PLs) catalyze hydrolysis of various phospholipids, depending on tyrosine phosphorylation. Thus, activated PLC-g1 phospholipase cleaves the membrane phospholipid phosphatidylinositol 4,5-bisphosphate, generating two second messengers, diacylglycerol (DAG) and inositol 1,4,5-trisphosphate (IP3). Numerous works demonstrate the involvement of PLs in various enzymatic pathways depending on the generation or mediation by ROS, but it seems that in most of them phospholipases participate indirectly by catalyzing protein kinases responsible for ROS signaling.

Nonetheless, as early as 1988, Shasby et al. [219] suggested that exogenous oxidants (H_2O_2) initiated hydrolysis of inositol phospholipids catalyzed by phospholipases A and C in endothelial cells. In recent work, Hong et al. [220] investigated the effect of H_2O_2 on the regulation of intracellular calcium signaling in rat cortical astrocytes. They found that H_2O_2 triggered the generation of oscillations of intracellular calcium concentration and activated phospholipase C-γ1 (PLC-γ1) in a dose-dependent manner. It was concluded that low concentrations of exogenously applied H_2O_2 generated calcium oscillations by activating PLC-γ1 through sulfhydryl oxidation-dependent mechanisms. These works demonstrate the possible involvement of phospholipases in ROS signaling.

4.3 GTPASES

GTPases are a large family of hydrolases that bind and hydrolyze guanosine triphosphate (GTP) into guanosine diphosphate (GDP). It is believed that the hydrolysis of GTP into GDP and inorganic phosphate (Pi) occurs by the S_N2 nucleophilic mechanism (Figure 4.4).

$$GTP + HO^- \rightarrow GDP + Pi^- \qquad (4.1)$$

Therefore, signaling by $O_2\cdot^-$ (a "supernucleophile") in the reactions catalyzed by GTPases seems to be quite possible.

The participation of GTPases in catalytic cascades mediated by $O_2\cdot^-$ and H_2O_2 was already discussed in the sections considering ROS signaling in the reactions of protein kinases. Thus, Rygiel et al. [63] reported that Tiam1, an activator of the small GTPase Rac, influenced Nox-mediated ROS production and regulation ERK phosphorylation in keratinocytes. Sabbatini et al. [119] studied the effects of Ras/ERK1/2 signaling inhibition on oxidative damage in cultured renal and endothelial cells and renal ischemia-reperfusion injury in the rat.

Adachi et al. [165] suggested that H_2O_2 generated by Nox catalyzed glutathiolation of p21ras Cys118 to form the reactive intermediate GSS-p21ras, which activated p38 and Akt phosphorylation in VSMCs. These authors also concluded that the formation of GSS-p21ras by the $ONOO^-$-initiated reaction of glutathione with p21ras resulted in the activation of ERK and Akt but not p38 kinase in endothelial cells [166]. Joneson and Bar-Sagi [180] demonstrated that Rac1-mediated $O_2\cdot^-$ production induced JNK activation in quiescent fibroblasts.

Actually, as early as 1995, Lander et al. [221] proposed that redox modulators such as H_2O_2 can directly promote guanine nucleotide exchange on p21ras. They also pointed out that direct activation of p21ras might be a central mechanism by which a variety of redox stress stimuli (ROS) transmitted their signal to the nucleus. For example, Irani et al. [222] demonstrated that fibroblasts stably transformed with a constitutively active isoform of p21Ras, H-RasV12 produced large amounts of $O_2\cdot^-$, which formation was inhibited by N-acetyl-L-cysteine. In already cited work, Bhunia et al. [60] suggested that $O_2\cdot^-$ stimulated ERK (p44) activation through the p21^{rasz}GTP loading. This hypothesis is vital for understanding $O_2\cdot^-$ signaling mechanisms because loading of p21^{rasz}GTP suggests hydrolysis of GP to GDP (Figure 4.4). Yang et al. [223] suggested that $O_2\cdot^-$ mediated ras-induced transformation in v-Ha-ras-transfected rat kidney epithelial cells; CuZnSOD and MnSOD inhibited ras-induced transformation.

Campbell and coworkers [224–226] investigated the mechanisms of of ROS and RNS signaling in processes catalyzed by the ras superfamily GTPase. They suggested that $O_2\cdot^-$ reacts with the redox-active GTPases to release GDP as an unstable oxygenated adduct (5-oxo-GDP). They also suggested that Cys-118 and Phe-28 residues are involved in $O_2\cdot^-$-mediated Ras guanine nucleotide dissociation. In the case of RNS-initiated GTPase activation, these authors suggested that NO is converted into the reactive radical $\cdot NO_2$ capable of reacting with the thiol groups and that free radicals $O_2\cdot^-/\cdot NO_2$ are the actual active species in GTP activation by H_2O_2 in the presence of iron ions.

A rather complicated cascade of chemical reactions mediated by ROS, RNS, and their derivatives was suggested, which unfortunately included some doubtful stages. (For example, it is proposed [224–226] that ROS/RNS are able to abstract a hydrogen atom from the sulfhydryl group, although it is now known that in the case of $O_2\cdot^-$ this reaction proceeds through addition to the S atom. Some peculiarities of these mechanisms will be considered later.)

In recent years, it has been discovered that ROS induced vascular smooth muscle contraction via the activation of small GTPase and its downstream target Rho/Rho kinase. Thus, Jin et al. [227] have shown that $O_2\cdot^-$ and H_2O_2 activated Rho/Rho kinase and mediated vascular contraction in rat aorta rings. Kajimoto et al. [228] demonstrated that molecular oxygen activated the Rho/Rho-kinase pathway in human and rabbit ductus arteriosus by increasing mitochondria-derived ROS.

4.4 PHOSPHATASES

Phosphatases are a group of enzymes that catalyze removal of phosphate anion from phosphorylated biomolecules by hydrolysis of the P-O bond. Phosphatase catalytic activity is an important regulatory factor of protein kinase activation, and their competition is responsible for proper enzymatic action. Phosphatases consist of two main groups: the cysteine-dependent protein tyrosine phosphatases (PTP) and metallophosphatases. The active sites of these enzymes are cysteine residues and transition metal ions, respectively.

ROS signaling is an important factor of activation/deactivation of all phosphatases. It was shown that $O_2\cdot^-$ and H_2O_2 inhibited phosphatases by the oxidation of cysteines or through the change of the iron ion valence. Therefore, it is believed that in many cases the activation of protein kinases by ROS is actually due to the deactivation of some phosphatases. Therefore, ROS opposite effects on protein kinases and phosphatases have been studied in numerous works. Mechanisms of ROS signaling in the reactions of phosphatases are probably studied much better than for other enzymes, but they will be discussed later in Chapter 7. Now, we will consider the most important studies concerning ROS signaling in catalysis by phosphatases.

Denu and Tanner [229] suggested that phosphatases can be regulated by reversible reduction/oxidation with ROS such as H_2O_2. They studied the effect of H_2O_2 on PTPs and PP (serine/threonine protein phosphatase). H_2O_2 had no apparent effect on PP activity, while PTPs were relatively rapidly inactivated (the rate constant of inactivation was equal to $10-20$ $M^{-1}s^{-1}$) even with low micromolar concentrations of H_2O_2. PTP inactivation was fully reversible in the presence of glutathione and other thiols. It was found that H_2O_2 oxidized the catalytic cysteine thiolate of PTPs to the cysteine sulfenic acid intermediate (Cys-SOH). The mechanism of reversible inactivation involving a cysteine sulfenic acid intermediate has been discussed (see Chapter 7).

Caselli et al. [230] showed that low-molecular-weight phosphotyrosine-protein phosphatase (LMW-PTP), which shares no general sequence homology with other PTPs, was also inhibited by H_2O_2. In this case, H_2O_2 oxidized two cysteines (Cys12 and Cys17) out of the eight LMW-PTP cysteines to form a disulfide bond. Because a physiological concentration of H_2O_2 induced enzyme inactivation and considering that its activity was restored by reduction by low-molecular-weight thiols, the authors suggested that H_2O_2 was able to regulate LMW-PTP in the cell. Reversible oxidation and inactivation of multiple PTPs by H_2O_2 was then demonstrated under in vivo conditions [231].

Groen et al. [232] found that the different PTPs are oxidized by H_2O_2 in different ways. Thus, the membrane-proximal RPTPα-D1 was catalytically active but not readily oxidized, while the membrane-distal RPTPα-D2 was a poor PTP, but was easily oxidized by H_2O_2. On the other hand, H_2O_2 readily oxidized the catalytically active PTP1B and LAR-D1 phosphatases. It was concluded that PTPs are oxidized at physiological pH and H_2O_2 concentrations in a different way. In addition, the PTP loop arginine is apparently an important determinant for susceptibility to oxidation. Meng et al. [233] studied the effect of reversible PTP oxidation in the signaling response to insulin. These authors found that the stimulation of cells with insulin resulted in the rapid oxidation and inhibition of two distinct PTPs (PTP1B and TC45, the 45-kDa spliced variant of the T cell protein-tyrosine phosphatase). It was also shown that TC45 is an inhibitor of insulin signaling.

The aforementioned works demonstrate the importance of H_2O_2 signaling in the inactivation of PTPs through the reaction with PTP's cysteines. However, there are some unresolved problems concerning the mechanism of reaction of H_2O_2 with sulfhydryl groups of cysteines, which make doubtful the direct interaction between these molecules. (Full discussion of these questions is in Chapter 7.)

Barrett et al. [234] concluded that $O_2 \cdot^-$ signaling should be a more reliable mechanism of these reactions. These authors point out that in the case of PTP-1B phosphatase, Cys215 is surrounded by main-chain amides, which constitute the phosphate-binding loop, and the positively charged amino acid residues Arg-45, Arg-47, Lys-116, Lys-120, Arg-112, and Arg-221. Therefore, it is reasonable to propose that this site will be a favorable target for the negatively charged $O_2 \cdot^-$. In order to identify a reactive species (H_2O_2 or $O_2 \cdot^-$) in PTP-1B catalysis, the kinetics of PTP-1B inactivation by each of these ROS was examined.

It was found [234] that $O_2 \cdot^-$ is kinetically a more efficient and more specific ROS than H_2O_2 for inactivating PTP-1B phosphatase. The rate constants for the $O_2 \cdot^-$ and H_2O_2-stimulated inactivation are equal to 334 ± 45 M^{-1}s^{-1} and 42.8 ± 3.8 M^{-1}s^{-1}, respectively. Furthermore, it is important that after oxidation by H_2O_2 PTP-1B exhibited significantly more oxidized methionine residues and showed a lower degree of reversibility. Further oxidation of the Cys-215 sulfenic derivative (an initial oxidative product) into irreversible sulfinic and sulfonic derivatives was prevented by glutathionylation of the sulfenic derivative to form a glutathionylated PTP-1B, which can be reactivated by dithiothreitol or thioltransferase. Thus, $O_2 \cdot^-$ signaling in RTP inactivation might be a more efficient pathway for the direct reaction of H_2O_2 with sulfhydryl groups.

This important conclusion by Barrett et al. [234] about $O_2 \cdot^-$ signaling in the reactions catalyzed by PTPs was confirmed in subsequent work. Wang et al. [235] reported that polyaromatic quinones, such as the environmental pollutants 9,10-phenanthrene-diones, are ROS-mediated inactivators of PTPs. Under aerobic conditions, PTP inactivation by 9,10-phenanthrenediones proceeded with rate constants of 4300, 387, and 5200 M^{-1}s^{-1} at pH 7.2 for CD45, PTP1B, and LAR phosphatases, respectively. As these rate constants are about 40 times higher than the rate constant for PTP inhibition by H_2O_2 [234], it was concluded that ROS-mediated PTP inhibition by 9,10-phenanthrenediones depended on $O_2 \cdot^-$ and possibly semiquinones, but not H_2O_2.

RNS can probably also participate in the regulation of protein tyrosine phosphatases. For example, Barrett et al. [236] demonstrated that the stimulation of NO generation leading to S-nitrosylation and inhibition of PTPs was an important factor of cellular signal transduction pathways.

Serine/threonine protein phosphatases (PPs) are another class of phosphatases that are regulated by ROS, but in a different way. One of the most studied PPs is calcineurin (also called protein phosphatase 2B), a major calmodulin-binding protein in the brain and the only serine/threonine PP under the control of calcium/calmodulin [237].

In 1996, Wang et al. [238] showed that SOD protected calcineurin from inactivation, supposedly owing to the suppression of oxidation of the Fe–Zn active center of enzyme by $O_2{}^{\cdot-}$. In subsequent works, the mechanism of calcineurin inactivation by ROS and RNS was widely discussed. Sommer et al. [239] suggested that $O_2{}^{\cdot-}$ and H_2O_2 can inhibit calcineurin activity through the oxidation of the ferrous center or the thiol groups. However, in the same year, Bogumil et al. [240] proposed that H_2O_2 inactivated calcineurin by the oxidation of cysteine residues to disulfide and that iron was not oxidized in the active enzyme. However, later on, these authors agreed that the catalytically active ferrous–zinc binuclear center at the calcineurin active site plays an important role in calcineurin inhibition by $O_2{}^{\cdot-}$ and H_2O_2, in which it is oxidized to a ferric–zinc complex. Overall, the inactivation of calcineurin by $O_2{}^{\cdot-}$ is apparently the complex redox process, which is modified by calcium, NO, and superoxide dismutase [241]. Although $O_2{}^{\cdot-}$ is capable of targeting the ferrous–zinc active complex in calcineurin, the possibility of the interaction of H_2O_2 with the thiol groups cannot be excluded [242].

Sommer et al. [243] compared the effects of $O_2{}^{\cdot-}$, NO, and H_2O_2 on the activities of three highly homologous serine/threonine phosphatases: protein phosphatase type 1 (PP1), protein phosphatase type 2A (PP2A), and calcineurin (protein phosphatase type 2B). It was found that only calcineurin was a sensitive target for inhibition by $O_2{}^{\cdot-}$ and NO and that none of the phosphatases were sensitive to inhibition by H_2O_2. The authors suggested that ROS and RNS inhibited calcineurin by oxidizing both a catalytic metal and a critical thiol.

The unique role of $O_2{}^{\cdot-}$ in calcineurin inhibition was confirmed in the study of the inhibitory effects of a redox cycler 2,3-dimethoxy-1,4-naphthoquinone (DMNQ) on calcineurin activity in endothelial cells. Namgaladze et al. [244] demonstrated that DMNQ caused inhibition of cellular calcineurin activity and its inhibitory effect was reversible upon DMNQ removal. Inhibition was suppressed by CuZnSOD.

4.5   THE OTHER ENZYMES

ROS signaling has been demonstrated in the reactions catalyzed by the other enzymes; some works are cited in the following subsections.

4.5.1   GLUTATHIONE TRANSFERASE, SOLUBLE GUANYLATE CYCLASE

Murata et al. [245] found that xanthine oxidase-generated $O_2{}^{\cdot-}$ activated rat glutathione transferases (GSTs), GST activation being suppressed by SOD. Brune et al. [246] reported that $O_2{}^{\cdot-}$ was found to be a potent and reversible inhibitor of soluble guanylate cyclase.

4.5.2 Tryptophan Hydroxylase

It has already been noted earlier that tryptophan hydroxylase (TH), the initial and rate-limiting enzyme in serotonin biosynthesis, is inactivated by $ONOO^-$ [131]. Inactivation was not inhibited by $O_2\cdot^-$ and hydroxyl radical inhibitors SOD, catalase, DMSO, and mannitol. The rate constant for inhibition of tryptophan hydroxylase by $ONOO^-$ was estimated as 3.4×10^4 $M^{-1}s^{-1}$ at 25°C and pH 7.4. These findings suggested that $ONOO^-$ inactivated tryptophan hydroxylase through sulfhydryl oxidation. Modification of tyrosyl residues by $ONOO^-$ played a relatively minor role in the inhibition of TH catalytic activity.

Contradictions in the mechanism of inhibition of TH were discussed in the following works. Although Kuhn and Geddes [131] concluded that cysteine oxidation rather than tyrosine nitration was responsible for TH inactivation by $ONOO^-$, reexamination of the reaction of $ONOO^-$ with purified TH failed to produce cysteine oxidation but resulted in an increase in tyrosine nitration and inactivation [247]. Tyrosine residue 423 and, to a lesser extent tyrosine residues 428 and 432, were modified by nitration. The rate constant of $(3.8 \pm 0.9) \times 10^3$ $M^{-1}s^{-1}$ for the reaction of $ONOO^-$ with TH was determined at pH 7.4 and 25°C in stopped-flow experiments to be 10 times less than Kuhn and Geddes [131]. It was concluded that $ONOO^-$ reacted with the metal center of the enzyme with subsequent nitration of tyrosine residue 423, which is responsible for the inactivation of TH. However, in their 2002 work, Kuhn et al. [248] determined the site of $ONOO^-$ attack on TH. They identified the residues Tyr-423, Tyr-428, and Tyr-432 as the sites of $ONOO^-$-induced nitration in TH. It was concluded that no single tyrosine residue appears to be critical for TH catalytic function, and tyrosine nitration is neither necessary nor sufficient for $ONOO^-$-induced inactivation. The loss of TH catalytic activity caused by $ONOO^-$ is associated with oxidation of cysteine residues.

4.5.3 Aldose Reductase

Recent studies have shown the importance of ROS and RNS signaling in the activation of aldose reductase (AR). AR reduces cytotoxic aldehydes and glutathione conjugates of aldehydes derived from lipid peroxidation. Its inhibition increases oxidative injury, but the mechanisms by which ischemia regulates AR activity remain unclear. Kaiserova et al. [249] found that rat hearts subjected to ischemia displayed an increase in AR activity. The AR activity can be prevented by the free radical scavengers (N-(2-mercaptoproprionyl) glycine, superoxide dismutase mimetic tiron, or dithiothreitol. It was suggested that ROS formed in the ischemic heart activated AR by modifying its cysteine residues (Cys-298 and Cys-303) to sulfenic acids. In a following work, these authors [250] studied the role of RNS in the regulation of AR. They concluded that AR is a cardioprotective enzyme and that its activation in the ischemic heart is due to $ONOO^-$-mediated oxidation of Cys-298 to sulfenic acid via the PI3K/Akt/endothelial NOS pathway.

4.5.4 Metalloproteinases (MMPs)

Metalloproteinases (MMPs) are a family of zinc-dependent endopeptidases capable of degrading extracellular matrix components, which participate in the atherosclerotic process by remodeling the extracellular matrix. MMPs play a pivotal role in angiogenesis, atherogenesis, vascular remodeling after vascular injury, and instability of atherosclerotic plaque. ROS signaling was identified in the processes catalyzed by MMPs in various cells. Inoue et al. [251] have shown that lysophosphatidycholine (LPC) induced generation of ROS in bovine aortic endothelial cells (BAECs). An increase in $O_2{\cdot}^-$ production stimulated the production and activation of MMP-2.

Grote et al. [252] suggested that mechanical stretch can enhance MMP expression and activity in a Nox-dependent manner. Accordingly, mouse VSMCs were exposed to cyclic mechanical stretch. It was found that mechanical stretch induced Nox-dependent ROS formation that enhanced MMP-2 mRNA expression and pro-MMP-2 release. Deem and Cook-Mills [253] have shown that the endothelial cell–associated activity of matrix MMP was increased within minutes after VCAM-1 cross-linking and that MMP activation depended on the endothelial cell Nox-derived ROS. Zalba et al. [254] investigated the association between phagocytic Nox and MMP-9 in human atherosclerosis. They found that enhanced Nox-dependent $O_2{\cdot}^-$ production stimulated MMP-9 in monocytes and that this relationship may be relevant in the atherosclerotic process.

4.5.5 ENDOTHELIN CONVERTING ENZYME

Endothelin converting enzyme (ECE) is a member of a larger zinc MMP family that includes neutral endopeptidase. Lopez-Ongil et al. [255] found that $O_2{\cdot}^-$ inhibited ECE by ejecting zinc from the enzyme. Importantly, only $O_2{\cdot}^-$ can inhibit ECE, while H_2O_2 and NO did not affect ECE activity. Superoxide inhibition was reversible, since the addition of zinc restored ECE activity. These authors considered different mechanisms of zinc ejection from ECE by $O_2{\cdot}^-$. One possible mechanism is the reduction of the residues that coordinate zinc binding, while the other one is Zn^{2+} reduction to Zn^+. Although the mechanism of ECE inhibition by $O_2{\cdot}^-$ remains uncertain, the second mechanism is certainly wrong because zinc is not a transition metal and, hence, cannot change its valence.

New important evidence regarding ROS signaling has been obtained recently by Deudero et al. [256]. These authors reported that $O_2{\cdot}^-$ produced by Nox was a critical signaling element in the previously described action of VEGF on the expression of its own transcription factor, HIF-1, and on VEGF itself.

REFERENCES

1. VJ Thannickal and BL Fanburg. Reactive oxygen species in cell signaling. *Am J Physiol Lung Cell Physiol* 279: L1005–L1028, 2000.
2. KK Griendling, D Sorescu, B Lassègue, and M Ushio-Fukai. Modulation of protein kinase activity and gene expression by reactive oxygen species and their role in vascular physiology and pathophysiology. *Arterioscler Thromb Vasc Biol* 20: 2175–2183, 2000.
3. GEN Kass, SK Duddy, and S Orrenius. Activation of hepatocyte protein kinase C by redox-cycling quinones. *Biochem J* 260: 499–507, 1989.
4. R Larsson and P Cerutti. Translocation and enhancement of phosphotransferase activity of protein kinase c following exposure in mouse epidermal cells to oxidants. *Cancer Res* 49: 5627–5632, 1989.

5. H Konishi, M Tanaka, Y Takemura, H Matsuzaki, Y Ono, U Kikkawa, and Y Nishizuka. Activation of protein kinase C by tyrosine phosphorylation in response to H_2O_2. *Proc Natl Acad Sci USA* 94: 11233–11237, 1997.

6. H Konishi, E Yamauchi, H Taniguchi, T Yamamoto, H Matsuzaki, Y Takemura, K Ohmae, U Kikkawa, and Y Nishizuka. Phosphorylation sites of protein kinase C δ in H_2O_2-treated cells and its activation by tyrosine in vitro. *Proc Natl Acad Sci USA* 98: 6587–6592, 2001.

7. E Klann, ED Roberson, LT Knapp, and JD Sweatt. A role for superoxide in protein kinase C activation and induction of long-term potentiation. *J Biol Chem* 273: 4516–4522, 1998.

8. LT Knapp and E Klann. Superoxide-induced stimulation of protein kinase C via thiol modification and modulation of zinc content. *J Biol Chem* 275, 24136–24145, 2000.

9. I Korichneva, B Hoyos, R Chua, E Levi, and U Hammerling. Zinc release from protein kinase c as the common event during activation by lipid second messenger or reactive oxygen. *J Biol Chem* 277: 44327–44331, 2002.

10. R Gopalakrishna and U Gundimeda. Antioxidant regulation of protein kinase c in cancer prevention. *J Nutr* 132: 3819S–3823S, 2002.

11. IB Afanas'ev, in *Superoxide Ion: Chemistry and Biological Implications*, vol. 1. Boca Raton, FL: CRC Press, 1989.

12. U Gundimeda, Z-H Chen, and R Gopalakrishna. Tamoxifen modulates protein kinase c via oxidative stress in estrogen receptor-negative breast cancer cells. *J Biol Chem* 271: 13504–13508, 1996.

13. R Gopalakrishna, U Gundimeda, JE Schiffman, and TH McNeil. A direct redox regulation of protein kinase C isoenzymes mediates oxidant-induced neuritogenesis in PC12 cells. *J Biol Chem* 283: 14430–14444, 2008.

14. J Hongpaisan, CA Winters, and SB Andrews. Strong calcium entry activates mitochondrial superoxide generation, upregulating kinase signaling in hippocampal neurons. *J Neurosci* 24: 10878–10887, 2004.

15. M Van Marwijk Kooy, JWN Akkerman, S Van Asbeck, L Borghnis, and HC Van Prooijen. UVB radiation exposes fibrinogen binding sites on platelets by activation protein kinase C via reactive oxygen species. *Br J Haematol* 83: 253–258,1993.

16. N Ogata, H Yamamoto, K Kugiyama, H Yasue, and E Miyamoto. Involvement of protein kinase C in superoxide anion-induced activation of nuclear factor-kappa B in human endothelial cells. *Cardiovasc Res* 45: 513–521, 2000.

17. AMN Kabir, JE Clark, M Tanno, X Cao, JS Hothersall, S Dashnyam, DA Gorog, M Bellahcene, MJ Shattock, and MS Marber. Cardioprotection initiated by reactive oxygen species is dependant on the activation of PKC{epsilon}. *Am J Physiol Heart Circ Physiol* 291: H1893–H1899, 2006.

18. W-S Wu, RK Tsai, CH Chang, S Wang, J-R Wu, and Y-X Chang. Reactive oxygen species mediated sustained activation of protein kinase C {alpha} and extracellular signal-regulated kinase for migration of human hepatoma cell Hepg2. *Mol Cancer Res* 4: 747–758, 2006.

19. K Kuribayashi, K Nakamura, M Tanaka, T Sato, J Kato, K Sasaki, R Takimoto, K Kogawa, T Terui, T Takayama, T Onuma, T Matsunaga, and Y Niitsu. Essential role of protein kinase C zeta in transducing a motility signal induced by superoxide and a chemotactic peptide, fMLP. *J Cell Biol* 176: 1049–1060, 2007.

20. F Chu, NE Ward, and CA O'Brian. PKC isozyme S-cysteinylation by cystine stimulates the pro-apoptotic isozyme PKC delta and inactivates the oncogenic isozyme PKC epsilon. *Carcinogenesis* 24: 317–325, 2003.

21. F Chu, LH Chen, and CA O'Brian. Cellular protein kinase C isozyme regulation by exogenously delivered physiological disulfides—implications of oxidative protein kinase C regulation to cancer prevention. *Carcinogenesis* 25: 585–596, 2004.

22. RT Waldron and E Rozengurt. Oxidative stress induces protein kinase D activation in intact cells. Involvement of Src and dependence on protein kinase C. *J Biol Chem* 275: 17114–17121, 2000.
23. RT Waldron, O Rey, E Zhukova, and E Rozengurt. Oxidative stress induces protein kinase C-mediated activation loop phosphorylation and nuclear redistribution of protein kinase D. *J Biol Chem* 279: 27482–27493, 2004.
24. Y Wang, JM Schattenberg, RM Rigoli, P Storz, and MJ Czaja. Hepatocyte resistance to oxidative stress is dependant on protein kinase C mediated down-regulation of c-Jun/AP-1. *J Biol Chem* 279: 31089–31097, 2004.
25. P Storz, H Doppler, and A Toker. Protein kinase C{delta} selectively regulates protein kinase D-dependant activation of NF-{kappa}b in oxidative stress signaling. *Mol Cell Biol* 24: 2614–2626, 2004.
26. H Doppler and P Storz. A novel tyrosine phosphorylation site in protein kinase D contributes to oxidative stress-mediated activation. *J Biol Chem* 282: 31873–31881, 2007.
27. MA Baxter, RG Leshe, and WG Reeves. The stimulation of superoxide anion production in guinea-pig peritoneal macrophages and neutrophils by phorbol myristate acetate opsonized zymosan and IgG2-containing soluble immune complexes. *Immunology* 48: 657–665, 1983.
28. T Matsubara and M Ziff. Superoxide anion release by human endothelial cells: Synergism between a phorbol ester and a calcium ionophore. *J Cell Physiol* 127: 207–210, 1986.
29. BP Salimath and G Sawtha. Mechanism of inhibition by cyclic AMP of protein kinase C-triggered respiratory burst in Ehrlich ascites tumor cells. *Cell Signal* 4: 651–653, 1992.
30. Y Ohara, TE Peterson, B Zheng, JF Kuo, and DG Harrison. Lysophosphatidylcholine increases vascular superoxide anion production via protein kinase C activation. *Arterioscler Thromb Vasc Biol* 14: 1007–1013, 1994.
31. Q Li, V Subbulakshmi, AP Fields, NR Murray, and MK Catheart. Protein kinase C regulates human monocyte O2.- production and low density lipoprotein lipid peroxidation. *J Biol Chem* 274: 3764–3771, 1999.
32. WM Armstead and WG Mayhan. Superoxide generation links protein kinase C activation to impaired atp-sensitive K+ channel function after brain injury. *Stroke* 30: 153–159, 1999.
33. Z Ungvari, A Csiszar, PM Kaminski, MS Wolin, and A Koller. Chronic high pressure-induced arterial oxidative stress: Involvement of protein kinase C-dependant NAD(P)H oxidase and local renin-angiotensin system. *Am J Pathol* 165: 219–226, 2004.
34. YL Siow, KK Au-Yeung, CW Woo, and K O. Homocysteine stimulates phosphorylation of NADPH oxidase p47phox and p67phox subunits in monocytes via protein kinase Cbeta activation. *Biochem J* 398: 73–82, 2006.
35. W Zhou, X-L Wang, KG Lamping, and H-C Lee. Inhibition of PKC{beta} protects against diabetes-induced impairment in arachidonic acid dilation of small coronary arteries. *J Pharmacol Exp Ther* 319: 199–207, 2006.
36. Q Li, V Subbulakshmi, CM Oldfield, R Aamir, CM Weyman, A Wolfman, and MK Cathcart. PKCalpha regulates phosphorylation and enzymatic activity of cPLA(2) in vitro and in activated human monocytes. *Cell Signal* 19: 359–366, 2007.
37. CH Kim, SI Han, SY Lee, HS Youk, JY Moon, HQ Duong, MJ Park, YM Joo, HG Park, YJ Kim, MA Yoo, SC Lim, and HS Kang. Protein kinase C-ERK1/2 signal pathway switches glucose depletion-induced necrosis to apoptosis by regulating superoxide dismutases and suppressing reactive oxygen species production in A549 lung cancer cells. *J Cell Physiol* 211: 371–385, 2007.

38. HM Korchak, LB Dorsey, H Li, D Mackie, and LE Kilpatrick. Selective roles for alpha-PKC in positive signaling for O-(2) generation and calcium mobilization but not elastase release in differentiated HL60 cells. *Biochim Biophys Acta* 1773: 440–449, 2007.

39. RL Miller, GY Sun, and AY Sun. Cytotoxicity of paraquat in microglial cells: Involvement of PKCdelta- and ERK1/2-dependant NADPH oxidase. *Brain Res* 1167: 129–139, 2007.

40. A Goel, Y Zhang, L Anderson, and R Rahimian. Gender difference in rat aorta vasodilation after acute exposure to high glucose: Involvement of protein kinase C beta and superoxide but not of rho kinase. *Cardiovasc Res* 76: 351–360, 2007.

41. L Xia, H Wang, S Munk, H Frecker, HJ Goldberg, IG Fantus, and CI Whiteside. Reactive oxygen species, PKC-{beta}1, and PKC-{zeta} mediate high glucose-induced vascular endothelial growth factor expression in mesangial cells. *Am J Physiol Endocrinol Metab* 293: E1280–E1288, 2007.

42. M Shaw, P Cohen, and DR Alessi. The activation of protein kinase B by H2O2 or heat shock is mediated by phosphoinositide 3-kinase and not by mitogen-activated protein kinase-activated protein kinase-2. *Biochem J* 336: 241–246, 1998.

43. M Ushio-Fukai, RW Alexander, M Akers, QQ Yin, Y Fujio, K Walsh, and KK Griendling. Reactive oxygen species mediate the activation of Akt/protein kinase B by angiotensin II in vascular smooth muscle cells. *J Biol Chem* 274: 22699–22704, 1999.

44. D-Y Shi, Y-R Deng, S-L Liu, Y-D Zhang, and L Wei. Redox stress regulates cell proliferation and apoptosis of human hepatoma through Akt protein phosphorylation. *FEBS Lett* 542: 60–64, 2003.

45. X Wang, KD McCullough, TF Franke, and NJ Holbrook. Epidermal growth factor receptor-dependant Akt activation by oxidative stress enhances cell survival. *J Biol Chem* 275: 14624–14631, 2000.

46. Y Gorin, N-H Kim, D Feliers, B Bhandari, GG Choudhury, and HE Abboud. Angiotensin II activates Akt/protein kinase B by an arachidonic acid/redox-dependant pathway and independent of phosphoinositide 3-kinase. *FASEB J* 15: 1909–1920, 2001.

47. Y Gorin, JM Ricono, N-H Kim, B Bhandari, GG Choudhury, and HE Abboud. Nox4 mediates angiotensin II-induced activation of Akt/protein kinase B in mesangial cells. *Am J Physiol Renal Physiol* 285: F219–F229, 2003.

48. SD Hingtgen, X Tian, J Yang, SM Dunlay, AS Peek, Y Wu, RV Sharma, JF Engelhardt, and RL Davisson. Nox2-containing NADPH oxidase and Akt activation play a key role in angiotensin II-induced cardiomyocyte hypertrophy. *Physiol Genomics* 26: 180–191, 2006.

49. F Li and KU Malik. Angiotensin II-induced Akt activation is mediated by metabolites of arachidonic acid generated by CaMKII-stimulated Ca2+-dependant phospholipase A2. *Am J Physiol Heart Circ Physiol* 288: H2306–H2316, 2005.

50. S Lim and MV Clement. Phosphorylation of the survival kinase Akt by superoxide is dependant on an ascorbate-reversible oxidation of PTEN. *Free Radic Biol Med* 42: 1178–1192, 2007.

51. CR Hoyal, A Gutierrez, BM Young, SD Catz, J-H Lin, PN Tsichlis, and BM Babior. Modulation of p47phos activity by site-specific phosphorylation: Akt-dependant activation of the NADPH oxidase. *Proc Natl Acad Sci USA* 100: 5130–5135, 2003.

52. JX Chen, H Zeng, QH Tuo, H Yu, B Meyrick, and JL Aschner. NADPH oxidase modulates myocardial Akt, ERK1/2 activation, and angiogenesis after hypoxia-reoxygenation. *Am J Physiol Heart Circ Physiol* 292: H1664–H1674, 2007.

53. Y Wang, MM Zeigler, GK Lam, MG Hunter, TD Eubank, VV Khramtsov, S Tridandapani, CK Sen, and CB Marsh. The role of the NADPH oxidase complex, p38 MAPK, and Akt in regulating human monocyte/macrophage survival. *Am J Respir Cell Mol Biol* 36: 68–77, 2007.

54. MH Li, YN Cha, and YJ Surh. Peroxynitrite induces HO-1 expression via PI3K/Akt-dependant activation of NF-E2-related factor 2 in PC12 cells. *Free Radic Biol Med* 41: 1079–1091, 2006.

55. D Martin, M Salinas, N Fujita, T Tsuruo, and A Cuadrado. Ceramide and reactive oxygen species generated by H2O2 induce caspase-3-independant degradation of Akt/protein kinase B. *J Biol Chem* 277: 42943–42952, 2002.

56. B Kou, J Ni, M Vatish, and DF Singer. Xanthine oxidase interaction with vascular endothelial growth factor in human endothelial cell angiogenesis. *Microcirculation* 15: 251–267, 2008.

57. B Govindarajan, JE Sligh, BJ Vincent, M Li, JA Canter, BJ Nickoloff, RJ Rodenburg, JA Smeitink, L Oberley, Y Zhang, J Slingerland, RS Arnold, JD Lambeth, C Cohen, L Hilenski, K Griendling, M Martínez-Diez, JM Cuezva, and JL Arbiser. Overexpression of Akt converts radial growth melanoma to vertical growth melanoma. *J Clin Invest* 117: 719–729, 2007.

58. S Ichihara, Y Yamada, Y Kawai, T Osawa, K Furuhashi, Z Duan, and G Ichihara. Roles of oxidative stress and Akt signaling in doxorubicin cardiotoxicity. *Biochem Biophys Res Commun* 359: 27–33, 2007.

59. AS Baas and BC Berk. Differential activation of mitogen-activated protein kinases by H2O2 and O2.- in vascular smooth muscle cells. *Circ Res* 77: 29–36, 1995.

60. AK Bhunia, H Han, A Snowden, and S Chatterjee. Redox-regulated signaling by acto-sylceramide in the proliferation of human aortic smooth muscle cells. *J Biol Chem* 272: 15642–15649, 1997.

61. YM Janssen-Heininger, I Macara, and BT Mossman. Cooperativity between oxidants and tumor necrosis factor in the activation of nuclear factor (NF)-kappaB. Requirement of ras/mitogen-activated protein kinases in the activation of nf-kappab by oxidants. *Am J Respir Cell Mol Biol* 20: 942–952, 1999.

62. S Svegliati, R Cancello, P Sambo, M Luchetti, P Paroncini, G Orlandini, G Discepoli, R Paterno, M Santillo, C Cuozzo, S Cassano, EV Avvedimento, and A Gabrielli. Platelet-derived growth factor and reactive oxygen species (ROS) regulate ras protein levels in primary human fibroblasts via ERK1/2: Amplification of ROS and ras in systemic sclerosis fibroblasts. *J Biol Chem* 280: 36474–36482, 2005.

63. TP Rygiel, AE Mertens, K Strumane, R van der Kammen, and JG Collard. The Rac activator Tiam1 prevents keratinocyte apoptosis by controlling ROS-mediated ERK phosphorylation. *J Cell Sci* 121: 1183–1192, 2008.

64. KZ Guyton, Y Liu, M Gorospe, Q Xu, and NJ Holbrook. Activation of mitogen-activated protein kinase by H2O2. Role in cell survival following oxidant injury. *J Biol Chem* 271: 4138–4142, 1996.

65. SD Goldstone and NH Hunt. Redox regulation of the mitogen-activated protein kinase pathway during lymphocyte activation. *Biochim Biophys Acta* 1355: 353–360, 1997.

66. MV Gurjar, J Deleon, RV Sharma, and RC Bhalla. Role of reactive oxygen species in IL-1beta-stimulated sustained ERK activation and MMP-9 induction. *Am J Physiol Heart Circ Physiol* 281: H2568–H2574, 2001.

67. NR Madamanchi, SK Moon, ZS Hakim, S Clark, M Ali, C Patterson, and MS Runge. Differential activation of mitogenic signaling pathways in aortic smooth muscle cells deficient in superoxide dismutase isoforms. *Arterioscler Thromb Vasc Biol* 25: 950–956, 2005.

68. HY Yu, T Inoguchi, M Nakayama, H Tsubouchi, N Sato, N Sonoda, S Sasaki, K Kobayashi, and H Nawata. Statin attenuates high glucose-induced and angiotensin II-induced MAP kinase activity through inhibition of NAD(P)H oxidase activity in cultured mesangial cells. *Med Chem* 1: 461–466, 2005.

69. M Ortego, A Gomez-Hernandez, C Vidal, E Sanchez-Galan, LM Blanco-Colio, JL Martin-Ventura, J Tunon, C Diaz, G Hernandez, and J Egido. HMG-coa reductase inhibitors reduce Ikappab kinase activity induced by oxidative stress in monocytes and vascular smooth muscle cells. *J Cardiovasc Pharmacol* 45: 468–475, 2005.

70. Z Xie, DR Pimental, S Lohan, A Vasertriger, C Pligavko, WS Colucci, and K Singh. Regulation of angiotensin II-stimulated osteopontin expression in cardiac microvascular endothelial cells: Role of p42/44 mitogen-activated protein kinase and reactive oxygen species. *J Cell Physiol* 188: 132–138, 2001.

71. EL Greene, V Velarde, and AA Jaffe. Role of reactive oxygen species in bradykinin-induced mitogen-activated protein kinase and c-fos induction in vascular cells. *Hypertension* 35: 942–947, 2000.

72. A Nishiyama, M Yoshizumi, H Hitomi, S Kagami, S Kondo, A Miyatake, M Fukunaga, T Tamaki, H Kiyomoto, M Kohno, T Shokoji, S Kimura, and Y Abe. The SOD mimetic tempol ameliorates glomerular injury and reduces mitogen-activated protein kinase activity in Dahl salt-sensitive rats. *J Am Soc Nephrol* 15: 306–315, 2004.

73. Z Xie, M Singh, and K Singh. ERK1/2 and JNKs, but not p38 kinase, are involved in reactive oxygen species-mediated induction of osteopontin gene expression by angiotensin II and interleukin-1beta in adult rat cardiac fibroblasts. *J Cell Physiol* 198: 399–407, 2004.

74. KZ Gong, ZG Zhang, AH Li, YF Huang, P Bu, F Dong, and J Liu. ROS-mediated ERK activation in delayed protection from anoxic preconditioning in neonatal rat cardiomyocytes. *Chin Med J* (Engl) 117: 395–400, 2004.

75. YV Mukhin, MN Garnovskaya, G Collinsworth, JS Grewal, D Pendergrass, T Nagai, S Pinckney, EL Greene, and JR Raymond. 5-Hydroxytryptamine1A receptor/gibeta-gamma stimulates mitogen-activated protein kinase via NAD(P)H oxidase and reactive oxygen species upstream of src in chinese hamster ovary fibroblasts. *Biochem J* 347: 61–67, 2000.

76. AR Simon, M Severgnini, S Takahashi, L Rozo, B Andrahbi, A Agyeman, BH Cochran, RM Day, and BL Fanburg. 5-HT induction of c-fos gene expression requires reactive oxygen species and Rac1 and Ras GTPases. *Cell Biochem Biophys* 42: 263–276, 2005.

77. YJ Gao, K Takemori, LY Su, WS An, C Lu, AM Sharma, and RM Lee. Perivascular adipose tissue promotes vasoconstriction: The role of superoxide anion. *Cardiovasc Res* 71: 363–373, 2006.

78. F-S Wang, C-J Wang, Y-J Chen, P-R Chang, Y-T Huang, Y-C Sun, H-C Huang, Y-J Yang, and KD Yang. Ras induction of superoxide activates ERK-dependant angiogenic transcription factor HIF-1{alpha} and VEGF-A expression in shock wave-stimulated osteoblasts. *J Biol Chem* 279: 10331–10337, 2004.

79 SA Milligan, MW Owens, and MB Grisham. Differential regulation of extracellular signal-regulated kinase and nuclear factor-kappa B signal transduction pathways by hydrogen peroxide and tumor necrosis factor. *Arch Biochem Biophys* 352: 255–262, 1998.

80. RL de la Conde, MH Schoemaker, TE Vrenken, M Buist-Homan, R Havinga, PL Jansen, and H Moshage. Superoxide anions and hydrogen peroxide induce hepatocyte death by different mechanisms: Involvement of JNK and ERK MAP kinases. *J Hepatol* 44: 918–929, 2006.

81. T Yamakawa, S Tanaka, Y Yamakawa, J Kamei, K Numaguchi, ED Motley, T Inagami, and S Eguchi. Lysophosphatidylcholine activates extracellular signal-regulated kinases 1/2 through reactive oxygen species in rat vascular smooth muscle cells. *Arterioscler Thromb Vasc Biol* 22: 752–758, 2002.

82. E Kefaloyianni, C Gaitanaki, and I Beis. ERK1/2 and p38-MAPK signalling pathways, through MSK1, are involved in NF-kappaB transactivation during oxidative stress in skeletal myoblasts. *Cell Signal* 18: 2238–2251, 2006.

83. SL Lee, WW Wang, GA Finlay, and BL Fanburg. Serotonin stimulates mitogen-activated protein kinase activity through the formation of superoxide anion. *Am J Physiol Lung Cell Mol Physiol* 277: L282–L291, 1999.

84. SL Lee, AR Simon, WW Wang, and BL Fanburg. H(2)O(2) signals 5-HT-induced ERK MAP kinase activation and mitogenesis of smooth muscle cells. *Am J Physiol Lung Cell Mol Physiol* 281: L646–L652, 2001.

85. S Devadas, L Zaritskaya, SG Rhee, L Oberlay, and MS Williams. Discrete generation of superoxide and hydrogen peroxide by T cell receptor stimulation: Selective regulation of mitogen-activated protein kinase activation and Fas ligand expression. *J Exp Med* 195: 59–70, 2002.

86. DY Rhyu, Y Yang, H Ha, GT Lee, JS Song, S Uh, and HB Lee. Role of reactive oxygen species in TGF-β1-induced mitogen-activated protein kinase activation and epithelial-mesenchymal transition in renal tubular epithelial cells. *J Am Soc Nephrol* 16: 667–675, 2005.

87. CL Lin, FS Wang, YR Kuo, YT Huang, HC Huang, YC Sun, and YH Kuo. Ras modulation of superoxide activates ERK-dependant fibronectin expression in diabetes-induced renal injuries. *Kidney Int* 69: 1593–600, 2006.

88. R Locher, RP Brandes, W Vetter, and M Baron. Native LDL induces proliferation of human vascular smooth muscle cells via redox-mediated activation of ERK 1/2 mitogen-activated protein kinases. *Hypertension* 39 (Part 2): 645–650, 2002.

89. F-S Wang, C-J Wang, S-M Sheen-Chen, Y-R Kuo, R-F Chen, and KD Yang. Superoxide mediates shock wave induction of ERK-dependant osteogenic transcription factor (CBFA1) and mesenchymal cell differentiation toward osteoprogenitors. *J Biol Chem* 277: 10931–10937, 2002.

90. WC Huang, CC Chio, KH Chi, HM Wu, and WW Lin. Superoxide anion-dependant Raf/MEK/ERK activation by peroxisome proliferator activated receptor gamma agonists 15-deoxy-delta(12,14)-prostaglandin J(2), ciglitazone, and GW1929. *Exp Cell Res* 277: 192–200, 2002.

91. S Susa and I Wakabayashi. Extracellular alkalosis activates ERK mitogen-activated protein kinase of vascular smooth muscle cells through NADPH-mediated formation of reactive oxygen species. *FEBS Lett* 554: 399–402, 2003.

92. X Zhang, P Shan, M Sasidhar, GL Chupp, RA Flavell, AM Choi, and PJ Lee. Reactive oxygen species and extracellular signal-regulated kinase 1/2 mitogen-activated protein kinase mediate hyperoxia-induced cell death in lung epithelium. *Am J Respir Cell Mol Biol* 28: 305–315, 2003.

93. SM Kulich and CT Chu. Role of reactive oxygen species in extracellular signal-regulated protein kinase phosphorylation and 6-hydroxydopamine cytotoxicity. *J Biosci* 28: 83–89, 2003.

94. B Badrian, TM Casey, MC Lai, PE Rakoczy, PG Arthur, and MA Bogoyevitch. Contrasting actions of prolonged mitogen-activated protein kinase activation on cell survival. *Biochem Biophys Res Commun* 345: 843–850, 2006.

95. M Sano, K Fukuda, T Sato, H Kawaguchi, M Suematsu, S Matsuda, S Koyasu, H Matsui, K Yamauchi-Takihara, M Harada, Y Saito, and S Ogawa. ERK and p38 MAPK, but not NF-kappaB, are critically involved in reactive oxygen species-mediated induction of IL-6 by angiotensin II in cardiac fibroblasts. *Circ Res* 89: 661–669, 2001.

96. GD Frank, S Eguchi, T Yamakawa, S Tanaka, T Inagami, and ED Motley. Involvement of reactive oxygen species in the activation of tyrosine kinase and extracellular signal-regulated kinase by angiotensin II. *Endocrinology* 141: 3120–3126, 2000.

97. M Kyaw, M Yoshizumi, K Tsuchiya, K Kirima, and T Tamaki. Antioxidants inhibit JNK and p38 MAPK activation but not ERK 1/2 activation by angiotensin II in rat aortic smooth muscle cells. *Hypertens Res* 24: 251–261, 2001.

98. MA Laplante, R Wu, A El Midaoui, and J De Champain. NAD(P)H oxidase activation by angiotensin II is dependant on p42/44 ERK-MAPK pathway activation in rat's vascular smooth muscle cells. *J Hypertens* 21: 927–936, 2003.
99. I Papparella, G Ceolotto, L Lenzini, M Mazzoni, L Franco, M Sartori, L Ciccariello, and A Semplicini. Angiotensin II-induced over-activation of p47phox in fibroblasts from hypertensives: Which role in the enhanced ERK1/2 responsiveness to angiotensin II? *J Hypertens* 23: 793–800, 2005.
100. L Ding, A Chapman, R Boyd, and HD Wang. ERK activation contributes to regulation of spontaneous contractile tone via superoxide anion in isolated rat aorta of angiotensin II-induced hypertension. *Am J Physiol Heart Circ Physiol* 292: H2997–H3005, 2007.
101. J-M Li, S Wheatcroft, LM Fan, MT Kearney, and AM Shah. Opposing roles of p47phox in basal versus angiotensin II–stimulated alterations in vascular O2⁻ production, vascular tone, and mitogen-activated protein kinase activation. *Circulation* 109: 1307–1313, 2004.
102. J Ross and WM Armstead. Differential role of PTK and ERK MAPK in superoxide impairment of KATP and KCa channel cerebrovasodilation. *Am J Physiol Regul Integr Comp Physiol* 285: R149–R154, 2003.
103. S Philip and WM Armstead. Differential role of PTK, ERK and p38 MAPK in superoxide impairment of NMDA cerebrovasodilation. *Brain Res* 979: 98–103, 2003.
104. RA Oeckler, PM Kaminski, and MS Wolin. Stretch enhances contraction of bovine coronary arteries via an NAD(P)H oxidase–mediated activation of the extracellular signal–regulated kinase mitogen-activated protein kinase cascade. *Circ Res* 92: 23–31, 2003.
105. Q Chen, W Li, Z Quan, and BE Sumpio. Modulation of vascular smooth muscle cell alignment by cyclic strain is dependant on reactive oxygen species and P38 mitogen-activated protein kinase. *J Vasc Surg* 37: 660–668, 2003.
106. TM Guest, G Vlastos, FM Alameddine, and WR Taylor. Mechanoregulation of monocyte chemoattractant protein-1 expression in rat vascular smooth muscle cells. *Antioxid Redox Signal* 8: 1461–1471, 2006.
107. MJ Czaja, H Liu, and Y Wang. Oxidant-induced hepatocyte injury from menadione is regulated by ERK and AP-1 signaling. *Hepatology* 37: 1405–1413, 2003.
108. G Ceolotto, M Bevilacqua, I Papparella, E Baritono, L Franco, C Corvaja, M Mazzoni, A Semplicini, and A Avogaro. Insulin generates free radicals by an NAD(P)H, phosphatidylinositol 3'-kinase-dependant mechanism in human skin fibroblasts ex vivo. *Diabetes* 53: 1344–1351, 2004.
109. DP Gelain, M Cammarota, A Zanotto-Filho, RB de Oliveira, F Dal-Pizzol, I Izquierdo, LR Bevilaqua, and JC Moreira. Retinol induces the ERK1/2-dependant phosphorylation of CREB through a pathway involving the generation of reactive oxygen species in cultured Sertoli cells. *Cell Signal* 18: 1685–1694, 1996.
110. SK Moon, SK Kang, and CH Kim. Reactive oxygen species mediates disialoganglioside GD3-induced inhibition of ERK1/2 and matrix metalloproteinase-9 expression in vascular smooth muscle cells. *FASEB J* 20: 1387–1395, 2006.
111. RN Hasan and AI Schafer. Hemin upregulates Egr-1 expression in vascular smooth muscle cells via ROS ERK-1/2 Elk-1 and NF-{kappa}B. *Circ Res* 2007, Oct 25 [Epub ahead of print].
112. C Wakade, MM Khan, LM De Sevilla, Q-G Zhang, VB Mahesh, and DW Brann. Tamoxifen neuroprotection in cerebral ischemia involves attenuation of kinase activation and superoxide production and potentiation of mitochondrial superoxide dismutase. *Endocrinology* 149: 367–379, 2008.
113. R Navarro, I Busnadiego, MB Ruiz-Larrea, and JI Ruiz-Sanz. Superoxide anions are involved in doxorubicin-induced ERK activation in hepatocyte cultures. *Ann NY Acad Sci* 1090: 419–428, 2006.

114. GA Finlay, VJ Thannickal, BL Fanburg, and DJ Kwiatkowski. Platelet-derived growth factor-induced p42/44 mitogen-activated protein kinase activation and cellular growth is mediated by reactive oxygen species in the absence of TSC2/tuberin. *Cancer Res* 65: 10881–10890, 2005.

115. K Traore, R Sharma, RK Thimmulappa, WH Watson, S Biswal, and MA Trush. Redox-regulation of ERK1/2-directed phosphatase by reactive oxygen species: Role in signaling TPA-induced growth arrest in ML-1 cells. *J Cell Physiol* 216: 276–285, 2008.

116. LA Tephly and AB Carter. Constitutive NADPH oxidase and increased mitochondrial respiratory chain activity regulate chemokine gene expression. *Am J Physiol Lung Cell Mol Physiol* 293: L1143–L1155, 2007.

117. SR Datla, H Peshavariya, GJ Dusting, K Mahadev, BJ Goldstein, and F Jiang. Important role of Nox4 type NADPH oxidase in angiogenic responses in human microvascular endothelial cells in vitro. *Arterioscler Thromb Vasc Biol* 27: 2319–2324, 2007.

118. M Inaba, T Takahashi, Y Kumeda, T Kato, F Hato, Y Yutani, H Goto, Y Nishizawa, and S Kitagawa. Increased basal phosphorylation of mitogen-activated protein kinases and reduced responsiveness to inflammatory cytokines in neutrophils from patients with rheumatoid arthritis. *Clin Exp Rheumatol* 26: 52–60, 2008.

119. M Sabbatini, M Santillo, A Pisani, R Paterno, F Uccello, R Serù, G Matrone, G Spagnuolo, M Andreucci, V Serio, P Esposito, B Cianciaruso, G Fuiano, and EV Avvedimento. Inhibition of Ras/ERK1/2 signaling protects against postischemic renal injury. *Am J Physiol Renal Physiol* 290: F1408–F1415, 2006.

120. ED van Deel, Z Lu, X Xu, G Zhu, X Hu, TD Oury, RJ Bache, DJ Duncker, and Y Chen Extracellular superoxide dimutase protects the heart against oxidative stress and hypertrophy after myocardial infarction. *Free Radic Biol Med* 44: 1305–1313, 2008.

121. G Wojcicka, A Jamroz-Wisniewska, S Widomska, M Ksiazek, and J Beltowski. Role of extracellular signal-regulated kinases (ERK) in leptin-induced hypertension. *Life Sci* 82: 402–412, 2008.

122. AT Huddleston, W Tang, H Takeshima, SL Hamilton, and E Klann. Superoxide-induced potentiation in the hippocampus requires activation of ryanodine receptor type 3 and ERK. *J Neurophysiol* 99: 1565–1571, 2008.

123. FM Lyng, P Maguire, B McClean, C Seymour, and C Mothersill. The involvement of calcium and MAP kinase signaling pathways in the production of radiation-induced bystander effects. *Radiat Res* 165: 400–409, 2006.

124. S Pervin, R Singh, WA Freije, and G Chaudhuri. MKP-1-induced dephosphorylation of extracellular signal-regulated kinase is essential for triggering nitric oxide-induced apoptosis in human breast cancer cell lines: Implications in breast cancer. *Cancer Res* 63: 8853–8860, 2003.

125. DD Thomas, MG Espey, LA Ridnour, LJ Hofseth, D Mancard, CC Harris, and DA Wink. Hypoxic inducible factor 1α, extracellular signal-regulated kinase, and p53 are regulated by distinct threshold concentrations of nitric oxide. *Proc Nat Acad Sci USA* 101, 8894–8899, 2004.

126. DD Thomas, LA Ridnour, MG Espey, S Donzelli, S Ambs, SP Hussain, CC Harris, W DeGraff, DD Roberts, JB Mitchell, and DA Wink. Superoxide fluxes limit nitric oxide-induced signaling. *J Biol Chem* 281: 25984–25993, 2006.

127. A Scorziello, M Santillo, A Adornetto, C Dell'aversano, R Sirabella, S Damiano, LM Canzoniero, GF Renzo, and L Annunziato. NO-induced neuroprotection in ischemic preconditioning stimulates mitochondrial Mn-SOD activity and expression via RAS/ERK1/2 pathway. *J Neurochem* 103: 1472–1480, 2007.

128. W Wang, S Wang, EV Nishanian, A Del Pilar Cintron, RA Wesley, and RL Danner. Signaling by eNOS through a superoxide-dependant p42/44 mitogen-activated protein kinase pathway. *Am J Physiol Cell Physiol* 281: C544–C554, 2001.

129. S Mizuno, M Kadowaki, Y Demura, S Ameshima, I Miyamori, and T Ishizaki. p42/44 Mitogen-activated protein kinase regulated by p53 and nitric oxide in human pulmonary arterial smooth muscle cells. *Am J Respir Cell Mol Biol* 31: 184–192, 2004.

130. G-X Zhang, Y Nagai, T Nakagawa, H Miyanaka, Y Fujisawa, A Nishiyama, K Izuishi, K Ohmori, and S Kimura. Involvement of endogenous nitric oxide in angiotensin II-induced activation of vascular mitogen-activated protein kinases. *Am J Physiol Heart Circ Physiol* 293: H2403–H2408, 2007.

131. DM Kuhn and TJ Geddes. Peroxynitrite inactivates tryptophan hydroxylase via sulf-hydryl oxidation. Coincident nitration of enzyme tyrosyl residues has minimal impact on catalytic activity. *J Biol Chem* 274: 29726–29732, 1999.

132. C Zouki, S-L Zhang, JSD Chan, and JG Filep. Peroxynitrite induces integrin-dependant adhesion of human neutrophils to endothelial cells via activation of the Raf-1/MEK/Erk pathway. *FASEB J* 10.1096/fj.00–0521fje, 2000.

133. RK Upmacis, RS Deeb, MJ Resnick, R Lindenbaum, C Gamss, D Mittar, and DP Hajjar. Involvement of the mitogen-activated protein kinase cascade in peroxynitrite-mediated arachidonic acid release in vascular smooth muscle cells. *Am J Physiol Cell Physiol* 286: C1271–C1280, 2004.

134. Y Seko, N Takahashi, K Tobe, T Kadowaki, and Y Yazaki. Hypoxia and hypoxia/reoxygenation activate p65PAK, p38 mitogen-activated protein kinase (MAPK), and stress-activated protein kinase (SAPK) in cultured rat cardiac myocytes. *Biochem Biophys Res Commun* 239: 840–844, 1997.

135. J Huot, F Houle, F Marceau, and J Landry. Oxidative stress–induced actin reorganiza-tion mediated by the p38 mitogen-activated protein kinase/heat shock protein 27 path-way in vascular endothelial cells. *Circ Res* 80: 383–392, 1997.

136. A Clerk, SJ Fuller, A Michael, and PH Sugden. Stimulation of "stress-regulated" mito-gen-activated protein kinases (stress-activated protein kinases/c-Jun n-terminal kinases and p38-mitogen-activated protein kinases) in perfused rat hearts by oxidative and other stresses. *J Biol Chem* 273: 7228–7232, 1998.

137. M Ushio-Fukai, RW Alexander, M Akers, and KK Griendling. P38 Mitogen-activated protein kinase is a critical component of the redox-sensitive signaling pathways acti-vated by angiotensin II. Role in vascular smooth muscle cell hyperthrophy. *J Biol Chem* 273: 15022–15029, 1998.

138. C Chiu, DA Maddock, Q Zhang, KP Souza, AR Townsend, and Y Wan. TGF-beta-induced p38 activation is mediated by Rac1-regulated generation of reactive oxygen species in cultured human keratinocytes. *Int J Mol Med* 8: 251–255, 2001.

139. JJ Haddad. The involvement of L-gamma-glutamyl-L-cysteinyl-glycine (glutathione/GSH) in the mechanism of redox signaling mediating MAPK (p38)-dependant regulation of pro-inflammatory cytokine production. *Biochem Pharmacol* 63: 305–320, 2002.

140. B Herrera, M Fernandez, C Roncero, JJ Ventura, A Porras, A Valladares, M Benito, and I Fabregat. Activation of p38MARK by TGF-β in fetal rat hepatocytes requires radical oxygen production, but is dispensable for cell death. *FEBS Lett* 499: 225–229, 2001.

141. Q Wang and CM Doerschuk. The p38 mitogen-activated protein kinase mediates cytoskeletal remodeling in pulmonary microvascular endothelial cells upon intracellular adhesion molecule-1 ligation. *J Immunol* 166: 6877–6884, 2001.

142. E Babilonia, D Li, Z Wang, P Sun, D-H Lin, Y Jin, and W-H Wang. Mitogen-activated protein kinases inhibit the ROMK (Kir 1.1)-like small conductance k channels in the cortical collecting duct. *J Am Soc Nephrol* 17: 2687–2696, 2006.

143. Y Wei, Z Wang, E Babilonia, H Sterling, P Sun, and WH Wang. Effect of hydrogen peroxide on ROMK channels in the cortical collecting duct. *Am J Physiol Renal Physiol* 292: F1151–F1156, 2007.

144. P Becuwe, A Bianchi, C Didelot, M Barberi-Heyob, and M Dauca. Arachidonic acid activates a functional AP-1 and an inactive NF-kappaB complex in human HepG2 hepatoma cells. *Free Radic Biol Med* 35: 636–647, 2003.

145. C Gaitanaki, M Papatriantafyllou, K Stathopoulou, and I Beis. Effects of various oxidants and antioxidants on the p38-MAPK signalling pathway in the perfused amphibian heart. *Mol Cell Biochem* 291: 107–117, 2006.

146. TJ Hsieh, SL Zhang, JG Filep, SS Tang, JR Ingelfinger, and JS Chan. High glucose stimulates angiotensinogen gene expression via reactive oxygen species generation in rat kidney proximal tubular cells. *Endocrinology* 143: 2975–2985, 2002.

147. TJ Hsieh, P Fustier, CC Wei, SL Zhang, JG Filep, SS Tang, JP Ingelfinger, IG Fantus, P Hamet, and JS Chan. Reactive oxygen species blockade and action of insulin on expression of angiotensinogen gene in proximal tubular cells. *J Endocrinol* 183: 535–550, 2004.

148. A Kulisz, N Chen, NS Chandel, Z Shao, and PT Schumacker. Mitochondrial ROS initiate phosphorylation of p38 MAP kinase during hypoxia in cardiomyocytes. *Am J Physiol Lung Cell Mol Physiol* 282: L1324–L1329, 2002.

149. BM Emerling, LC Platanias, E Black, AR Nebreda, RJ Davis, and NS Chandel. Mitochondrial reactive oxygen species activation of p38 mitogen-activated protein kinase is required for hypoxia signaling. *Mol Cell Biol* 25: 4853–4862, 2005.

150. S Pawate, Q Shen, F Fan, and NR Bhat. Redox regulation of glial inflammatory response to lipopolysaccharide and interferongamma. *J Neurosci Res* 77: 540–551, 2004.

151. R Harfouche, NA Abdel-Malak, RP Brandes, A Karsan, K Irani, and SN Hussain. Roles of reactive oxygen species in angiopoietin-1/tie-2 receptor signaling. *FASEB J* Jul 2005; 10.1096/fj.04–3621fje.

152. L Xu, PS Pathak, and D Fukumura. Hypoxia-induced activation of p38 mitogen-activated protein kinase and phosphatidylinositol 3'-kinase signaling pathways contributes to expression of interleukin 8 in human ovarian carcinoma cells. *Clin Cancer Res* 10: 701–707, 2004.

153. F Amersi, XD Shen, D Anselmo, J Melinek, S Iyer, DJ Southard, M Katori, HD Volk, RW Busuttil, R Buelow, and JW Kupiec-Weglinski. Ex vivo exposure to carbon monoxide prevents hepatic ischemia/reperfusion injury through p38 MAP kinase pathway. *Hepatology* 35: 815–823, 2002.

154. BS Zuckerbraun, BY Chin, M Bilban, J de Costa d'Avila, J Rao, TR Billiar, and LE Otterbein. Carbon monoxide signals via inhibition of cytochrome c oxidase and generation of mitochondrial reactive oxygen species. *FASEB J* 21: 1099–1106, 2007.

155. W-S Choi, D-S Eom, BS Han, WK Kim, BH Han, E-J Choi, TH Oh, GJ Markelonis, JW Cho, and YJ Oh. Phosphorylation of p38 MAPK induced by oxidative stress is linked to activation of both caspase-8- and -9-mediated apoptotic pathways in dopaminergic neurons. *J Biol Chem* 279: 20451–20460, 2004.

156. W Bao, DJ Behm, SS Nerurkar, Z Ao, R Bentley, RC Mirabile, DG Johns, TN Woods, CP Doe, RW Coatney, JF Ohlstein, SA Douglas, RN Willette, and TL Yue. Effects of p38 MAPK inhibitor on angiotensin II-dependant hypertension, organ damage, and superoxide anion production. *J Cardiovasc Pharmacol* 49: 362–368, 2007.

157. ML Brezniceanu, CC Wei, SL Zhang, TJ Hsieh, DF Guo, MJ Hebert, JR Ingelfinger, JG Filep, and JS Chan. Transforming growth factor-beta 1 stimulates angiotensinogen gene expression in kidney proximal tubular cells. *Kidney Int* 69: 1977–1985, 2006.

158. J Widder, T Behr, D Fraccarollo, K Hu, P Galuppo, P Tas, CE Angermann, G Ertl, and J Bauersachs. Vascular endothelial dysfunction and superoxide anion production in heart failure are p38 MAP kinase-dependant. *Cardiovasc Res* 63: 161–167, 2004.

159. M Alvarez-Maqueda, R El Bekay, J Monteseirin, G Alba, P Chacon, A Vega, C Santa Maria, JR Tejedo, J Martin-Nieto, FJ Bedoya, E Pintado, and F Sobrino. Homocysteine enhances superoxide anion release and NADPH oxidase assembly by human neutrophils. Effects on MAPK activation and neutrophil migration. *Atherosclerosis* 172: 229–238, 2004.

160. T Peng, X Lu, and Q Feng. Pivotal role of gp91phox-containing NADH oxidase in lipopolysaccharide-induced tumor necrosis factor-alpha expression and myocardial depression. *Circulation* 111: 1637–1644, 2005.

161. C Nediani, E Borchi, C Giordano, S Baruzzo, V Ponziani, M Sebastiani, P Nassi, A Mugelli, G d'Amati, and E Cerbai. NADPH oxidase-dependant redox signaling in human heart failure: Relationship between the left and right ventricle. *J Mol Cell Cardiol* 42: 826–834, 2007.

162. P Rocic, C Kolz, R Reed, B Potter, and WM Chilian. Optimal reactive oxygen species concentration and p38 MAP kinase are required for coronary collateral growth. *Am J Physiol Heart Circ Physiol* 292: H2729–H2736, 2007.

163. S Mitra and E Abraham. Participation of superoxide in neutrophil activation and cytokine production. *Biochim Biophys Acta* 1762: 732–741, 2006.

164. C Nito, H Kamada, H Endo, K Niizuma, DJ Myer, and PH Chan. Role of the p38 mitogen-activated protein kinase/cytosolic phospholipase A(2) signaling pathway in blood-brain barrier disruption after focal cerebral ischemia and reperfusion. *J Cereb Blood Flow Metab* 2008 Jun 11. [Epub ahead of print].

165. T Adachi, DR Pimentel, T Heibeck, X Hou, YJ Lee, B Jiang, Y Ido, and RA Cohen. S-glutathiolation of Ras mediates redox-sensitive signaling by angiotensin II in vascular smooth muscle cells. *J. Biol. Chem* 279: 857–862, 2004.

166. N Clavreul, T Adachi, DR Pimental, Y Ido, C Schoneich, and RA Cohen. S-glutathiolation by peroxynitrite of p21ras at cysteine-118 mediates its direct activation and downstream signaling in endothelial cells. *FASEB J* 10.1096/fj.05–4875fje. (2007).

167. S Azhar, P Abidi, S Leers-Sucheta, Y Cortez, J Han, H Zhang, S Zaidi, and W Shen. Oxidative stress-induced inhibition of adrenal steroidogenesis requires participation of p38 MAPK signaling pathway. *J Endocrinol* 2008; 10.1677/JOE–07–0570.

168. K Suzuki, M Hino, H Kutsuna, F Hato, C Sakamoto, T Takahashi, N Tatsumi, and S Kitagawa. Selective activation of p38 mitogen-activated protein kinase cascade in human neutrophils stimulated by IL-1beta. *J Immunol* 167: 5940–5947, 2001.

169. T Saito, H Takahashi, H Doken, H Koyama, and Y Aratani. Phorbol myristate acetate induces neutrophil death through activation of p38 mitogen-activated protein kinase that requires endogenous reactive oxygen species other than HOCl. *Biosci Biotechnol Biochem* 69: 2207–2212, 2005.

170. A Huwiler and J Pfeilschifter. Nitric oxide stimulates the stress-activated protein kinase p38 in rat renal mesangial cells. *J Exp Biol* 202: 655–660, 1999.

171. DD Browning, ND Windes, and RD Ye. Activation of p38 mitogen-activated protein kinase by lipopolysaccharide in human neutrophils requires nitric oxide-dependant cGMP accumulation. *J Biol Chem* 274: 537–542, 1999.

172. DD Browning, MP McShane, C Marty, and RD Ye. Nitric oxide activation of p38 mitogen-activated protein kinase in 293T fibroblasts requires cGMP-dependant protein kinase. *J Biol Chem* 275: 2811–2816, 2000.

173. BK Yoo, JW Choi, CY Shin, SJ Jeon, SJ Park, JH Cheong, SY Han, JR Ryu, MR Song, and KH Ko. Activation of p38 MAPK induced peroxynitrite generation in LPS plus IFN-gamma-stimulated rat primary astrocytes via activation of iNOS and NADPH oxidase. *Neurochem Int* 52: 1188–1197, 2008.

174. YYC Lo, JMS Wong, and TF Cruz. Reactive oxygen species mediate cytokine activation of c-Jun NH2-terminal kinases. *J Biol Chem* 271: 15703–15707, 1996.

175. KR Laderoute and KA Webster. Hypoxia/reoxygenation stimulates Jun kinase activity through redox signaling in cardiac myocytes. *Circ Res* 80: 336–344, 1997.
176. JR Lee and GA Koretzky. Production of reactive oxygen intermediates following CD40 ligation correlates with c-Jun N-terminal kinase activation and IL-6 secretion in murine B lymphocytes. *Eur J Immunol* 28: 4188–4197, 1998.
177. XL Cui and JG Douglas. Arachidonic acid activates c-jun N-terminal kinase through NADPH oxidase in rabbit proximal tubular epithelial cells. *Proc Natl Acad Sci USA* 94: 3771–3776, 1997.
178. H Peng, T Takano, J Papillon, K Bijian, A Khadir, and AV Cybulsky. Complement activates the c-Jun N-terminal kinase/stress-activated protein kinase in glomerular epithelial cells. *J Immunol* 169: 2594–2601, 2002.
179. G Pani, R Colavitti, S Borrello, and T Galeotti. Endogenous oxygen radicals modulate protein tyrosine phosphorylation and JNK-1 activation in lectin-stimulated thymocotes. *Biochem J* 347: 173–181, 2000.
180. T Joneson and D Bar-Sagi. A Rac1 effector site controlling mitogenesis through superoxide production. *J Biol Chem* 273: 17991–17994, 1998.
181. M Benhar, I Dalyot, D Engelberg, and A Levitzki. Enhanced ROS production in oncogenically transformed cells potentiates c-Jun n-terminal kinase and p38 mitogen-activated protein kinase activation and sensitization to genotoxic stress. *Mol Cell Biol* 21: 6913–6926, 2001.
182. Y-M Go, AL Levonen, D Moellering, A Ramachandran, RP Patel, H Jo, and VM Darley-Usmar. Endothelial Nos-dependant activation of c-jun NH_2-terminal kinase by oxidized low-density lipoprotein. *Am J Physiol Heart Circ Physiol* 281: H2705–H2713, 2001.
183. N Matesanz, N Lafuente, V Azcutia, D Martin, A Cuadrado, J Nevado, L Rodriguez-Manas, CF Sanchez-Ferrer, and C Peiro. Xanthine oxidase-derived extracellular superoxide anions stimulate activator protein 1 activity and hypertrophy in human vascular smooth muscle via c-Jun N-terminal kinase and p38 mitogen-activated protein kinases. *J Hypertens* 25: 609–618, 2007.
184. C Viedt, HI Krieger-Brauer, J Fei, C Elsing, W Kubler, and J Kreuzer. Differential activation of mitogen-activated protein kinases in smooth muscle cells by angiotensin II: Involvement of p22phox and reactive oxygen species. *Arterioscler Thromb Vasc Biol* 20: 940–949, 2000.
185. G Ding, A Zhang, S Huang, X Pan, G Zhen, R Chen, and T Yang. ANG II induces c-Jun NH2-terminal kinase activation and proliferation of human mesangial cells via redox-sensitive transactivation of the EGFR. *Am J Physiol Renal Physiol* 293: F1889–F1897, 2007.
186. PK Vayalil, KE Iles, J Choi, A-K Yi, EM Postlethwait, and R-M Liu. Glutathione suppresses TGF-{beta}-induced PAI-1 expression by inhibiting p38 and JNK MAPK and the binding of AP-1, SP-1, and Smad to the PAI-1 promoter. *Am J Physiol Lung Cell Mol Physiol* 293: L1281–L1292, 2007.
187. SK Mantena and SK Katiyar. Grape seed proanthocyanidins inhibit UV-radiation-induced oxidative stress and activation of MAPK and NF-kappaB signaling in human epidermal keratinocytes. *Free Radic Biol Med* 40: 1603–1614, 2006.
188. SJ Lin, SK Shyue, MC Shih, TH Chu, YH Chen, HH Ku, JW Chen, KB Tam, and YL Chen. Superoxide dismutase and catalase inhibit oxidized low-density lipoprotein-induced human aortic smooth muscle cell proliferation: Role of cell-cycle regulation, mitogen-activated protein kinases, and transcription factors. *Atherosclerosis* 190, 124–134, 2007.
189. A Pechtelidou, I Beis, and C Gaitanaki. Transient and sustained oxidative stress differentially activate the JNK1/2 pathway and apoptotic phenotype in H9c2 cells. *Mol Cell Biochem* 309: 177–189, 2008.
190. J Pfeilschifter and A Huwiler. Nitric oxide stimulates stress-activated protein kinases in glomerular endothelial and mesangial cells. *FEBS Lett* 396: 67–70, 1996.

191. H-S Park, S-H Huh, M-S Kim, SH Lee, and E-J Choi. Nitric oxide negatively regulates c-Jun N-terminal kinase/stress-activated protein kinase by means of S-nitrosylation. *Proc Natl Acad Sci USA* 97: 14382–14387, 2000.

192. YM Go, RP Patel, MC Maland, H Park, JS Beckman, VM Darley-Usmar, and H Jo. Evidence for peroxynitrite as a signaling molecule in flow-dependant activation of c-Jun NH(2)-terminal kinase. *Am J Physiol* 277: H1647–H1653, 1999.

193. Y-M Go, YC Boo, H Park, MC Maland, R Patel, KA Pritchard Jr, Y Fujio, K Walsh, V Darley-Usmar, and H Jo. Protein kinase B/Akt activates c-Jun NH2-terminal kinase by increasing NO production in response to shear stress. *J Appl Physiol* 91: 1574–1581, 2001.

194. J-J Ventura, P Cogswell, RA Flavell, AS Baldwin, Jr., and RJ Davis. JNK potentiates TNF-stimulated necrosis by increasing the production of cytotoxic reactive oxygen species. *Genes & Dev* 18: 2905–2915, 2004.

195. J Antosiewicz, A Herman-Antosiewicz, SW Marynowski, and SV Singh. c-Jun NH2-terminal kinase signaling axis regulates diallyl trisulfide-induced generation of reactive oxygen species and cell cycle arrest in human prostate cancer cells. *Cancer Res* 66: 5379–5386, 2006.

196. A Nakajima, S Komazawa-Sakon, M Takekawa, T Sasazuki, WC Yeh, H Yagita, K Okumura, and H Nakano. An antiapoptotic protein, c-FLIPL, directly binds to MKK7 and inhibits the JNK pathway. *EMBO J* 25: 5549–5559, 2006.

197. S Dimon-Gadal, P Gerbaud, G Keryer, W Anderson, D Evain-Brion, and F Raynaud. In vitro effects of oxygen-derived free radicals on type I and type II cAMP-dependant protein kinases. *J Biol Chem* 273: 22833–22840, 1998.

198. T Toyoda, T Hayashi, L Miyamoto, S Yonemitsu, M Nakano, S Tanaka, K Ebihara, H Masuzaki, K Hosoda, G Inoue, A Otaka, K Sato, T Fushiki, and K Nakao. Possible involvement of the α1 isoform of 5'AMP-activated protein kinase in oxidative stress-stimulated glucose transport in skeletal muscle. *Am J Physiol Endocrinol Metab* 287: E166–E173, 2004.

199. ME Sandström, S-J Zhang, J Bruton, JP Silva, MB Reid, H Westerblad, and A Katz. Role of reactive oxygen species in contraction-mediated glucose transport in mouse skeletal muscle. *J Physiol* 575: 251–262, 2006.

200. Z An, H Wang, P Song, M Zhang, X Geng, and M-H Zou. Nicotine-induced activation of AMP-activated protein kinase inhibits fatty acid synthase in 3T3-L1 adipocytes: A role for oxidant stress. *J Biol Chem* 282: 26793–26801, 2007.

201. M Zhang, Y Dong, J Xu, Z Xie, Y Wu, P Song, M Guzman, J Wu, and M-H Zou. Thromboxane receptor activates the AMP-activated protein kinase in vascular smooth muscle cells via hydrogen peroxide. *Circ Res* 102: 328–337, 2008.

202. GL Schieven, JM Kirihara, DE Myers, JA Ledbetter, and FM Uckun. Reactive oxygen intermediates activate NF-kB in a tyrosine kinase-dependant mechanism and in combination with vanadate activate the p56lck and p59fyn tyrosine kinases in human lymphocytes. *Blood* 82: 1212–1220, 1993.

203. JH Brumell, AL Burkhardt, JB Bolen, and S Grinstein. Endogenous reactive oxygen intermediates activate tyrosine kinases in human neutrophils. *J Biol Chem* 271: 1455–1461, 1996.

204. JS Hardwick and BM Sefton. Activation of the Lck tyrosine protein kinase by hydrogen peroxide requires the phosphorylation of Tyr-394. *Proc Natl Acad Sci USA* 92: 4527–4531, 1995.

205. SR Yan and G Berton. Regulation of Src family tyrosine kinase activities in adherent human neutrophils. Evidence that reactive oxygen intermediates produced by adherent neutrophils increase the activity of the p58c-fgr and p53/56lyn tyrosine kinases. *J Biol Chem* 271: 23464–23471, 1996.

206. L Wu and J de Champlain. Superoxide anion-induced formation of inositol phosphates involves tyrosine kinase activation in smooth muscle cells from rat mesenteric artery. *Biochem Biophys Res Commun* 259: 239–243, 1999.

207. H Tang, Q Hao, SA Rutherford, B Low, and ZJ Zhao. Inactivation of Src family tyrosine kinase by reactive oxygen species *in vivo. J Biol Chem* 280, 23918–23925, 2005.

208. E Babilonia, Y Wei, H Sterling, P Kaminski, M Wolin, and W-H Wang. Superoxide anions are involved in mediating the effect of low K intake on c-Src expression and renal k secretion in the cortical collecting duct. *J Biol Chem* 280: 10790–10796, 2005.

209. Md Ruhul Abid, KC Spokes, S-C Shih, and WC Aird. NADPH oxidase activity selectively modulates vascular endothelial growth factor signaling pathways. *J Biol Chem* 282: 35373–35385, 2007.

210. M Bouaouina, E Blouin, L Halbwachs-Mecarelli, P Lesavre, and P Rieu. TNF-induced β₂ integrin activation involves src kinases and a redox-regulated activation of p38 MAPK. *J Immunol* 173: 1313–1320, 2004.

211. YK Kim, YY Jang, DH Kim, HH Ko, ES Han, and CS Lee. Differential regulation of protein tyrosine kinase on free radical production, granule enzyme release, and cytokine synthesis by activated murine peritoneal macrophages. *Biochem Pharmacol* 61: 87–96, 2001.

212. B Schieffer, M Luchtefeld, S Braun, A Hilfiker, D Hilfiker-Kleiner, and H Drexler. Role of NAD(P)H oxidase in angiotensin II-induced JAK/STAT signaling and cytokine induction. *Circ Res* 87: 1195–1201, 2000.

213. K Moriwaki, H Kiyomoto, H Hitomi, G Ihara, K Kaifu, K Matsubara, T Hara, N Kondo, K Ohmori, A Nishiyama, T Fukui, and M Kohno. Interferon-gamma enhances superoxide production in human mesangial cells via the JAK-STAT pathway. *Kidney Int* 70: 788–793, 2006.

214. Y Gotoh and JA Cooper. Reactive oxygen species- and dimerization-induced activation of apoptosis signal-regulating kinase 1 in tumor necrosis factor-α signal transduction. *J Biol Chem* 273: 17477–17482, 1998.

215. J Du, H Suzuki, F Nagase, AA Akhand, XY Ma, T Yokoyama, T Miyata, and I Nakashima. Superoxide-mediated early oxidation and activation of ASK1 are important for initiating methylglyoxal-induced apoptosis process. *Free Radic Biol Med* 31: 469–478, 2001.

216. EH Goldman, L Chen, and H Fu. Activation of apoptosis signal-regulating kinase 1 by reactive oxygen species through dephosphorylation at serine 967 and 14-3-3 dissociation. *J Biol Chem* 279: 10442–10449, 2004.

217. C Zhou, C Ziegler, LA Birder, AF Stewart, and FS Levitan. Angiotensin II and stretch activate NADPH oxidase to destabilize cardiac Kv4.3 channel mRNA. *Circ Res* 98: 1040–1047, 2006.

218. PJ Nadeau, SJ Charette, MB Toledano, and J Landry. Disulfide bond-mediated multimerization of Ask1 and its reduction by thioredoxin-1 regulate H2O2-induced c-Jun NH2-terminal kinase activation and apoptosis. *Mol Biol Cell* 18: 3903–3913, 2007.

219. DM Shasby, M Yorek, and SS Shasby. Exogenous oxidants initiate hydrolysis of endothelial cell inositol phospholipids. *Blood* 72: 491–499, 1988.

220. JH Hong, SJ Moon, HM Byun, MS Kim, H Jo, YS Bae, S-I Lee, MD Bootman, HL Roderick, DM Shin, and JT Seo. Critical role of PLCgamma 1 in the generation of H2O2-evoked [Ca2+]i oscillations in cultured rat cortical astrocytes. *J Biol Chem* 281: 13057–13067, 2006.

221. HM Lander, JS Ogiste, and KK Teng, A Novogrodsky. p21ras as a common signaling target of reactive free radicals and cellular redox stress. *J Biol Chem* 270: 21195–21198, 1995.

222. K Irani, Y Xia, JL Zweier, SJ Sollott, CJ Der, ER Fearon, M Sundaresan, T Finkel, and PJ Goldschmidt-Clermont. Mitogenic signaling mediated by oxidants in Ras-transformed fibroblasts. *Science* 275: 1567–1568, 1997.

223. JQ Yang, GR Buettner, FE Domann, Q Li, JF Engelhardt, CD Weydert, and LW Oberley. v-Ha-ras mitogenic signaling through superoxide and derived reactive oxygen species. *Mol Carcinog* 33: 206–218, 2002.

224. J Heo and SL Campbell. Superoxide anion radical modulates the activity of Ras and Ras-related GTPases by a radical-based mechanism similar to that of nitric oxide. *J Biol Chem* 280: 12438–12445, 2005.

225. J Heo and SL Campbell. Ras regulation by reactive oxygen and nitrogen species. *Biochemistry* 45: 2200–2210, 2006.

226. KW Raines, MG Bonini, and SL Campbell. Nitric oxide cell signaling: S-nitrosation of Ras superfamily GTPases. *Cardiovasc Res* 75: 229–239, 2007.

227. L Jin, Z Ying, and RC Webb. Activation of Rho/Rho kinase signaling pathway by reactive oxygen species in rat aorta. *Am J Physiol Heart Circ Physiol* 287: H1495–1500, 2004.

228. H Kajimoto, K Hashimoto, SN Bonnet, A Haromy, G Harry, R Moudgil, T Nakanishi, I Rebeyka, B Thébaud, ED Michelakis, and SL Archer. Oxygen activates the Rho/Rho-kinase pathway and induces RhoB and ROCK-1 expression in human and rabbit ductus arteriosus by increasing mitochondria-derived reactive oxygen species: A newly recognized mechanism for sustaining ductal constriction. *Circulation* 115: 1777–1788, 2007.

229. JM Denu and KG Tanner. Specific and reversible inactivation of protein tyrosine phosphatases by hydrogen peroxide: Evidence for a sulfenic acid intermediate and implications for redox regulation. *Biochemistry* 37: 5633–5642, 1998.

230. A Caselli, R Marzocchini, G Camici, G Manao, G Moneti, G Pieraccini, and G Ramponi. The inactivation mechanism of low molecular weight phosphotyrosine-protein phosphatase by H2O2. *J Biol Chem* 273: 32554–32560, 1998.

231. T-C Meng, T Fukada, and NK Tonks. Reversible oxidation and inactivation of protein tyrosine phosphotases in vivo. *Molec Cell* 9: 387–399, 2002.

232. A Groen, S Lemeer, T van der Wijk, J Overvoorde, AJ R Heck, A Ostman, D Barford, M Slijper, and J den Hertog. Differential oxidation of protein-tyrosine phosphatases. *J Biol Chem* 280: 10298–10304, 2005.

233. T-C Meng, DA Buckley, S Galic, S Tiganis, and NK Tonks. Regulation of insulin signaling through reversible oxidation of the protein-tyrosine phosphatases TC45 and PTP1B. *J Biol Chem* 279: 37716–37725, 2004.

234. WC Barrett, JP DeGnore, YF Keng, ZY Zhang, MB Yim, and PB Chock. Roles of superoxide radical anion in signal transduction mediated by reversible regulation of protein-tyrosine phosphatase 1B. *J Biol Chem* 274: 34543–34546, 1999.

235. Q Wang, D Dube, RW Friesen, TG LeRiche, KP Bateman, L Trimble, J Sanghara, R Pollex, C Ramachandran, MJ Gresser, and S Huang. Catalytic inactivation of protein-tyrosine phosphatase CD4 and protein tyrosine phosphatase 1B by polyaromatic quinones. *Biochemistry* 43: 4294–4303, 2004.

236. DM Barrett, SM Black, H Todor, RK Schmidt-Ullrich, KS Dawson, and RB Mikkelsen. Inhibition of protein-tyrosine phosphatases by mild oxidative stresses is dependant on S-nitrosylation. *J Biol Chem* 280: 14453–14461, 2005.

237. CB Klee, H Ren, and X Wang. Regulation of the calmodulin-stimulated protein phosphatase, calcineurin. *J Biol Chem* 273: 13367–13370, 1998.

238. X Wang, VC Culotta, and CB Klee. Superoxide dismutase protects calcineurin from inactivation. *Nature* 383: 434–437, 1996.

239. D Sommer, KL Fakata, SA Swanson, and PM Stemmer. Modulation of the phosphatase activity of calcineurin by oxidants and antioxidants in vitro. *Eur J Biochem* 267: 2312–2322, 2000.

240. R Bogumil, D Namgaladze, D Schaarschmidt, T Schmachtel, S Hellstern, R Mutzel, and V Ullrich. Inactivation of calcineurin by hydrogen peroxide and phenylarsine oxide: Evidence for a dithiol–disulfide equilibrium and implications for redox regulation. *Eur J Biochem* 267: 1407–1415, 2000.

241. D Namgaladze, HW Hofer, and V Ullrich. Redox control of calcineurin by targeting the binuclear Fe^{2+}-Zn^{2+} center at the enzyme active site. *J Biol Chem* 277: 5962–5969, 2002.

242. V Ullrich, D Namgaladze, and A Frein. Superoxide as inhibitor of calcineurin and mediator of redox regulation. *Toxicol Lett* 139: 107–110, 2003.

243. D Sommer, S Coleman, SA Swanson, and PM Stemmer. Differential susceptibilities of serine/threonine phosphatases to oxidative and nitrosative stress. *Arch Biochem Biophys* 404: 271–278, 2002.

244. D Namgaladze, I Shcherbyna, J Kienhofer, HW Hofer, and V Ullrich. Superoxide targets calcineurin signaling in vascular endothelium. *Biochem Biophys Res Commun* 334: 1061–1067, 2005.

245. T Murata, I Hatayama, K Satoh, S Tsuchida, and K Sato. Activation of rat glutathione transferases in class mu by active oxygen species. *Biochem Biophys Res Commun* 171: 845–851, 1998.

246. B Brune, KU Schmidt, and V Ullrich. Activation of soluble guanylate cyclase by carbon monoxide and inhibition by superoxide anion. *Eur J Biochem* 192: 683–688, 1990.

247. B Blanchard-Fillion, JM Souza, T Friel, GCT Jiang, K Vrana, V Sharov, L Barrón, C Schöneich, C Quijano, B Alvarez, R Radi, S Przedborski, GS Fernando, J Horwitz, and H Ischiropoulos. Nitration and inactivation of tyrosine hydroxylase by peroxynitrite. *J Biol Chem* 276: 46017–46023, 2001.

248. DM Kuhn, M Sadidi, X Liu, C Kreipke, T Geddes, C Borges, and JT Watson. Peroxynitrite-induced nitration of tyrosine hydroxylase. Identification of tyrosines 423, 428, and 432 as sites of modification by matrix-assisted laser desorption ionization time-of-flight mass spectrometry and tyrosine-scanning mutagenesis. *J Biol Chem* 277: 14336–14342, 2002.

249. K Kaiserova, S Srivastava, JD Hoetker, SO Awe, X-L Tang, J Cai, and A Bhatnagar. Redox activation of aldose reductase in the ischemic heart. *J Biol Chem* 281: 15110–15120, 2006.

250. K Kaiserova, X-L Tang, S Srivastava, and A Bhatnagar. Role of nitric oxide in regulating aldose reductase activation in the ischemic heart. *J Biol Chem* 283: 9101–9112, 2008.

251. N Inoue, S Takeshita, D Gao, T Ishida, S Kawashima, H Akita, R Tawa, H Sakurai, and M Yokoyama. Lysophosphatidylcholine increases the secretion of matrix metalloproteinase 2 through the activation of NADH/NADPH oxidase in cultured aortic endothelial cells. *Atherosclerosis* 155: 45–52, 2001.

252. K Grote, I Flach, M Luchtefeld, E Akin, SM Holland, H Drexler, and B Schieffer. Mechanical stretch enhances mRNA expression and proenzyme release of matrix metalloproteinase-2 (MMP-2) via NAD(P)H oxidase–derived reactive oxygen species. *Circ Res* 92: e80–e86, 2003.

253. TL Deem and JM Cook-Mills. Vascular cell adhesion molecule 1 (VCAM-1) activation of endothelial cell matrix metalloproteinases: Role of reactive oxygen species. *Blood* 104: 2385–2393, 2004.

254. G Zalba, A Fortuño, J Orbe, G San José, MU Moreno, M Belzunce, JA Rodríguez, O Beloqui, JA Páramo, and J Díez. Phagocytic NADPH oxidase-dependant superoxide production stimulates matrix metalloproteinase-9: Implications for human atherosclerosis. *Arterioscler Thromb Vasc Biol* 27: 587–593, 2007.

255. S Lopez-Ongil, V Senchak, M Saura, C Zaragoza, M Ames, BJ Ballermann, M Rodriguez-Puyol, D Rodriguez-Puyol, and CJ Lowenstein. Superoxide regulation of endothelin converting enzyme. *J Biol Chem* 275: 26423–26427, 2000.

256. JJP Deudero, C Caramelo, MC Castellanos, F Neria, R Fernández-Sánchez, O Calabia, S Peñate, and FR González-Pacheco. Induction of hypoxia-inducible factor 1alpha gene expression by vascular endothelial growth factor. Role of a superoxide-mediated mechanism. *J Biol Chem* 283: 11435–11444, 2008.

5 ROS and RNS Signaling in Apoptosis

Apoptosis is a form of programmed cell death in multicellular organisms character-ized by biochemical events that lead to a variety of morphological changes, including blebbing and changes to the cell membranes. These changes depend on many sig-naling processes, including important reactive oxygen species (ROS)- and reactive nitrogen species (RNS)-mediated cascades. In contrast to apoptosis, necrosis is an accidental death of cells, which also depends on ROS, but in this case, free radicals play mainly a damaging role, and their signaling functions may not be as important as in apoptosis. Therefore, this chapter will be dedicated to the study of the signaling role of ROS and RNS in apoptosis.

5.1 STIMULATION AND INHIBITION OF ROS-INDUCED APOPTOSIS

ROS (superoxide [$O_2\cdot^-$] and hydrogen peroxide [H_2O_2]) signaling is an important ini-tiation stage of apoptosis in mitochondria. In many cases, the start of apoptosis is associated with a leak of $O_2\cdot^-$ from mitochondrial respiratory chain and cytochrome c release (CCR). Numerous factors have been identified that regulate ROS stimulation of apoptosis in mitochondria. Among them is Bcl-2 protein, a general inhibitor of apoptosis in mammalian cells [1,2]. Bcl-2 is localized to intracellular sites of ROS generation, including mitochondria, endoplasmic reticula, and nuclear membranes. Bcl-2 protected cells from H_2O_2- and menadione-induced oxidative deaths. Following an apoptotic signal, cells stimulate lipid peroxidation, and overexpression of Bcl-2 suppresses lipid peroxidation.

The role of Bcl-2 in apoptosis has been studied in various works. Esposti et al. [3] suggest that Bcl-2 lets the cells adapt to ROS overproduction, increasing cellular antioxidant defenses and counteracting ROS-mediated damaging effects by different cell-death stimuli. Ling et al. [4] demonstrated that ROS generation played an important role in the initiation of the apoptosis cascade by the proteasome inhibitor bortezomib and that overexpression of Bcl-2 in prostate carcinoma cells diminished bortezomib-induced apoptosis. Hildeman et al. [5] found that the superoxide dismutase (SOD) mimetic MnTBAP decreased intracellular ROS and prevented apoptosis of activated T cells in vitro. They showed that MnTBAP increased the expression of Bcl-2, which was normally decreased by T-cell activation. There was a tight inverse correlation between the levels of Bcl-2 and ROS within T cells. In vivo, ROS production in acti-vated T cells occurred before Bcl-2 downregulation. In general, these findings sug-gested that ROS sensitized T cells to apoptosis by decreasing the expression of Bcl-2.

Wang et al. [6] reported that H_2O_2 was a major mediator of Bcl-2 downregulation and apoptosis induction by cisplatin in human lung cancer cells.

The p53 tumor suppressor gene can induce either apoptosis or a permanent growth arrest (also termed senescence) phenotype in response to cellular stresses. It has been shown that ROS are the potent activators of p53. Jonson et al. [7] investigated whether p53 enhanced ROS levels during the induction of p53-dependent apoptosis. They found that the cells sensitive to p53-mediated apoptosis produced ROS concomitantly with p53 overexpression, whereas cells resistant to p53 failed to produce ROS. In sensitive cells, both ROS production and apoptosis were inhibited by antioxidant treatment. These findings suggest that ROS mediate p53-depended apoptosis.

Macip et al. [8] showed that an increase in intracellular ROS that was associated with the magnitude of p53 protein expression correlated with the induction of either senescence or apoptosis in both normal and cancer cells. ROS inhibitors ameliorated both p53-dependent cell fates, implicating ROS accumulation as an effector in each case. Moreover, physiological p53 levels in combination with an exogenous ROS source were able to convert a p53 senescence response into apoptosis. Fujioka et al. [9] demonstrated that NFκB activated the wild-type p53 tumor suppressor to initiate apoptosis in response to overproduction of $O_2\cdot^-$ in human pancreatic tumor cells. Hu et al. [10] reported that selenite induced a rapid $O_2\cdot^-$ burst and p53 activation in prostate cancer cells, leading to a synergistic caspase-9/3 cascade-mediated apoptosis.

It has been demonstrated that apoptosis is a main event in cancer chemoprevention by selenium and that ROS induce apoptosis by selenium compounds. Xiang et al. [11] investigated the role of $O_2\cdot^-$ and mitochondria in selenite-induced apoptosis in human prostate cancer cells. They found that selenite induced cancer cell death and apoptosis by producing $O_2\cdot^-$ and that $O_2\cdot^-$ production, cell death, and apoptosis were inhibited by overexpression of MnSOD but not CuZnSOD (SOD1), or catalase. Furthermore, selenite treatment resulted in a decrease in mitochondrial membrane potential, the release of cytochrome c into the cytosol, and activation of caspases 9 and 3. Taken together, these findings suggest that mitochondrial $O_2\cdot^-$ production is, at least in part, a key event in selenium-induced apoptosis in prostate cancer cells.

It has been shown that mtDNA damage can initiate apoptosis, although the mechanism involved remains unclear. Ricci et al. [12] investigated angiotensin II (Ang II)-mediated apoptosis in cells that were transduced with a lentiviral vector to overexpress the DNA repair enzyme 8-oxoguanine glycosylase, or treated with inhibitors that blocked Ang II-induced mtDNA damage. It was found that mtDNA-damaged cells showed two phases of $O_2\cdot^-$ generation that originated from NADPH oxidase (Nox) and mitochondria, respectively, while control cells exhibited only the first one. During the second phase of mtDNA-damaged cells, the mitochondrial membrane potential collapsed, cytochrome c was released, and the cells underwent apoptosis. The authors propose a novel mechanism of apoptosis based on the stimulation of mitochondrial $O_2\cdot^-$ generation by mtDNA damage. Pajusto et al. [13] have shown that $O_2\cdot^-$ induced efficient apoptotic DNA degradation after the initial formation of DNA strand breaks in primary CD4+ T cells.

It is believed that neutrophils, which have a short half-life, die by apoptosis both in vivo and in vitro. Fadeel et al. [14] have shown that caspases, important members of the apoptotic cascade, are activated in a time-dependent manner in neutrophils undergoing spontaneous apoptosis. However, ROS produced by NOX actually prevented the activation of caspases in neutrophils. Therefore, caspases were apparently suppressed in activated neutrophils. Fay et al. [15] demonstrated that apoptosis in neutrophils mediated by $O_2\cdot^-$ and H_2O_2 can be initiated by 1-ethyl-2-benzimidazolinone (1-EBIO), an activator of Ca(2+)-activated potassium channels of small conductance (SC) and intermediate conductance (IC). The proapoptotic effect of 1-EBIO was independent of the NOX pathway and probably depended on mitochondrial ROS production.

To determine the mechanisms of ROS-induced apoptosis, caspase activation and nuclear factor-κB (NFκB) signaling have been studied by Jones et al. [16]. They found that the treatment of hepatocytes with H_2O_2 or the $O_2\cdot^-$ generator menadione caused 26% and 33% cell death, respectively. Death from ROS occurred by apoptosis as indicated by morphology, induction of caspase activation, and DNA fragmentation. Importantly, caspase inhibition blocked H_2O_2- but not menadione-induced apoptosis. At the same time, inhibition of NFκB activation decreased apoptosis in both cases. Thus, H_2O_2 and $O_2\cdot^-$ signaling stimulated distinct apoptotic pathways in hepatocytes by both caspase-dependent and independent ways. In contrast to a known protective effect of NFκB activation in TNF-α-induced hepatocyte apoptosis, NFκB promoted death from ROS in hepatocytes.

Hirpara et al. [17] showed that H_2O_2 formed by dismutation of $O_2\cdot^-$ in mitochondria induced intracellular acidification and release of cytochrome c independently of the inner membrane pore, resulting in caspase activation and apoptosis of leukemia cells. Valencia and Moran [18] also showed that $O_2\cdot^-$ and H_2O_2 differently affected neuronal apoptotic death. $O_2\cdot^-$-initiated apoptotic death of cultured cerebellar granules occurred through caspase activation, nuclear condensation, phosphatidylserine translocation, and a decrease in intracellular calcium levels. On the other hand, H_2O_2 led to a necrosis-like cell death that did not induce caspase activation, phosphatidylserine translocation, or changes in calcium levels. It was concluded that $O_2\cdot^-$ and H_2O_2 participated in different mechanisms of neuronal cell death through necrosis and apoptosis. (It should be noted that the difference in signaling functions of $O_2\cdot^-$ and H_2O_2 in apoptosis had also been shown earlier by Li et al. [19] who found that, in vascular smooth-muscle cells, $O_2\cdot^-$ induced proliferation H_2O_2 caused apoptosis. The mechanism of superoxide-stimulated proliferation is unknown.)

Numerous biomolecules can stimulate apoptotic processes in different cells. Kogure et al. [20] have shown that α-tocopheryl hemisuccinate (TS)-induced apoptosis of rat vascular smooth-muscle cells (VSMC) was caused by exogenous $O_2\cdot^-$ generated by TS-activated NOX. Herdener et al. [21] demonstrated that ROS signaling mediated intercellularly the induction of apoptosis of transformed fibroblasts by their nontransformed neighbors. Intercellular signaling was inhibited by SOD, taurine (a scavenger of HOCl), DMSO (a hydroxyl radical scavenger), and two inhibitors of NO synthase (NOS). It was suggested that apoptosis was initiated by $O_2\cdot^-$ and its products H_2O_2 or HOCl.

Schimmel and Bauer [22] have shown that, after cellular glutathione depletion, extracellular ROS generated by transformed fibroblasts exhibited a strong apoptosis-inducing potential. In contrast to extracellular ROS, intracellular ROS did not induce apoptosis in transformed fibroblasts. Nontransformed glutathione-depleted fibroblasts did not generate substantial extracellular ROS at all, but apoptosis was efficiently induced in these cells by intracellular ROS.

Nilakantan et al. [23] showed that 20-HETE mediated cytotoxicity and apoptosis in ischemic kidney epithelial cells by H_2O_2 signaling. Mao et al. [24] reported that norepinephrine induced cell apoptosis by endoplasmic reticulum (ER) stress, and mitochondrial death was mediated via oxidative stress and the inhibition of the phosphatidylinositol 3-kinase survival pathway. Fu et al. [25] have shown that high concentrations of norepinephrine induced cardiomyocyte apoptosis in vitro via the ROS–TNF-α–caspases signaling pathway. Antioxidants and anti-TNF treatments drastically blocked apoptosis.

Moreno-Manzano et al. [26] investigated the roles of $O_2 \cdot^-$, H_2O_2, and peroxynitrite ($ONOO^-$) in TNF-α-induced apoptosis of mesangial cells. TNF-α-stimulated mesangial cells produced $O_2 \cdot^-$ and underwent apoptosis. Apoptosis was inhibited by transfection with MnSOD or treatment with the $O_2 \cdot^-$ scavenger tiron. In contrast, although exogenous H_2O_2 and $ONOO^-$ are well-known stimulants of apoptosis, TNF-α-triggered apoptosis is not affected by their inhibitors or scavengers. These results suggested that, in mesangial cells, TNF-α selectively induced apoptosis through $O_2 \cdot^-$ signaling, which was not mediated by H_2O_2 or $ONOO^-$.

Han et al. [27] found that BCA (2′-benzoyl-oxycinnamaldehyde) induced growth arrest and apoptosis of tumor cells. BCA-induced apoptosis was mediated by ROS and inhibited by the pretreatment of cells with glutathione or N-acetyl-cysteine. Kamp et al. [28] demonstrated that asbestos altered mitochondrial function of alveolar epithelial cells in part by ROS generation, which in turn resulted in apoptosis. These findings suggest that the mitochondrial death pathway is important in regulating pulmonary toxicity from asbestos.

α-Lipoic acid (LA) is a natural compound possessing antioxidative properties. However, it has been shown that LA also possesses anticancer activity due to its ability to induce apoptosis and inhibit proliferation of cancer cells relative to normal cells. Moungjaroen et al. [29] found that LA induced ROS formation and an increase in apoptosis of human lung epithelial cancer H460 cells, which was inhibited by glutathione peroxidase and SOD. Taken together, these findings demonstrate a novel prooxidant role of LA in apoptosis induction and its regulation by Bcl-2, which may be exploited for the treatment of cancer and related apoptosis disorders.

Rio et al. [30] demonstrated that paraquat-induced apoptosis in lymphocytes depended on ROS, mitochondrial dysfunction, transcriptional factors, and caspase-3 activation. Shibayama-Imazu et al. [31] have shown that, during the treatment of ovarian cancer cells with vitamin K2 (a promoter of differentiation and apoptosis in many human cancer cell lines), $O_2 \cdot^-$ was produced, which was followed shortly thereafter by release of mitochondrial cytochrome c. These data suggest that $O_2 \cdot^-$ production might cause damage to mitochondrial membranes, open permeability transition pores, and result in disruption of mitochondrial potential with subsequent release of cytochrome c. It was proposed that the loss of mitochondrial potential caused by $O_2 \cdot^-$ might be the major cause of vitamin-K2-stimulated apoptosis.

Criddle et al. [32] have shown that menadione-induced ROS generation promoted apoptosis of murine pancreatic acinar cells. ROS generation was prevented by N-acetyl-L-cysteine but not by the NOX inhibitor diphenyliodonium. These results suggest that acute generation of ROS by menadione occurs via redox cycling with subsequent induction of apoptosis. In a subsequent work [33], these authors showed that menadione-induced ROS initiated two independent apoptotic pathways within pancreatic acinar cells: the classical mitochondrial calcium-dependent pathway and a slower, caspase-8-mediated pathway that depended on the lysosomal activities of cathepsins that was used when the caspase-9 pathway was disabled.

Han et al. [34] investigated the involvement of glutathione and ROS in pyrogallol-induced HeLa cell death. $O_2{}^{.-}$ production significantly increased, while SOD activity and intracellular GSH content decreased by the action of pyrogallol. Treatment of cells with NAC or SOD caused the recovery of GSH depletion and significantly rescued cells from pyrogallol-induced apoptosis. Taken together, these findings show that pyrogallol potently increased intracellular $O_2{}^{.-}$ production and decreased GSH content in HeLa cells.

Juan et al. [35] have shown the activity of *trans*-resveratrol on the mitochondria apoptosis pathway, which was evidenced by $O_2{}^{.-}$ production in the mitochondria of cells undergoing apoptosis. *trans*-Resveratrol inhibited cell proliferation without cytotoxicity and stimulated apoptosis in HT-29. These findings provide evidence demonstrating the antitumor effect of *trans*-resveratrol via an ROS-dependent apoptosis pathway in colorectal carcinoma.

Khawaja et al. [36] demonstrated that patupilone (epothilone B) induced mito-chondrial membrane potential collapse, mitochondrial morphological changes, and CCR, leading to apoptosis. Patupilone increased the mitochondrial formation of $O_2{}^{.-}$ and H_2O_2. Collectively, these findings highlighted the importance of mitochondria in the activation of apoptotic signals triggered by patupilone.

Ricci et al. [37] investigated the effect of insulin withdrawal on apoptosis of cultured neonatal cardiomyocytes. Removal of insulin caused a significant increase in ROS production and resulted in oxidative mitochondrial DNA damage. It is important that the effects of insulin withdrawal can be decreased by treatment with the antioxidant tiron. Thus, these findings demonstrate that insulin deficiency leads to apoptosis and oxidative mitochondrial DNA damage.

Kongkaneramit et al. [38] showed that cationic liposomes such as Lipofectamine stimulated $O_2{}^{.-}$-mediated apoptosis in human lung epithelial cells.

Wartenberg et al. [39] investigated the mechanism of apoptosis of oral mucosa cancer cells in response to electromagnetic fields. Electrical field treatment (4 V/m, 24 h) increased apoptosis as it was followed from the analysis of cleaved caspase-3 and poly-(ADP-ribose)-polymerase-1 (PARP-1). Furthermore, Nox-enhanced ROS production was observed. Pretreatment with *N*-acetyl cysteine and the SOD mimetic EUK-8 abolished caspase-3 and PARP-1 induction, suggesting that electric field-stimulated apoptosis in oral mucosa cancer cells was mediated by ROS.

Hayashi et al. [40] have shown that radiation-induced apoptosis of stem/progenitor cells in human umbilical cord blood depended on ROS formation and following changes in intracellular pH. Ge et al. [41] have studied Ce4+-induced apoptosis of cultured *Taxus cuspidata* cells. Ce4+ induced $O_2{}^{.-}$ formation and cell apoptosis, which was inhibited by the Nox inhibitor diphenyl iodonium. Wenzel et al. [42] studied the effects of ascorbic acid on apoptosis in HT-29 human colon carcinoma cells induced by two apoptosis stimuli: camptothecin or the flavonoid flavone. In both cases, ascorbic acid inhibited apoptosis supposedly via reaction with $O_2{\cdot}^-$.

5.2 MITOCHONDRIAL ROS AS APOPTOSIS INITIATORS

There are various pathways of ROS interfering with apoptotic processes in mitochondria. Various biomolecules are able to stimulate apoptosis through the inhibition of mitochondrial Complex I and following $O_2{\cdot}^-$ overproduction. Li et al. [43] studied the mechanism of apoptosis initiated by rotenone, an inhibitor of Complex I. As should be expected, the inhibition of Complex I by rotenone-induced ROS production, both in isolated mitochondria and the cells. Rotenone-induced apoptosis was confirmed by DNA fragmentation, CCR, and caspase 3 activity. There was a quantitative correlation between rotenone-induced apoptosis and rotenone-induced ROS production. Furthermore, rotenone-induced apoptosis was inhibited by the antioxidants glutathione or N-acetylcysteine.

Pelicano et al. [44] developed a new strategy to enhance ROS-mediated apoptosis induced by anticancer agents in human leukemia cells via hindrance of mitochondrial electron transport. They showed that As_2O_3, a clinically active antileukemia agent, inhibited mitochondrial respiratory activity and increased $O_2{\cdot}^-$ production in leukemia cells. Marella et al. [45] used the expression of a rotenone-insensitive yeast NADH quinone oxidoreductase (Ndi1) for the suppression of the Complex I dysfunction in mammalian cells. They found that Complex I inhibition provoked the activation of specific kinase pathways and the release of mitochondrial proapoptotic factors, apoptosis-inducing factor, and endonuclease G. Prevention of apoptotic cell death by Ndi1 confirmed that the presence of Ndi1 suppressed rotenone-induced ROS generation from Complex I.

Chang et al. [46] investigated the role and metabolism of mitochondrial H_2O_2 during apoptosis. They found that the depletion of peroxiredoxin (Prx III), a mitochondrion-specific H_2O_2 scavenging enzyme, increased H_2O_2 levels and stimulated staurosporine- or TNF-α-dependent apoptosis in HeLa cells. These findings suggest that Prx III is a critical regulator of H_2O_2 formation in mitochondria, which promotes apoptosis together with the other mediators of apoptotic signaling.

Chandel et al. [47] showed that TNF-receptor (TRAF2)-mediated signaling pathways control ROS generation in mitochondria. They suggested that ROS production in mitochondria play a central role in regulating cell survival and apoptosis in TRAF2 signaling pathways. Gardai et al. [48] found that ROS-induced mitochondrial injury was a major factor of eosinophil apoptosis, which was enhanced by glucocorticosteroids-induced prolonged JNK activation that was in turn inhibited by GM-CSF.

Both mitochondrial MnSOD and SOD1 influence mitochondrial apoptosis. French et al. [49] suggested that exercise-induced increase in MnSOD provided cardioprotection by reducing the ischemia-reperfusion (IR)-induced oxidative modification to critical calcium-handling proteins, thereby decreasing their calpain-mediated cleavage and cardiomyocyte death. After IR, myocardial apoptosis and infarct size were significantly reduced in hearts of exercised animals. These findings suggest that MnSOD provides cardioprotection by the suppression of IR-induced oxidation and prevention of myocardial apoptosis and necrosis. Inarrea et al. [50] have shown that the regulation of intermembrane CuZn–SOD dismutase activity by mitochondrial respiratory chain and thioredoxin reductase influenced mitochondrial energy metabolism and apoptosis.

McInnis et al. [51] investigated the role of superoxide in N-methyl-D-aspartate (NMDA)-receptor-mediated apoptosis. They proposed that mitochondrial superoxide generation occurred upstream of NFκB activation and acted as an intracellular signal by changing the redox state of the cell. Li et al. [52] studied the $O_2\cdot^-$-mediated apoptosis in VSMC. Diethyldithiocarbamic acid (DDC) was used to inhibit SOD1 for increasing intracellular $O_2\cdot^-$ levels. It was found that DDC was able to induce VSMC apoptosis. In the apoptotic process, the mitochondrial membrane potential was decreased, and caspase-3, -8 and -9 were activated. Surprisingly, CCR was not observed. Oncogenic transformed fibroblasts induced extracellular $O_2\cdot^-$ generation through a membrane-associated Nox.

Kajitani et al. [53] have shown that the enhancement of intracellular calcium in A23187-treated HL-60 cells was associated with the formation of intracellular and extracellular ROS and induction of apoptosis. A23187-induced apoptosis was prevented by cyclosporin A, an inhibitor of mitochondrial permeability transition (MPT). The generation of extracellular ROS was suppressed by diphenylene iodonium and SOD, but these inhibitors did not affect A23187-induced apoptosis. However, the suppression of intracellular ROS completely inhibited the induction of MPT and apoptosis. These results point out that only intracellular ROS produced by A23187 depend on the opening of MPT pores and the promotion of apoptotic cell death.

Madesh et al. [54] found that ROS generation by activated macrophages stimulated an intracellular calcium transient in endothelial cells that was inhibited by the combination of SOD and an anion channel blocker. $O_2\cdot^-$-induced calcium mobilization was preceded by a decrease in mitochondrial membrane potential that was independent of other mitochondrially derived ROS. $O_2\cdot^-$ selectively activated apoptosis. Present findings suggest that $O_2\cdot^-$ induced an InsP(3)R-linked apoptotic cascade and exhibits a critical function in I/R injury and inflammation.

Piskernik et al. [55] found that antimycin A and lipopolysaccharide induced the leakage of $O_2\cdot^-$ and apoptosis in rat liver mitochondria. Quagliaro et al. [56] have shown that mitochondrially produced $O_2\cdot^-$ initiated a cascade of events leading to hyperglycemia-generated apoptosis in high-glucose-treated human umbilical vein endothelial cells. Hawkins et al. [57] demonstrated that superoxide flux across the endothelial cell plasma membrane occurred through ClC-3 channels and induced intracellular calcium release, and activated mitochondrial $O_2\cdot^-$ generation and cellular apoptosis.

5.3 MECHANISM OF SUPEROXIDE SIGNALING IN APOPTOSIS

As follows from the aforementioned data, $O_2\cdot^-$ is a very important initiator of mitochondrial apoptosis. Furthermore, the mechanism of $O_2\cdot^-$ signaling in apoptosis was probably studied better than other major ROS and RNS species (H_2O_2, nitric oxide [NO], and ONOO$^-$). Numerous works (for example, References [58,59] as well as some works already quoted earlier) demonstrate that $O_2\cdot^-$ begins apoptotic cascades by CCR from mitochondria.

In an important work, Madesh and Hajnoczky [60] have shown that exposure of permeabilized HepG2 cells to $O_2\cdot^-$ resulted in rapid CCR, whereas H_2O_2 failed to stimulate it. Both $O_2\cdot^-$ and H_2O_2 promoted activation of the mitochondrial permeability transition pore by calcium, but a calcium-dependent pore opening was not required for $O_2\cdot^-$-induced CCR. Furthermore, only $O_2\cdot^-$ stimulated CCR without damage to the inner mitochondrial membrane barrier. Importantly, a block of the voltage-dependent anion channel (VDAC) prevented $O_2\cdot^-$-induced CCR. The proapoptotic protein Bak was not detected in HepG2 cells and $O_2\cdot^-$-induced CCR; the latter was also independent of Bax translocation to mitochondria. $O_2\cdot^-$-induced CCR was followed by caspase activation and apoptosis development. Taken together, these results suggest that $O_2\cdot^-$ signaling starts apoptosis via VDAC-dependent permeabilization of the mitochondrial outer membrane without the apparent contribution of proapoptotic Bcl-2 family proteins. Furthermore, these findings demonstrate that there are different signaling pathways for apoptosis initiation by $O_2\cdot^-$ and H_2O_2.

It has been shown that cytochrome c exists in mitochondria as a complex with cardiolipin; therefore, its release must depend on the dissociation of this complex by some specific agents. Based on the experiments with isolated liver mitochondria, Ott et al. [61] proposed that CCR occurred by a two-step process. Because cytochrome c is present as loosely and tightly bound complexes with cardiolipin attached to the inner membrane, these complexes must first be disrupted to generate a soluble cytochrome. It was also suggested that the formed free cytochrome c is able to penetrate the outer mitochondrial membrane and enter the extramitochondrial environment.

Correspondingly, Petrosillo et al. [62] found that the dissociation of cytochrome c from the inner mitochondrial membrane (IMM), where it is bound to cardiolipin, was the first step of CCR by ROS. ROS were produced by rat liver mitochondria oxidizing succinate in the nonphosphorylating state. It was found that succinate-supported ROS production resulted in the release of cytochrome c from mitochondria and a parallel loss of cardiolipin content. ROS-induced CCR was independent from MPT but apparently involved VDAC. It was also suggested that mitochondrial-induced ROS promoted CCR from mitochondria by a two-step process, consisting of its dissociation from cardiolipin and permeabilization of the outer membrane.

Dussmann et al. [63] investigated apoptosis of human breast carcinoma cells stimulated by the proapoptotic agent staurosporine. They found that an increase in $O_2\cdot^-$ production resulted in CCR and a new enhancement of $O_2\cdot^-$ formation. Thirunavukkarasu et al. [64] showed that $O_2\cdot^-$-induced apoptosis of hepatic stellate cells (HSCs) involved the release of cytochrome c, augmented Bax expression, and increased caspase-3 activity and hydrolysis of PARP. Antioxidants N-acetylcysteine, vitamin E, and SOD inhibited apoptosis. It is interesting that SOD1, a cytosolic enzyme that is also localized in mitochondria in various types of cells, might also be released together with cytochrome c during apoptosis [65].

It has been proposed that the destruction of cytochrome c–cardiolipin complexes might be initiated by peroxidation of cardiolipin. However, $O_2\cdot^-$ is not an antioxidant; therefore, it cannot directly initiate lipid peroxidation during the $O_2\cdot^-$-induced release of cytochrome c. Recently, Kagan and coworkers [66] investigated the possibility of modification of cytochrome c during its binding to cardiolipin. They suggested that the bound cytochrome c might change its tertiary structure and acquire peroxidase activity. These changes in cytochrome structure may suppress its reduction by mitochondrial Complex III. These authors also suggested that, in this case, cytochrome–cardiolipin complexes will lose their ability to react with $O_2 \cdot^-$. In a subsequent work [67], they proposed that Bax/Bak activation and cardiolipin peroxidation are essential for CCR during apoptosis. They found that exposure of mouse embryonic fibroblast (MEF) cells to actinomycin D resulted in apoptosis development in the order Bax translocation–superoxide production–cardiolipin peroxidation.

Some time ago, we proposed another mechanism for $O_2 \cdot^-$-induced CCR from mitochondria as a first step of the apoptotic cascade [68,69]. The problem with the peroxidase model of CCR from mitochondria to the cytosol by $O_2 \cdot^-$ is the inability of $O_2 \cdot^-$ to oxidize lipids. (The inability of $O_2 \cdot^-$ to abstract a hydrogen atom even from the most reactive C-H bonds has been shown theoretically and experimentally and is now a well-established fact; see, for example, References [70,71].) The only way for $O_2 \cdot^-$ to initiate cytochrome c–mediated cardiolipin peroxidation is MnSOD-catalyzed dismutation into H_2O_2. Burkitt et al. [72] studied the oxidation of cytochrome c by H_2O_2 and showed that, in this reaction, a peroxidase compound, and not hydroxyl radicals, was formed. Thus, it is possible that $O_2 \cdot^-$ can indirectly stimulate CCR through dismutation into H_2O_2.

We proposed that $O_2 \cdot^-$ can directly initiate CCR through the reduction of $Fe^{3+}cyt$ to $Fe^{2+}cyt$. It is believed that cytochrome c attached to the inner mitochondrial membrane by two ways: (1) electrostatically by ion–ion interaction with positively charged lysine residues of cytochrome and negatively charged phosphate groups of cardiolipin and (2) through the formation of a tight complex with partial embedding of cytochrome into the membrane. It seems reasonable to suggest that the release of cytochrome c from the inner membrane takes place by the reduction of electrostatically bound cytochrome by superoxide. (The rate constant for the reduction of a bound oxidized cytochrome is sufficiently high [about $10^5 \, M^{-1}s^{-1}$] [73]; that is, the binding of cytochrome c to lipid membranes insignificantly affects the rate of cytochrome c reduction.)

Such a mechanism of CCR is possible if the bound cytochrome exists in the oxidized ferric form. There are contradictory literature data concerning the valence of mitochondrial cytochrome c. Earlier, Iwase et al. [74] reported that cytochrome c is bound to the membrane in the ferric form. However, Nantes et al. [75] did not agree with this conclusion because they found that the absorption and MCD spectra of bound cytochrome are similar to the spectra of high-spin iron complexes. However, Tuominen et al. [76] convincingly demonstrated that the lipid binding of cytochrome c changed its ferrous absorption spectrum to the ferric one. Therefore, it can be accepted that cytochrome c mainly bound to the inner mitochondrial membrane in its oxidative ferric form.

It is reasonable to suggest that, from two forms of cytochrome c–cardiolipin complexes, a loosely electrostatically bound complex, is a major target for the attack by $O_2 \cdot^-$. If there is equilibrium between this complex and a tight complex with partial embedding of cytochrome into the membrane, then $O_2 \cdot^-$ will be able to shift this equilibrium and release all mitochondrial cytochrome c. Previously, we have shown

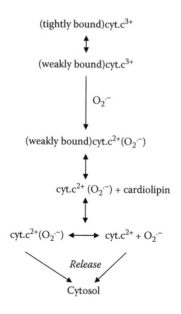

FIGURE 5.1 Mechanism of $O_2\cdot^-$-initiated CCR from mitochondria.

[77,78] that $O_2\cdot^-$ readily reacts with ferric and ferrous hemes to form the (heme)Fe(II) $(O_2\cdot^-)$ complexes. Therefore, the following mechanism for $O_2\cdot^-$-induced CCR might be proposed.

$$O_2\cdot^- + \text{(bound)cyt.c}^{3+} \Rightarrow O_2 + \text{(bound)cyt.c}^{2+} \qquad (5.1)$$

$$O_2\cdot^- + \text{(bound)cyt.c}^{2+} \Rightarrow \text{(bound)cyt.c}^{2+}(O_2\cdot^-) \qquad (5.2)$$

$$\text{(bound)cyt.c}^{2+}(O_2\cdot) \overset{\text{release}}{\Rightarrow} \text{cyt.c}^{2+}(O_2\cdot^-) \qquad (5.3)$$

At present, it is impossible to say whether the cyt c^{2+} $(O_2\cdot^-)$ complex or cyt c^{2+} is released to the cytosol through the outer mitochondrial membrane. The whole mechanism of CCR by $O_2\cdot^-$ can be presented as in Figure 5.1.

5.4   ENZYMATIC APOPTOTIC CASCADES MEDIATED BY ROS

Numerous studies demonstrate various enzymatic cascades that mediate ROS signaling in apoptosis. In 1999, Li et al. [79] investigated the role of protein kinase C (PKC) in H_2O_2-initiated apoptosis of VSMC. It was found that exposure of VSMCs to H_2O_2 led to PKC activation and apoptosis when PKC was activated by PMA. The authors also concluded that PKC converted the ROS-induced signals from necrotic cell death to the activation of apoptosis in VSMC.

Bergamo et al. [80] investigated the conjugated linoleic-acid (CLA)-mediated apoptosis in Jurkat T cells. Pretreatment of cells with antioxidants (trolox, quercetin, catalase, SOD, and N-acetyl-L-cysteine) and the inhibitors of Nox or PKC suppressed CLA-mediated caspase-3 activation. These results suggested that CLA can initiate ROS-mediated apoptosis by the PKC/Nox pathway.

Byun et al. [81] investigated the role of rottlerin, a specific PKCδ inhibitor, in the stimulation of death-receptor-mediated apoptosis through cytochrome c-dependent- or independent pathways. It was found that treatment with rottlerin protected the murine fibrosarcoma L929 cells against TNF-α-induced necrosis, but it stimulated apoptosis in these cells promoted by cotreatment with Hsp90 inhibitor geldanamycin and TNF-α. TNF-α treatment induced rapid generation of mitochondrial $O_2 \cdot^-$ through Nox1 NADPH oxidase when cells undergo necrosis. This study suggests that Nox1 NADPH oxidase is a new molecular target for the antinecrotic activity of rottlerin.

Methylglyoxal (MG) is a physiological metabolite capable of inducing stress and causing apoptosis. Du et al. [82] found that MG-induced apoptosis depended on rapid production of $O_2 \cdot^-$, which was followed by the activation of apoptosis-signal-regulating kinase 1 (ASK1). Antioxidants completely inhibited the activation of ASK1 and MG-induced apoptosis, while the SOD inhibitor DDC enhanced MG-mediated ASK1 activation. These results suggest that ASK1 is an important mediator of $O_2 \cdot^-$ initiation of apoptosis in MG-treated Jurkat cells.

Fujino et al. [83] also studied the role of ASK 1 in ROS-induced cellular responses. Upon ROS stimulation, ASK1 dissociated from thioredoxin (Trx) and formed a fully activated higher-molecular-mass complex by recruitment of TRAF2 and TRAF6. These findings demonstrated that Trx, TRAF2, and TRAF6 regulate ASK1 activity during H_2O_2-induced apoptosis. Lim et al. [84] found that troglitazone induced mitochondrial generation of $O_2 \cdot^-$, which activated the (Trx2)/Ask1 signaling pathway, leading to cell injury to human hepatocytes.

Serine–threonine kinase Akt/protein kinase B (PKB) is an important regulator of cell-surviving pathways. It has been shown that Akt phosphorylation protects cells from apoptosis induced by various stimuli. Shi et al. [85] showed that exogenous and endogenous ROS or antioxidants regulated Akt activation through the inhibition of Akt phosphorylation and the reduction of c-Fos/c-Jun expression, inducing proliferation or apoptosis of hepatoma cells.

Inoue et al. [86] demonstrated that diclofenac, a member of the nonsteroidal anti-inflammatory drugs (NSAID) family, induced growth inhibition and apoptosis of HL-60 cells through the cascade of mitochondrial ROS, Akt, caspase-8, and Bid. ROS generation preceded CCR, caspase activation, and DNA fragmentation. N-Acetyl-L-cysteine suppressed ROS generation, Akt inactivation, caspase-8 activation, and DNA fragmentation.

Gao et al. [87] have shown that 2-methoxyestradiol (2ME) induced apoptosis in human leukemia cells (U937 and Jurkat). 2ME-induced apoptosis occurred through the release of mitochondrial cytochrome c, ROS generation, downregulation of the Mcl-1 gene and X-linked inhibitor of apoptosis protein (XIAP) and Akt dephosphorylation accompanied by activation of JNK. In U937 cells, catalase and a SOD mimetic TBAP inhibited ROS formation and Akt inactivation. Together, these findings suggest that 2ME-related apoptosis in human leukemia cells occurred through oxidative injury, Akt inactivation, JNK activation, XIAP and Mcl-1 downregulation, mitochondrial injury, and apoptosis.

Fujita et al. [88] studied the molecular mechanism of the 6-hydroxydopamine (6-OHDA)-induced apoptosis of pheochromocytoma cells. 6-OHDA induced $O_2{}^{\cdot-}$ generation, Bid cleavage, and mitochondrial membrane depolarization. Akt phosphorylation enhanced cell survival, which was decreased by 6-OHDA, while p38-MAPK (mitogen-activated protein kinase) phosphorylation was increased. Thus, the 6-OHDA-induced apoptosis was initiated by $O_2{}^{\cdot-}$ generation, followed by caspase cascade activation, suppressed Akt phosphorylation, and increased p38 phosphorylation.

Mochizuki et al. [89] suggested that ROS generated by Nox4, at least in part, transmit cell survival signals through the Akt/ASK1 pathway in pancreatic cancer cells and that their depletion leads to apoptosis. Yu et al. [90] showed that the tyrosine kinase and histone deacetylase (HDAC) inhibitors induced apoptosis in non-small-cell lung cancer, ovarian cancer, and leukemia cells, and in cisplatin-resistant human ovarian cancer cells, through the inactivation of MAPK and Akt cascades. Increased apoptosis depended on the enhanced reactive ROS generation. N-acetyl-L-cysteine suppressed both ROS formation and the induction of apoptosis. Collectively, these findings suggest that MAPK and Akt inactivation, together with ROS generation, contribute to the synergistic cytotoxicity and apoptosis development of the studied inhibitors in a variety of human cancer cell types.

Yokouchi et al. [91] studied cadmium-induced apoptosis occurring through the induction of ER stress. Cadmium caused ROS generation, while cadmium-induced ER stress was inhibited by antioxidants. Apoptosis was mediated by $O_2{}^{\cdot-}$, H_2O_2, and $ONOO^-$ and inhibited by MnSOD. In addition, $O_2{}^{\cdot-}$ induced phosphorylation of proapoptotic c-Jun N-terminal kinase (JNK).

Liu et al. [92] investigated the regulation of myocyte survival by Hsp27. It was found that Hsp27 overexpression in myocytes treated with H_2O_2 suppressed apoptosis, ROS generation, and the loss of mitochondrial membrane potential. Furthermore, augmented Akt activation was observed in the Hsp27-overexpressed cells following H_2O_2 exposure. These findings show that Hsp27 inhibits apoptosis and reduces ROS generation and the augmentation of Akt activation.

In contrast to the survival effects of Akt/PKB kinase, p38-MAPK mainly exhibits proapoptotic activity. Choi et al. [93] demonstrated that p38 resulted in dopaminergic neuronal death. Phosphorylation of p38 preceded apoptosis. The SOD mimetic and the NO chelator blocked 6-OHDA-induced phosphorylation of p38, indicating $O_2{}^{\cdot-}$ and NO signaling in p38 activation. Taken together, these findings suggest that $O_2{}^{\cdot-}$ and NO induce the 6-OHDA-stimulated p38 signal pathway, leading to activation of both mitochondrial and extramitochondrial apoptotic pathways in a new developed culture model of Parkinson's disease.

Sun et al. [94] examined whether the prevention of cardiomyocyte apoptosis by postconditioning might be mediated by MAPK pathways. Primary cultured neonatal rat cardiomyocytes were postconditioned after hypoxia and reoxygenation. It was found that H/R stimulated cardiomyocyte apoptosis through the activation of JNK and p38 kinases, release of TNF-α, and the activation of caspases, while postconditioning diminished apoptosis mediated by inhibiting JNKs/p-38 signaling pathways and reducing TNF-α release and caspase expression.

Dai et al. [95] have shown that the interruption of the NFκB pathway in leukemia cells exposed to HDAC inhibitors resulted in a cascade of events, including ROS generation, downregulation of SOD2 and XIAP, and JNK1 activation following an increase in mitochondrial dysfunction and apoptosis.

Dhingra et al. [96] investigated the role of p38 and ERK 1/2 MAPK in the regulation by TNF-α and IL-10 of oxidative stress and cardiac myocyte apoptosis in Sprague–Dawley male rats. Exposure to TNF-α significantly increased ROS production, caused cell injury, and increased the number of apoptotic cells and Bax/Bcl-xl ratio. Correspondingly, there was an increase in phosphor-p38 and a decrease in phosphor-ERK1/2 kinases. Thus, p38 and ERK1/2 MAPKs play an essential role in TNF-α and IL-10 simulated cardiac myocyte apoptosis.

She et al. [97] have shown that manumycin A, a farnesyltransferase inhibitor, induced apoptosis of anaplastic thyroid cancer cells via the induction of ROS, which mediated DNA damage. Suppression of ROS with N-acetyl-L-cysteine prevented CCR by manumycin. Manumycin induced phosphorylation of p38-MAPK, which was also blocked by N-acetyl-L-cysteine. Therefore, ROS-mediated activation of p38-MAPK may be an important signaling factor of the intrinsic apoptotic pathway by manumycin.

Kou et al. [98] investigated the effects of xanthine oxidase (XO) on endothelial survival and apoptotic signaling by PKB/Akt and p38-MAPK in vascular endothelial growth factor (VEGF)-simulated human umbilical vein cells. Exogenous XO increased cellular ROS production and caused $O_2^{\cdot-}$-dependent inhibition of Akt phosphorylation and enhancement of p38-MAPK phosphorylation. Exogenous XO also reduced cell viability, proliferation, and vascular tube formation by p38-MAPK-dependent, phosphoinositide 3-kinase (PI3-K) reversible mechanisms.

Gomez-Lazaro et al. [99] demonstrated that malonate, an inhibitor of mitochondrial Complex II, increased ROS production in human neuroblastoma cells, resulting in CCR and apoptotic cell death. It was also shown that malonate-induced ROS production initiated the subsequent p38-MAPK activation and the activation of the proapoptotic Bax protein to induce mitochondrial membrane permeabilization and neuronal apoptosis.

c-Jun NH2-terminal kinase is another proapoptotic MAPK kinase. It is known that high glucose can induce ROS and apoptosis in endothelial cells, which might be a causal reason of diabetes-mellitus-induced multiple cardiovascular complications. Ho et al. [100] studied the signaling pathway of high glucose–induced apoptosis in human umbilical vein endothelial cells (HUVECs). High glucose activated c-Jun (JNK) kinase and caspase-3 but not extracellular signal–regulated kinase (ERK) 1/2 or p38-MAPK. Thus, these findings indicate that ROS induced by high glucose may be involved in JNK activation, which in turn triggers the apoptosis of HUVECs.

Fu et al. [101] studied the mechanism of norepinephrine (NE)-induced apoptosis in endothelial cells cultured from neonatal rat hearts. Treatment with norepinephrine increased intracellular ROS level and ERK, JNK, and p38 phosphorylation and decreased Akt phosphorylation. Vitamin C and N-acetyl L-cysteine inhibited NE-induced ROS production, JNK phosphorylation, caspase activation, and apoptosis. These findings show that norepinephrine induces apoptosis in neonatal rat endothelial cells via an ROS-dependent JNK activation pathway.

Huc et al. [102] investigated the role of pH(i) regulator Na(+)/H(+) exchanger isoform 1 (NHE1) in carcinogenic polycyclic aromatic hydrocarbon (PAH)-induced apoptosis. NHE1 was activated by an early cytochrome-P450 (CYP1)-dependent H_2O_2 production, leading to mitochondrial dysfunction and apoptosis. The mechanism of NHE1 activation by benzo(a)pyrene (BaP) was studied. It was found that the MAPK kinase 4/c-Jun NH(2)-terminal kinase (MKK4/JNK) pathway was a link between BaP-induced hydrogen production and NHE1 activation that finally resulted in mitochondrial $O_2\cdot^-$ overproduction.

Singh et al. [103] demonstrated that guggulsterone, a constituent of Indian ayurvedic medicinal plant *Commiphora mukul,* induced apoptosis in human cancer cells through ROS-dependent activation of JNK and p38-MAPK, which was inhibited by overexpression of catalase and SOD.

Ishikawa and Kitamura [104] have shown that the activation of JNK and ERK kinases, but not p38 kinase, is required for H_2O_2-initiated apoptosis of mesangial cells and that the flavonoid quercetin is capable of suppressing the JNK and ERK apoptotic cascade. It is interesting that it had been shown earlier [105] that quercetin induced apoptosis in tumor cells through the inhibition of synthesis and expression of a heat shock protein.

Zhang et al. [106] studied the mechanisms of TNF-α- and H_2O_2-induced apoptotic cascades in endothelial cells. It was found that H_2O_2, but not TNF-α, stimulated the phosphorylation of protein kinase D (PKD) and translocation of PKD from the endothelial cell membrane to the cytoplasm, activating the JNK upstream activator, ASK1. Thus, PKD is a critical mediator in H_2O_2- but not TNF-induced ASK1-JNK signaling (Figure 5.2).

On the contrary, Kamata et al. [107] demonstrated that TNF-α induced ROS formation in mouse fibroblasts. ROS formation, which was inhibited by mitochondrial SOD, caused the oxidation and inhibition of JNK-inactivating phosphatases by oxidizing their catalytic cysteines to sulfenic acid. Inactivation of phosphatases led to JNK activation, CCR, and caspase 3 cleavage, as well as necrotic cell death. Treatment of cells of experimental animals with antioxidants prevented H_2O_2 accumulation and JNK-induced phosphatase oxidation, maintained JNK activity, and both forms of cell death.

Conde de la Rosa et al. [108] studied protection of primary hepatocytes against oxidative-stress-induced apoptosis by heme oxygenase-1 (HO-1) and carbon monoxide (CO). It was found that HO-1 overexpression inhibited $O_2\cdot^-$-induced apoptosis, while CO blocked $O_2\cdot^-$-induced JNK phosphorylation and caspase-9, -6, -3 activation and abolished apoptosis but did not increase necrosis. These findings suggest that HO-1 and CO protect primary hepatocytes against $O_2\cdot^-$-induced apoptosis partially through the inhibition of JNK activity.

Hong and Kim [109] demonstrated that genipin, the aglycone of geniposide, induced apoptotic cell death in human hepatoma cells via an Nox–ROS–JNK-dependent cascade. It was also found that the proapoptotic activity of genipin was mediated by mixed-lineage kinase 3 (MLK3) (Figure 5.2).

Suzumi et al. [110] investigated the effect of cyclic stretch on apoptosis of retinal vascular pericyte cells. Cyclic stretch increased intracellular ROS generation and enhanced c-Jun NH_2-terminal kinase phosphorylation, which were reduced by the Nox inhibitor diphenylene iodonium. These findings suggest that cyclic stretch

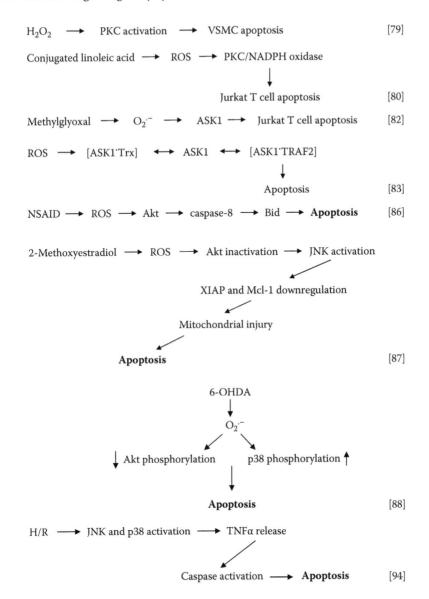

$H_2O_2 \longrightarrow$ PKC activation $\longrightarrow$ VSMC apoptosis   [79]

Conjugated linoleic acid $\longrightarrow$ ROS $\longrightarrow$ PKC/NADPH oxidase

Jurkat T cell apoptosis   [80]

Methylglyoxal $\longrightarrow O_2^{\cdot-} \longrightarrow$ ASK1 $\longrightarrow$ Jurkat T cell apoptosis   [82]

ROS $\longrightarrow$ [ASK1·Trx] $\longleftrightarrow$ ASK1 $\longleftrightarrow$ [ASK1·TRAF2]

Apoptosis   [83]

NSAID $\longrightarrow$ ROS $\longrightarrow$ Akt $\longrightarrow$ caspase-8 $\longrightarrow$ Bid $\longrightarrow$ **Apoptosis**   [86]

2-Methoxyestradiol $\longrightarrow$ ROS $\longrightarrow$ Akt inactivation $\longrightarrow$ JNK activation

XIAP and Mcl-1 downregulation

Mitochondrial injury

Apoptosis   [87]

6-OHDA

$O_2^{\cdot-}$

$\downarrow$ Akt phosphorylation   p38 phosphorylation $\uparrow$

Apoptosis   [88]

H/R $\longrightarrow$ JNK and p38 activation $\longrightarrow$ TNFα release

Caspase activation $\longrightarrow$ **Apoptosis**   [94]

FIGURE 5.2   ROS signaling in apoptosis [79,80,82,83,86–88,94].

induces apoptosis in porcine retinal pericytes by activation of the ROS–c-Jun NH_2-terminal kinase–caspase cascades.

Ham et al. [111] showed that calcium and ROS stimulate JNK1 activation during apoptosis induced by ginsenoside Rh2 (G-Rh2) in HeLa, MCF10A-*ras*, and MCF7 cells. Antioxidants *N*-acetyl-L-cysteine or catalase diminished G-Rh2-induced ROS generation, JNK1 activation, and apoptosis. These results suggest that ROS and calcium are important signaling intermediates that activate JNK MAPK kinase and depolarize the mitochondrial membrane potential in G-Rh2-induced apoptosis.

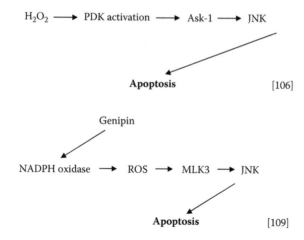

FIGURE 5.3   JNK signaling in apoptosis [106]. Mechanism genipin-mediated apoptosis [109].

Yokouchi et al. [112] showed that cadmium initiated apoptosis of porcine renal proximal tubular cells through the induction of ER stress and the generation of ROS. ROS formation and cadmium-induced ER stress were inhibited by antioxidants. Taken together, these data demonstrated that cadmium caused ER stress through the generation of ROS, and that $O_2 \cdot^-$ was selectively involved in cadmium-induced apoptosis and JNK phosphorylation.

Liu et al. [113] showed that proline oxidase (POX), localized on inner mitochondrial membranes, is encoded by a p53-induced gene. POX induced apoptosis in colorectal cancer cells, which was mediated by $O_2 \cdot^-$ production in mitochondria. MnSOD, but not SOD1 or CAT, inhibited this apoptosis.

AMP-activated protein kinase (AMPK) influences cellular metabolism, glucose-regulated gene expression, and insulin secretion of pancreatic beta cells. At low glucose concentrations or in the presence of 5-aminoimidazole-4-carboxamide riboside (AICAR), AMPK initiates apoptosis in beta cells. Cai et al. [114] showed that both low glucose- and AICAR-induced apoptosis depended on the increased formation of mitochondrial $O_2 \cdot^-$ and decreased mitochondrial activity. Mitochondrial dysfunction originated from an increased oxidized state of the mitochondrial flavins (FMN/FAD) but not NADPH. $O_2 \cdot^-$ formation was inhibited by vitamin E, N-acetylcysteine, or the SOD-mimetic MnTBAP. It was concluded that apoptosis induced by AMPK activation in beta cells resulted in the enhanced production of mitochondria-derived oxygen radicals.

Integrin-linked kinase (ILK) is a novel apoptotic regulator that plays an important role in the Akt signaling pathway. Saito et al. [115] demonstrated that the ILK cell-signaling pathway was activated after transient focal cerebral ischemia (FCI) in neurons, while SOD1 was able to inhibit the apoptosis induced by FCI through the reduction of $O_2 \cdot^-$ formation and ILK activation. The authors suggested that ILK might contribute to the neuroprotective role of SOD1 after cerebral ischemia.

The protooncogene Ras is a well-established modulator of apoptosis. Suppression of PKC activity can selectively induce apoptosis in cells expressing a constitutively activated Ras protein. Liou et al. [116] investigated whether ROS can participate in Ras-mediated apoptosis. It was found that ROS are downstream effectors of the Ras-mediated apoptotic response to PKC inhibition in Ras-transformed cells.

In order to study methylglyoxal (MGO)-inducing apoptosis in renal mesangial cells, Huang et al. [117] exposed MGO-treated rat mesangial cells to $O_2\cdot^-$ inhibitors SOD and diphenyliodonium. It was found that methylglyoxal rapidly enhanced Ras activation and progressively increased cytosolic p38 and nuclear c-Jun activation. Taken together, these findings show that methylglyoxal increases Ras modulation of $O_2\cdot^-$-mediated p38 activation and c-Jun activation, which resulted in enhanced apoptosis.

Wang et al. [118] investigated the mechanisms of $O_2\cdot^-$ signaling from Ras protein to the nucleus following apoptosis in Cr(VI)-stimulated cells. Ras-overexpressed human prostate tumor cell lines (Ras$^+$) exhibited higher susceptibility to apoptosis induced by Cr(VI) compared to control cells. Catalase, sodium formate, and deferoxamine inhibited Cr(VI)-induced apoptosis. Present results show that the Ras protein mediates $O_2\cdot^-$ production through the reduction of dioxygen by Nox in Cr(VI)-stimulated cells.

Lin et al. [119] have shown that high glucose induced Ras, Rac1 activation, $O_2\cdot^-$ burst, Wnt5a/beta-catenin destabilization, and subsequently promoted caspase-3 and PARP cleavage and apoptosis in mesangial cells. $O_2\cdot^-$ production was suppressed by SOD and diphenyloniodium, The Ras and Rac1 regulation of $O_2\cdot^-$ apparently raises apoptotic activity by activating glycogen synthase kinase GSK-3β and inhibiting Wnt5a/β-catenin signaling.

The cell surface receptor Fas and its ligand are the mediators of apoptosis and activation-induced T-cell death. It is believed that the generation of ROS and the activation of the caspase cascade are important factors of Fas-mediated apoptotic signaling. Using human T-cell leukemia Jurkat cells, Sato et al. [120] demonstrated that ROS are required for the formation of apoptosome. They showed that ROS derived from mitochondrial permeability transition MPT enhanced apoptotic events downstream of mitochondrial permeability transition. Furthermore, it was found that apoptosome formation in Fas-stimulated Jurkat cells was inhibited by N-acetyl-L-cysteine and MnSOD. Thus, these results suggest that ROS plays an important role in apoptosome formation by oxidizing apoptotic protease-activating factor 1 (Apaf-1) and the subsequent activation of caspase-9 and caspase-3.

Wang et al. [121] have studied the relationship between ROS generation and ICE-like protease FLICE in Fas-mediated apoptosis. They showed that ROS are required for FLIP downregulation and apoptosis induction by Fas ligand (FasL) in primary lung epithelial cells. Glutathione peroxidase and SOD effectively inhibited FLIP downregulation and FasL-induced apoptosis. It was found that H_2O_2 signaling was responsible for FLIP downregulation, whereas $O_2\cdot^-$ produced by Nox was a source of H_2O_2 and a scavenger of NO. Taken together, these results indicate that ROS signaling is an important pathway of apoptosis regulation in Fas-induced cell death and related apoptosis disorders.

Lin et al. [122] have shown that peroxisome proliferator-activated receptor-α (PPAR-α) is able to inhibit adriamycin-induced ROS and NF-κB activity and inhibit adriamycin-induced apoptosis of renal tubular cells. It is known that depriving sympathetic neurons of nerve growth factor (NGF) causes their apoptotic death through

Bax-dependent increase in mitochondrial ROS formation. Kirkland et al. [123] found that withdrawing NGF increased the levels of H_2O_2 in mitochondria probably formed by superoxide dismutation.

Emphysema development depends on ROS formation and alveolar cell apoptosis. Oxidative stress upregulates ceramides, proapoptotic signaling of sphingolipids that triggers further ROS formation, and alveolar space enlargement. Petrache et al. [124] proposed that ROS generated by ceramide are required for its pathogenetic effect on lung alveoli. $O_2\cdot^-$ formation could be a critical step in ceramide-mediated apoptosis and destruction in the lung.

Hua et al. [125] demonstrated that Granzyme M (GzmM) (an orphan granzyme) induced mitochondrial swelling and loss of mitochondrial transmembrane potential in natural killer (NK) cells. Furthermore, GzmM initiated ROS generation and CCR. The heat shock protein HSP75 suppressed ROS formation and protected cells from GzmM-mediated apoptosis. Martinvalet et al. [126] have studied the mechanism of proapoptotic activity of the killer lymphocyte protease granzyme A (GzmA) and shown that it cleaved an iron–sulfur (Fe–S) subunit of the NADH:ubiquinone oxidoreductase Complex I to interfere with NADH oxidation and stimulate $O_2\cdot^-$ production.

5.5 RNS IN APOPTOSIS

Both major reactive nitrogen species NO and $ONOO^-$ are signaling intermediates of apoptosis. They mainly exhibit opposite proapoptotic and antiapoptotic effects because NO signaling is usually favorable in most enzymatic processes, while $ONOO^-$ is principally a damaging species. However, in apoptosis, NO signaling might also be inhibitory, for example, due to nitration of caspases. Unfortunately, it is not always possible to divide NO and $ONOO^-$ effects in apoptosis because, as a rule, $O_2\cdot^-$ is formed in many biological processes together with NO, rapidly reacting with NO to form $ONOO^-$. Therefore, there is a competition between three signaling molecules, NO, $ONOO^-$, and $O_2\cdot^-$, which regulate pro- and antiapoptotic processes in cells.

In 1995, Bonfoco et al. [127] studied the role of NO, $O_2\cdot^-$, and $ONOO^-$ in NMDA-receptor-mediated neurotoxicity, which contributed to injury in many acute and chronic neurologic disorders, including focal ischemia, trauma, epilepsy, Huntington's disease, Alzheimer's disease, amyotrophic lateral sclerosis, AIDS dementia, and other neurodegenerative diseases. These authors found that exposure of cortical neurons to relatively short durations or low concentrations of NMDA or $ONOO^-$ induced an apoptotic form of neurotoxicity. SOD and catalase partially prevented $ONOO^-$-stimulated apoptosis. In contrast, the high concentrations of NMDA or $ONOO^-$ induced necrotic cell death. Thus, depending on NMDA or $NO/O_2\cdot^-$ concentrations, ROS and RNS can cause either apoptotic or necrotic neuronal cell damage.

The proapoptotic effect of NO has been widely demonstrated. For example, Hortelano et al. [128] found that NO efficiently activated mitochondrial permeability transition, causing the liberation of apoptotic factors from mitochondria, which induced nuclear apoptosis (DNA condensation and DNA fragmentation) in isolated nuclei of thymocytes. In addition, NO stimulated the disruption of the mitochondrial transmembrane potential, followed by ROS overproduction. Ward et al. [129] have shown that NO donors induced apoptosis in human neutrophils. Although this process was caspase dependent, it proceeded by a cGMP-independent mechanism and probably involved the formation of $ONOO^-$. Hortelano et al. [130] reported that treatment of elicited peritoneal macrophages with high concentrations of NO donors resulted in apoptotic cell death due to the changes in the mitochondrial transmembrane potential, which decreased in cells undergoing apoptosis through mitochondrial-dependent mechanisms. These findings also suggested that NO-dependent apoptosis in macrophages involved a chemical modification of cytochrome c that altered its structure and accelerated its release from the mitochondria. Pautz et al. [131] found that ceramide was an important mediator of apoptosis induction by both ROS and RNS in glomerular mesangial and endothelial cells. However, there are apparently cell-type-specific protective mechanisms that regulate the redox balance between ROS and RNS to initiate apoptosis or survival of cells.

Nazarewicz et al. [132] studied the mechanism of apoptosis induced by the anticancer drug tamoxifen. They showed that tamoxifen increased intramitochondrial calcium concentration and stimulated mitochondrial NO synthase (mtNOS) activity in the mitochondria from rat liver and human breast cancer cells. It was proposed that mtNOS activation was a major factor of apoptosis development through the inhibition of mitochondrial respiration and CCR.

Langer et al. [133] demonstrated that both NO donors and endothelial NO synthase overexpression promoted apoptosis in hepatic stellate cells (HSCs). NO-induced HSC death occurred through mitochondrial membrane depolarization by a caspase-independent pathway. It was proposed that NO induced HSC apoptosis through a signaling mechanism that was mediated by ROS and occurred independently of caspase activation.

Heigold et al. [134] have shown that NO mediated apoptosis induction in fibroblasts through constitutive Src kinase or induced *ras* oncogene expression but not in nontransformed parental cells. It was proposed that NO-mediated apoptosis depended on extracellular $O_2 \cdot^-$ production by transformed cells, which was inhibited by extracellular SOD, SOD mimetics, and the inhibitor of Nox apocynin. These results indicate that $ONOO^-$ seems to be an apoptosis inducer in NO-mediated apoptosis of transformed fibroblasts that was confirmed by apoptosis inhibition with the $ONOO^-$ scavengers ebselen and FeTPPS. Steinmann et al. [135] showed that NO-mediated apoptosis induction in transformed fibroblasts can be divided into two major phases: phase 1 apoptosis is mediated by $ONOO^-$ formed by the interaction of extracellular $O_2 \cdot^-$ with NO, and phase 2 apoptosis is mediated by NO and depends on intracellular $O_2 \cdot^-$ exclusively.

As shown earlier, $ONOO^-$-induced apoptosis occurs in many cells. In 1999, Ghafourifar et al. [136] studied the possible involvement of mitochondrial NO synthase in apoptosis. They showed that calcium uptake by mitochondria initiated mtNOS activity and stimulated the release of cytochrome c. It was proposed that calcium-induced mtNOS activation resulted in the formation of $ONOO^-$, which initiated apoptosis through the release of cytochrome c. Levrand et al. [137] studied apoptotic and necrotic cell death in cultured cardiomyocytes following a brief exposure to $ONOO^-$. In vitro, $ONOO^-$ killed cardiomyocytes mostly through apoptosis (DNA fragmentation, apoptotic nuclear alterations, caspase-3 activation, and PARP cleavage). In vivo, apoptosis was also a main origin of cardiomyocytes death in a rat model of myocardial I/R, where the effects of ONOO⁻ were suppressed by SOD mimetics and the ONOO⁻ scavenger Mn(III)-tetrakis(4-benzoic acid) porphyrin (MnTBAP).

Of great interest is the fact that, under certain conditions, ONOO⁻ can inhibit apoptosis. Such is a case when the carbon dioxide (CO_2) is present together with ONOO⁻. It is suggested that CO_2 is able to enhance the ONOO⁻-mediated nitration of aromatics and it partially impairs the oxidation of thiols. Ascenzi et al. [138] demonstrated that, in the presence of CO_2, ONOO⁻ inhibited the catalytic activity of human caspase-3 by oxidizing the Sγ atom of the cysteine catalytic residue.

Under conditions when ONOO⁻ formation is negligible, for example at low NO concentrations, it is possible that NO predominantly exhibits antiapoptotic activity. Mannick et al. [139] demonstrated that NOS inhibited Fas-induced apoptosis via a cGMP-independent mechanism in human leukocytes. In addition, NOS inhibited Fas-induced cleavage of PARP by cysteine proteases. Rossig et al. [140] showed that NO was an antiapoptotic regulator of caspase activity via S-nitrosation of the Cys63 residue of caspase-3 in HUVECS. Thus, NO can be a potent endogenous inhibitor of caspase-3-like protease activity. Kim et al. [141] found that NO inhibited the capacity of caspase-3 to cleave recombinant Bcl-2. Both Bcl-2 cleavage and CCR were inhibited in TNF-α- and actinomycin D–treated cells exposed to NO donors. Thus, NO is able to suppress the major step in the amplification of apoptosis by preventing Bcl-2 cleavage and CCR.

The antiapoptotic effect of NO on Hcy-stimulated caspase-dependent apoptosis in cultured HUVECS has also been shown by Lee et al. [142]. These authors demonstrated that Hcy induced apoptosis through ROS production, lipid peroxidation, p53 Nox expression, and mitochondrial CCR. The NO donor *S*-nitroso-*N*-acetylpenicillamine, and antioxidants (α-tocopherol, SOD, and catalase) and nitrate suppressed ROS production and Hcy-induced apoptosis. It was suggested that an increase in vascular NO production may prevent Hcy-induced endothelial dysfunction by S-nitrosylation.

It was noted earlier that low NO concentrations is the condition of the antiapoptotic effect of NO, but it does not seem to be a universal condition. Yuyama et al. [143] have shown that even low levels of NO are able to induce cell death with properties of apoptosis in undifferentiated PC12 pheochromocytoma cells. Such proapoptotic effect of NO was mediated by the inhibition of mitochondrial cytochrome *c* oxidase (COX) and the generation of ROS, whereas ONOO⁻ had an insignificant effect on cell viability. In addition, neither the release of cytochrome *c* from mitochondrial membranes nor the activation of caspase-3-like activities was observed. Thus, in this case, NO-stimulated apoptosis occurred by a caspase-independent mechanism. Collectively, these results suggest that apoptosis induced at physiologically low concentrations of NO is mediated by ROS production in mitochondria through the inhibition of COX, with ROS acting as an initiator of caspase-independent cell death.

Bobba et al. [144] investigated apoptosis in cerebellar granule cells, which was induced by lowering extracellular potassium. They found that NO initiated two opposite effects, depending on the time after induction of apoptosis. In an early phase, it supported survival of cells through a cGMP-dependent mechanism. After 3 h, nNOS expression and activity decreased, suppressing NO and cGMP production.

Residual NO participated in apoptosis by reacting with increased $O_2{}^{\cdot-}$ concentrations and forming $ONOO^-$. It was suggested that, while NO overproduction protects neurons from death in the early phase of neuronal damage, subsequent decrease in NO levels may contribute to neuronal cell death.

Protein kinases and MAPKs also mediate apoptotic RNS-depended processes. Thus, Cheng et al. [145] have shown that NO induced cell death by the activation of p38-MAP kinase, PARP, and caspase-3. NO caused the release of cytochrome c from mitochondria, while Bcl-2 protected neural progenitor cells against NO-induced apoptosis. Pervin et al. [146] studied the interplay between the extracellular signal-regulated kinase (ERK) and Akt pathways in NO-induced apoptosis of human breast cancer cells. NO inactivated ERK1/2, and its inactivation preceded the dephosphorylation of Akt kinase and apoptosis. Taken together, these results indicated that expression of MKP-1 transcripts by NO leading to dephosphorylation of ERK1/2 was the initial event of the apoptotic pathway in breast cancer cells.

Song et al. [147] demonstrated that exposure of cultured HUVECs to $ONOO^-$ or high glucose significantly inhibited both basal and insulin-stimulated Akt phosphorylation at Ser473 and Akt activity in parallel with increased apoptosis, phosphorylation, and the activity of phosphatase and tensin homologue deleted on chromosome 10 (PTEN). In addition, exposure of HUVECs to $ONOO^-$ or high glucose remarkably increased Ser428 phosphorylation of LKB1, a tumor suppressor. Finally, treatment with tempol, an SOD mimetic, and insulin, both of which reduced $ONOO^-$ formation, markedly reduced diabetes-enhanced LKB1-Ser428 phosphorylation, PTEN, and apoptosis in the endothelium of mouse aortas. These findings suggest that hyperglycemia triggers apoptosis by inhibiting Akt signaling by $ONOO^-$-mediated LKB1-dependent PTEN activation.

REFERENCES

1. DM Hockenbery, ZN Oltvai, XM Yin, CL Milliman, and SJ. Korsmeyer. Bcl-2 functions in an antioxidant pathway to prevent apoptosis. *Cell* 75: 241–251, 1993.
2. J Cai and DP Jones. Superoxide in apoptosi: Mitochondrial generation triggered by cytochrome c loss. *J Biol Chem* 273: 11401–11404, 1998.
3. MD Esposti, I Hatzinisiriou, H McLennan, and S Ralph. Bcl-2 and mitochondrial oxygen radicals: New approaches with reactive oxygen species-sensitive probes. *J Biol Chem* 274: 29831–29837, 1999.
4. Y-H Ling, L Liebes, Y Zou, and R Perez-Soler. Reactive oxygen species generation and mitochondrial dysfunction in the apoptotic response to bortezomib, a novel proteasome inhibitor, in human H460 non-small cell lung cancer cells. *J Biol Chem* 278: 33714–33723, 2003.
5. DA Hildeman, T Mitchell, B Aronow, S Wojciechowski, J Kappler, and P Marrack. Control of *Bcl-2* expression by reactive oxygen species. *Proc Nat Acad Sci USA* 100: 15035–15040, 2003.
6. L Wang, P Chanvorachote, D Toledo, C Stehlik, RR Mercer, V Castranova, and Y Rojanasakul. Peroxide is a key mediator of Bcl-2 downregulation and apoptosis induction by cisplatin in human lung cancer cells. *Mol Pharmacol* 73: 119–127, 2008.
7. TM Johnson, Z-X Yu, VJ Ferrans, RA Lowenstein, and T Finkel. Reactive oxygen species are downstream mediators of p53-dependent apoptosis. *Proc Natl Acad Sci USA* 93: 11848–11852, 1996.

8. S Macip, M Igarashi, P Berggren, J Yu, SW Lee, and SA Aaronson. Influence of induced reactive oxygen species in p53-mediated cell fate decisions. *Mol Cell Biol* 23: 8576–8585, 2003.
9. S Fujioka, C Schmidt, GM Sclabas, Z Li, H Pelicano, B Peng, A Yao, J Niu, W Zhang, DB Evans, JL Abbruzzese, P Huang, and PJ Chiao. Stabilization of p53: A novel mechanism for proapoptotic function of NF-kappa B. *J Biol Chem* 279: 27549–27559, 2004.
10. H Hu, C Jiang, T Schuster, GX Li, PT Daniel, and J Lu. Inorganic selenium sensitizes prostate cancer cells to TRAIL-induced apoptosis through superoxide/p53/Bax-mediated activation of mitochondrial pathway. *Mol Cancer Ther* 5: 1873–1883, 2006.
11. N Xiang, R Zhao, and W Zhong. Sodium selenite induces apoptosis by generation of superoxide via the mitochondrial-dependent pathway in human prostate cancer cells. *Cancer Chemother Pharmacol* 2008 Apr 1 [Epub ahead of print].
12. C Ricci, V Pastukh, J Leonard, J Turrens, G Wilson, D Schaffer, and SW Schaffer. Mitochondrial DNA damage triggers mitochondrial-superoxide generation and apoptosis. *Am J Physiol Cell Physiol* 294: C413–C422, 2008.
13. M Pajusto, TH Toivonen, J Tarkkanen, E Jokitalo, and PS Mattila. Reactive oxygen species induce signals that lead to apoptotic DNA degradation in primary CD4+ T cells. *Apoptosis* 10: 1433–1443, 2005.
14. B Fadeel, A Ahlin, J-I Henter, S Orrenius, and MB Hampton. Involvement of caspases in neutrophil apoptosis: Regulation by reactive oxygen species. *Blood* 92: 4808–4819, 1998.
15. AJ Fay, X Qian, YN Jan, and LY Jan. SK channels mediate NADPH oxidase-independent reactive oxygen species production and apoptosis in granulocytes. *Proc Natl Acad Sci USA* 103: 17548–17553, 2006.
16. BE Jones, CR Lo, H Liu, Z Pradhan, L Garcia, A Srinivasan, KL Valentino, and MJ Czaja. Role of caspases and NF-kappaB signaling in hydrogen peroxide- and superoxide-induced hepatocyte apoptosis. *Am J Physiol Gastrointest Liver Physiol* 278: G693–G699, 2000.
17. JL Hirpara, M-V Clement, and S Pervaiz. Intracellular acidification triggered by mitochondrial-derived hydrogen peroxide is an effector mechanism for drug-induced apoptosis in tumor cells. *J Biol Chem* 276: 514–521, 2001.
18. A Valencia and J Moran. Reactive oxygen species induce different cell death mechanisms in cultured neurons. *Free Radic Biol Med* 36: 1112–1125, 2004.
19. PF Li, R Dietz, and R von Harsdorf. Differential effect of hydrogen peroxide and superoxide anion on apoptosis and proliferation of vascular smooth muscle cells. *Circulation* 96: 3602–3609, 1997.
20. K Kogure, M Morita, S Nakashima, S Hama, A Tokumura, and K Fukuzawa. Superoxide is responsible for apoptosis in rat vascular smooth muscle cells induced by alpha-tocopheryl hemisuccinate. *Biochim Biophys Acta* 1528: 25–30, 2001.
21. M Herdener, S Heigold, M Saran, and G Bauer. Target cell-derived superoxide anions cause efficiency and selectivity of intercellular induction of apoptosis. *Free Radic Biol Med* 29: 1260–1271, 2000.
22. M Schimmel and G Bauer. Proapoptotic and redox state-related signaling of reactive oxygen species generated by transformed fibroblasts. *Oncogene* 21: 5886–5896, 2002.
23. V Nilakantan, C Maenpaa, G Jia, RJ Roman, and F Park. 20–HETE mediated cytotoxicity and apoptosis in ischemic kidney epithelial cells. *Am J Physiol Renal Physiol* 294: F562–F570, 2008.
24. W Mao, C Iwai, PC Keng, R Vulapalli, and CS Liang. Norepinephrine-induced oxidative stress causes PC-12 cell apoptosis by both endoplasmic reticulum stress and mitochondrial intrinsic pathway: inhibition of phosphatidylinositol 3-kinase survival pathway. *Am J Physiol Cell Physiol* 290: C1373–C1384, 2006.

25. Y-C Fu, C-S Chi, S-C Yin, B Hwang, Y-T Chiu, and S-L Hsu. Norepinephrine induces apoptosis in neonatal rat cardiomyocytes through a reactive oxygen species–TNFα–caspase signaling pathway. *Cardiovasc Res* 62: 558–567, 2004.

26. V Moreno-Manzano, Y Ishikawa, J Lucio-Cazana, and M Kitamura. Selective involvement of superoxide anion, but not downstream compounds hydrogen peroxide and peroxynitrite, in tumor necrosis factor-alpha-induced apoptosis of rat mesangial cells. *J Biol Chem* 275: 12684–12691, 2000.

27. DC Han, M-Y Lee, KD Shin, SB Jeon, JM Kim, K-H Son, H-C Kim, H-M Kim, and B-M Kwon. 2′-Benzoyloxycinnamaldehyde induces apoptosis in human carcinoma via reactive oxygen species. *J Biol Chem* 279: 6911–6920, 2004.

28. DW Kamp, V Panduri, SA Weitzman, and N Chandel. Asbestos-induced alveolar epithelial cell apoptosis: Role of mitochondrial dysfunction caused by iron-derived free radicals. *Mol Cell Biochem* 234–235: 153–160, 2002.

29. J Moungjaroen, U Nimmannit, PS Callery, L Wang, N Azad, V Lipipun, P Chanvorachote, and Y Rojanasakul. Reactive oxygen species mediate caspase activation and apoptosis induced by lipoic acid in human lung epithelial cancer cells through Bcl-2 downregulation. *J Pharmacol Exp Ther* 319: 1062–1069, 2006.

30. MJ Rio and C Velez-Pardo. Paraquat induces apoptosis in human lymphocytes: Protective and rescue effects of glucose, cannabinoids and insulin-like growth factor-1. *Growth Factors* 26: 49–60, 2008.

31. T Shibayama-Imazu, T Aiuchi, and K Nakaya. Vitamin K2-mediated apoptosis in cancer cells: Role of mitochondrial transmembrane potential. *Vitam Horn* 78: 211–226, 2008.

32. DN Criddle, S Gillies, HK Baumgartner-Wilson, M Jaffar, EC Chinje, S Passmore, M Chvanov, S Barrow, OV Gerasimenko, AV Tepikin, R Sutton, and OH Petersen. Menadione-induced reactive oxygen species generation via redox cycling promotes apoptosis of murine pancreatic acinar cells. *J Biol Chem* 281: 40485–40492, 2006.

33. HK Baumgartner, JV Gerasimenko, C Thorne, LH Ashurst, SL Barrow, MA Chvanov, S Gillies, DN Criddle, AV Tepikin, OH Petersen, R Sutton, AJM Watson, and OV Gerasimenko. Caspase-8-mediated apoptosis induced by oxidative stress is independent of the intrinsic pathway and dependent on cathepsins. *Am J Physiol Gastrointest Liver Physiol* 293: G296–G307, 2007.

34. YH Han, SZ Kim, SH Kim, and WH Park. Pyrogallol as a glutathione depletory induces apoptosis in HeLa cells. *Int J Mol Med* 21: 721–730, 2008.

35. ME Juan, U Wenzel, H Daniel, and JM Planas. Resveratrol induces apoptosis through ROS-dependent mitochondria pathway in HT-29 human colorectal carcinoma cells. *J Agric Food Chem* 56: 4813–4818, 2008.

36. NR Khawaja, M Carre, H Kovacic, M-A Esteve, and D Braguer. Patupilone-induced apoptosis is mediated by mitochondrial reactive oxygen species through Bim relocalization to mitochondria. *Mol. Pharmacol* 10.1124/mol.108.048405, 2008.

37. C Ricci, V Pastikh, M Mozaffari, and SW Schaffer. Insulin withdrawal induces apoptosis via a free radical-mediated mechanism. *Can J Physiol Pharmacol* 85: 455–464, 2007.

38. L Kongkaneramit, N Sarisuta, N Azad, Y Lu, AK Iyer, L Wang, and Y Rojanasakul. Dependence of reactive oxygen species and FLICE inhibitory protein on lipofectamine-induced apoptosis in human lung epithelial cells. *J Pharmacol Exp Ther* 325: 969–977, 2008.

39. M Wartenberg, N Wirtz, A Grob, W Niedermeier, J Hescheler, SC Peters, and H Sauer. Direct current electrical fields induce apoptosis in oral mucosa cancer cells by NADPH oxidase-derived reactive oxygen species. *Bioelectromagnetics* 29: 47–54, 2008.

40. T Hayashi, I Hayashi, T Shinohara, Y Morishita, H Nagamura, Y Kusunoki, S Kyoizumi, T Seyama, and K Nakachi. Radiation-induced apoptosis of stem/progenitor cells in human umbilical cord blood is associated with alterations in reactive oxygen and intracellular pH. *Mutat Res* 556: 83–91, 2004.
41. ZQ Ge, S Yang, JS Cheng, and YJ Yuan. Signal role for activation of caspase-3-like protease and burst of superoxide anions during Ce4+-induced apoptosis of cultured *Taxus cuspidata* cells. *Biometals* 18: 221–232, 2005.
42. U Wenzel, A Nickel, S Kuntz, and H Daniel. Ascorbic acid suppresses drug-induced apoptosis in human colon cancer cells by scavenging mitochondrial superoxide anions. *Carcinogenesis* 25: 703–712, 2004.
43. N Li, K Ragheb, G Lawler, J Sturgist, B Rajwa, JA Melendez, and JP Robinson. Mitochondrial complex I inhibitor rotenone induces apoptosis through enhancing mitochondrial reactive oxygen. *J Biol Chem* 278: 8516–8525, 2003.
44. H Pelicano, L Feng, Y Zhou, JS Carew, EO Hileman, W Plunkett, MJ Keating, and P Huang. Inhibition of mitochondrial repiration: A novel strategy to enhance drug-induced apoptosis in human leukemia cells by a reactive oxygen species-mediated mechanism. *J Biol Chem* 278: 37832–37839, 2003.
45. M Marella, BB Seo, A Matsuno-Yagi, and T Yagi. Mechanism of cell death caused by complex I defects in a rat dopaminergic cell line. *J Biol Chem* 282: 24146–24156, 2007.
46. T-S Chang, C-S Cho, S Park, S Yu, SW Kang, and SG Rhee. Peroxiredoxin III, a mitochondrion-specific peroxidase, regulates apoptotic signaling by mitochondria. *J Biol Chem* 279: 41975–41984, 2004.
47. NS Chandel, PT Schumacker, and RH Arch. Reactive oxygen species are downstream products of TRAF-mediated signal transduction. *J Biol Chem* 276: 42728–42736, 2001.
48. SJ Gardai, R Hoontrakoon, CD Goddard, BJ Day, LY Chang, PM Henson, and DL Bratton. Oxidant-mediated mitochondrial injury in eosinophil apoptosis: Enhancement by glucocorticoids and inhibition by granulocyte-macrophage colony-stimulating factor. *J Immunol* 170: 556–566, 2003.
49. JP French, KL Hamilton, JC Quindry, Y Lee, PA Upchurch, and SK Powers. Exercise-induced protection against myocardial apoptosis and necrosis: MnSOD, calcium-handling proteins, and calpain. *FASEB J* Apr 2008; 10.1096/fj.07–102541.
50. P Inarrea, H Moini, D Han, D Rettori, I Aguilo, MA Alava, M Iturralde, and E Cadenas. Mitochondrial respiratory chain and thioredoxin reductase regulate intermembrane CuZn-superoxide dismutase activity: Implications for mitochondrial energy metabolism and apoptosis. *Biochem J* 405: 173–179, 2007.
51. J McInnis, C Wang, N Anastasio, M Hultman, Y Ye, D Salvemini, and KM Johnson. The role of superoxide and nuclear factor-kappaB signaling in *N*-methyl-D-aspartate-induced necrosis and apoptosis. *J Pharmacol Exp Ther* 301: 478–487, 2002.
52. J Li, P-F Li, R Dietz, and R von Harsdorf. Intracellular superoxide induces apoptosis in VSMCs: Role of mitochondrial membrane potential, cytochrome c and caspases. *Apoptosis* 7: 511–517, 2002.
53. N Kajitani, H Kobuchi, H Fujita, H Yano, T Fujiwara, T Yasuda, and K Utsumi. Mechanism of A23187-induced apoptosis in HL-60 cells: Dependency on mitochondrial permeability transition but not on NADPH oxidase. *Biosci Biotechnol Biochem* 71: 2701–2711, 2007.
54. M Madesh, BJ Hawkins, T Milovanova, CD Bhanumathy, SK Joseph, SP Ramachandrarao, K Sharma, T Kurosaki, and AB Fisher. Selective role for superoxide in InsP3 receptor-mediated mitochondrial dysfunction and endothelial apoptosis. *J Cell Biol* 170: 1079–1090, 2005.
55. C Piskernik, S Haindl, T Behling, Z Gerald, I Kehrer, H Redl, and AV Kozlov. Antimycin A and lipopolysaccharide cause the leakage of superoxide radicals from rat liver mitochondria. *Biochim Biophys Acta* 1782: 280–285, 2008.

56. L Quagliaro, L Piconi, R Assaloni, R Da Ros, C Szabo, and A Ceriello. Primary role of superoxide anion generation in the cascade of events leading to endothelial dysfunction and damage in high glucose treated HUVEC. *Nutr Metab Cardiovasc Dis* 17: 257–267, 2007.

57. BJ Hawkins, M Madesh, CJ Kirkpatrick, and AB Fisher. Superoxide flux in endothelial cells via the chloride channel-3 mediates intracellular signaling. *Mol Biol Cell* 18: 2002–2012, 2007.

58. A Atlante, P Calissano, A Bobba, A Azzariti, E Marra, and S Passarella. Cytohcrome *c* is released from mitochondria in a reactive oxygen species (ROS)-dependent fashion and can operate as an ROS scavenger and as a respiratory substrate in cerebellar neurons undergoing excitotoxic death. *J Biol Chem* 275: 37159–37166, 2000.

59. CM Luetjens, NT Bui, B Sengpiel, G Munstermann, M Poppe, AJ Krohn, E Bauerbach, J Kriegistein, and HM Prehn. Delayed mitochondrial dysfunction in excitotoxic neuron death: CCR and a secondary increase in superoxide production. *J Neurosci* 20: 5715–5723, 2000.

60. M Madesh and G Hajnoczky. VDAC-dependent permeabilization of the outer mitochondrial membrane by superoxide induces rapid and massive cytochrome c release. *J Cell Biol* 156: 1003–1016, 2001.

61. M Ott, JD Robertson, V Gogvadze, B Zhvotovsky, and S Orrenius. CCR from mitochondria proceeds by a two-step process. *Proc Nat Acad Sci USA* 99: 1259–1263, 2002.

62. G Petrosillo, FM Ruggiero, and G Paradies. Role of reactive oxygen species and cardiolipin in the release of cytochrome c from mitochondria. *FASEB J* 17: 2202–2208, 2003.

63. H Dussmann, D Kogel, M Rehm, and LHM Prehn. Mitochondrial membrane permeabilization and superoxide production during apoptosis. A single-cell analysis. *J Biol Chem* 278: 12645–12649, 2003.

64. C Thirunavukkarasu, S Watkins, SA Harvey, and CR Gandhi. Superoxide-induced apoptosis of activated rat hepatic stellate cells. *J Hepatol* 41: 567–575, 2004.

65. Q Li, EF Sato, Y Kira, M Nishikawa, K Utsumi, and M Inoue. A possible cooperation of SOD1 and cytochrome c in mitochondria-dependent apoptosis. *Free Radic Biol Med* 40: 173–181, 2004.

66. LV Basova, IV Kurnikov, L Wang, VB Ritov, NA Belikova, II Vlasova, AA Pacheco, DE Winnica, J Peterson, H Bayir, DH Waldeck, and VE Kagan. Cardiolipin switch in mitochondria: Shutting off the reduction of cytochrome c and turning on the peroxidase activity. *Biochemistry* 46: 3423–3434, 2007.

67. J Jiang, Z Huang, Q Zhao, W Feng, NA Belikova, and VE Kagan. Interplay between bax, reactive oxygen species production, and cardiolipin oxidation during apoptosis. *Biochem Biophys Res Commun* 368: 145–150, 2008.

68. IB Afanas'ev. Superoxide and nitric oxide in pathological conditions associated with iron overload. The effects of antioxidants and chelators. *Curr Med Chem* 12: 2731–2739, 2005.

69. IB Afanas'ev. Free radical mechanisms of aging processes under physiological conditions. *Biogerontology* 6: 283–290, 2005.

70. IB Afanas'ev. *Superoxide Ion: Chemistry and Biological Implications*, Vol. I, II, Boca Raton, FL: CRC Press, 1989, 1990.

71. ET Denisov and IB Afanas'ev. *Oxidation and Antioxidants in Organic Chemistry and Biology*, Part III, Boca Raton, FL: CRC Press, 2005.

72. M Burkitt, C Jones, A Lawrence, and P Wardman. Activation of cytochrome c to a peroxidase compound I-type intermediate by H2O2: Relevance to redox signalling in apoptosis. *Biochem Soc Symp* 71: 97–106, 2004.

73. D Roos, CM Eckmann, M Yazdanbakhsh, MN Hamers, and M de Boer. Excretion of superoxide by phagocytes measured with cytochrome c entrapped in resealed erythrocyte ghosts. *J Biol Chem* 259: 1770–1775, 1984.

74. H Iwase, T Takatori, M Nagao, K Twadate, and M Nakajima. Monoepoxide production from linoleic acid by cytochrome c in the presence of cardiolipin. *Biochem Biophys Res Commun* 222: 83–89, 1996.

75. IL Nantes, MR Zucchi, OR Nascimento, and A Faljoni-Alario. Effect of heme iron valence state on the conformation of cytochrome c and its association with membrane interfaces: A CD and EPR investigation. *J Biol Chem* 276: 153–158, 2001.

76. EK Tuominen, JA Wallace, and PKJ Kinnunen. Phospholipid-cytochrome c interaction: Evidence for the extended lipid anchorage. *J Biol Chem* 277: 8822–8826, 2002.

77. IB Afanas'ev, SV Prigoda, AM Khenkin, and AA Sheiman. Interaction of oxygen radical anion O2.- with Fe(III)porphyrins. A new oxygenated heme complex. *Doklady USSR Acad Sci* 236: 641–645, 1977.

78. IB Afanas'ev. *Superoxide Ion: Chemistry and Biological Implications*, vol. I. Boca Raton, FL: CRC Press, 1989, p. 248.

79. P-F Li, C Maasch, H Haller, R Dietz, and R von Harsdorf. Requirment for protein kinase C in reactive oxygen species-induced apoptosis of vascular smooth muscle cells. *Circulation* 100: 967–973, 1999.

80. P Bergamo, D Luongo, and M Rossi. Conjugated linoleic acid-mediated apoptosis in jurkat T cells involves the production of reactive oxygen species. *Cell Physiol Biochem* 14: 57–64, 2004.

81. HS Byun, M Won, KA Park, YR Kim, BL Choi, H Lee, JH Hong, L Piao, J Park, JM Kim, GR Kweon, SH Kang, J Han, and GM Hur. Prevention of TNF-induced necrotic cell death by rottlerin through a Nox1 NADPH oxidase. *Exp Mol Med* 40: 186–195, 2008.

82. J Du, H Suzuki, F Nagase, AA Akhand, X Ma, T Yokoyama, T Miyata, and I Nakashima. Superoxide-mediated early oxidation and activation of ASK1 are important for initiating methylglyoxal-induced apoptosis process. *Free Radic Biol Med* 31: 469–478, 2001.

83. G Fujino, T Noguchi, A Matsuzawa, S Yamauchi, M Saitoh, K Takeda, and H Ichijo. Thioredoxin and TRAF family proteins regulate reactive oxygen species-dependent activation of ASK1 through reciprocal modulation of the N-terminal homophilic interaction of ASK1. *Mol Cell Biol* 27: 8152–8163, 2007.

84. PLK Lim, J Liu, ML Go, and UA Boelsterli. The mitochondrial superoxide/thioredoxin-2/Ask1 signaling pathway is critically involved in troglitazone-induced cell injury to human hepatocytes. *Toxicol Sci* 101: 341–349, 2008.

85. D-Y Shi, Y-R Deng, S-L Liu, Y-D Zhang, and L Wei. Redox stress regulates cell proliferation and apoptosis of human hepatoma through Akt protein phosphorylation. *FEBS Lett* 542: 60–64, 2003.

86. A Inoue, S Muranaka, H Fujita, T Kanno, H Tamai, and K Utsumi. Molecular mechanism of diclofenac-induced apoptosis of promyelocytic leukemia: Dependency on reactive oxygen species, akt, bid, cytochrome and caspase pathway. *Free Radic Biol Med* 37: 1290–1299, 2004.

87. N Gao, M Rahmani, P Dent, and S Grant. 2-Methoxyestradiol-induced apoptosis in human leukemia cells proceeds through a reactive oxygen species and Akt-dependent process. *Oncogene* 24: 3797–3809, 2005.

88. H Fujita, T Ogino, H Kobuchi, T Fujiwara, H Yano, J Akiyama, K Utsumi, and J Sasaki. Cell-permeable cAMP analog suppresses 6-hydroxydopamine-induced apoptosis in PC12 cells through the activation of the Akt pathway. *Brain Res* 1113: 10–23, 2006.

89. T Mochizuki, S Furuta, J Mitsushita, WH Shang, M Ito, Y Yokoo, M Yamaura, S Ishizone, J Nakayama, A Konagai, K Hirose, K Kiyosawa, and T Kamata. Inhibition of NADPH oxidase 4 activates apoptosis via the AKT/apoptosis signal-regulating kinase 1 pathway in pancreatic cancer PANC-1 cells. *Oncogene* 25: 3699–3707, 2006.

90. C Yu, BB Friday, J-P Lai, A McCollum, P Atadja, LR Roberts, and AA Adjei. Abrogation of MAPK and Akt signaling by AEE788 synergistically potentiates histone deacetylase inhibitor-induced apoptosis through reactive oxygen species generation. *Clin Cancer Res* 13: 1140–1148, 2007.

91. M Yokouchi, N Hiramatsu, K Hayakawa, M Okamura, S Du, A Kasai, Y Takano, A Shitamura, T Shimada, J Yao, and Mi Kitamura. Involvement of selective reactive oxygen species upstream of proapoptotic branches of unfolded protein response. *J Biol Chem* 283: 4252–4260, 2008.

92. L Liu, XJ Zhang, SR Jiang, ZN Ding, GX Ding, J Huang, and YL Cheng. Heat shock protein 27 regulates oxidative stress-induced apoptosis in cardiomyocytes: Mechanisms via reactive oxygen species generation and Akt activation. *Chin Med J (Engl)* 120: 2271–2277, 2007.

93. WS Choi, DS Eom, BS Han, WK Kim, BH Han, EJ Choi, TH Oh, GJ Markelonis, JW Cho, and YJ Oh. Phosphorylation of p38 MAPK induced by oxidative stress is linked to activation of both caspase-8- and 9-mediated apoptotic pathways in dopaminergic neurons. *J Biol Chem* 279: 20451–20460, 2004.

94. HY Sun, NP Wang, M Halkos, F Kerendi, H Kin, RA Guyton, J Vinten-Johansen, and ZQ Zhao. Postconditioning attenuates cardiomyocyte apoptosis via inhibition of JNK and p38 mitogen-activated protein kinase signaling pathways. *Apoptosis* 11: 1583–1593, 2006.

95. Y Dai, M Rahmani, P Dent, and S Grant. Blockade of histone deacetylase inhibitor-induced RelA/p65 acetylation and NF-{kappa}B activation potentiates apoptosis in leukemia cells through a process mediated by oxidative damage, XIAP downregulation, and c-Jun N-terminal kinase 1 activation. *Mol Cell Biol* 25: 5429–5444, 2005.

96. S Dhingra, AK Sharma, DK Singla, and PK Singal. p38 and ERK 1/2 MAP kinases mediate interplay of TNF- and IL-10 in regulating oxidative stress and cardiac myocyte apoptosis. *Am J Physiol Heart Circ Physiol* 293: H3524–H3531, 2007.

97. M She, H Yang, L Sun, and SC Yeung. Redox control of manumycin A-induced apoptosis in anaplastic thyroid cancer cells: Involvement of the xenobiotic apoptotic pathway. *Cancer Biol Ther* 5: 275–280, 2006.

98. B Kou, J Ni, M Vatish, and DR Singer. Xanthine oxidase interaction with vascular endothelial growth factor in human endothelial cell angiogenesis. *Microcirculation* 15: 251–267, 2008.

99. M Gomez-Lazaro, MF Galindo, RM Melero-Fernandez de Mera, FJ Fernandez-Gomez, CG Concannon, MF Segura, JX Comella, JHM Prehn, and J Jordan. Reactive oxygen species and p38 mitogen-activated protein kinase activate Bax to induce mitochondrial CCR and apoptosis in response to malonate. *Mol Pharmacol* 71: 736–743, 2007.

100. FM Ho, SH Liu, CS Liau, PJ Huang, and SY Lin-Shiau. High glucose-induced apoptosis in human endothelial cells is mediated by sequential activations of c-Jun NH2-terminal kinase and caspase-3. *Circulation* 101: 2618–2624, 2000.

101. YC Fu, SC Yin, CS Chi, B Hwang, and SL Hsu. Norepinephrine induces apoptosis in neonatal rat endothelial cells via a ROS-dependent JNK activation pathway. *Apoptosis* 11: 2053–2063, 2006.

102. L Huc, X Tekpli, JA Holme, M Rissel, A Solhaug, C Gardyn, G Le Moigne, M Gorria, M-T Dimanche-Boitrel, and D Lagadic-Gossmann. c-Jun NH2-terminal kinase-related Na^+/H^+ exchanger isoform 1 activation controls hexokinase II expression in benzo(a)pyrene-induced apoptosis. *Cancer Res* 67: 1696–1705, 2007.

103. SV Singh, S Choi, Y Zeng, E-R Hahm, and D Xiao. Guggulsterone-induced apoptosis in human prostate cancer cells is caused by reactive oxygen intermediate dependent activation of c-Jun NH2-terminal kinase. *Cancer Res* 67: 7439–7449, 2007.

104. I Ishikawa and M Kitamura. Anti-apoptotic effect of quercetin: Intervention in the JNK- and EPK-mediated apoptotic pathways. *Kidney Int* 58: 1078–1087, 2000.

105. Y Wei, X Zhao, Y Kariya, H Fukata, K Teshigawara, and A Uchida. Induction of apoptosis by quercetin: Involvement of heat shock protein. *Cancer Res* 54: 4952–4957, 1994.

106. W Zhang, S Zheng, P Storz, and W Min. Protein kinase D specifically mediates apoptosis signal-regulating kinase 1-JNK signaling induced by H2O2 but not tumor necrosis factor. *J Biol Chem* 280: 19036–19044, 2005.

107. H Kamata, SI Honda, S Maeda, L Chang, H Hirata, and M Karin. Reactive oxygen species promote TNFalpha-induced death and sustained JNK activation by inhibiting MAP kinase phosphatases. *Cell* 120: 649–661, 2005.

108. L Conde de la Rosa, TE Vrenken, RA Hannivoort, M Buist-Homan, R Havinga, DJ Slebos, HF Kauffman, KN Faber, PL Jansen, and H Moshage. Carbon monoxide blocks oxidative stress-induced hepatocyte apoptosis via inhibition of the p54 JNK isoform. *Free Radic Biol Med* 44: 1323–1333, 2008.

109. HY Hong and BC Kim. Mixed lineage kinase 3 connects reactive oxygen species to c-Jun NH2-terminal kinase-induced mitochondrial apoptosis in genipin-treated PC3 human prostate cancer cells. *Biochem Biophys Res Commun* 362: 307–312, 2007.

110. I Suzumi, T Murakami, K Suzuma, H Kaneto, D Watanabe, T Ojima, Y Honda, H Takagi, and N Yoshimura. Cyclic stretch-induced reactive oxygen species generation enhances apoptosis in retinal pericytes through c-Jun NH2-terminal kinase activation. *Hypertension* 49: 347–354, 2007.

111. Y-M Ham, J-H Lim, H-K Na, J-S Choi, B-D Park, H Yim, and S-K Lee. Ginsenoside-Rh2-induced mitochondrial depolarization and apoptosis are associated with reactive oxygen species- and Ca2+-mediated c-Jun NH2-terminal kinase 1 activation in HeLa cells. *J Pharmacol Exp Ther* 319: 1276–1285, 2006.

112. M Yokouchi, N Hiramatsu, K Hayakawa, M Okamura, S Du, A Kasai, Y Takano, A Shitamura, T Shimada, J Yao, and M Kitamura. Involvement of selective reactive oxygen species upstream of proapoptotic branches of unfolded protein response. *J Biol Chem* 283: 4252–4260, 2008.

113. Y Liu, GL Borchert, SP Donald, A Surazynski, C-A Hu, CJ Weydert, LW Oberley, and JM Phang. MnSOD inhibits proline oxidase-induced apoptosis in colorectal cancer cells. *Carcinogenesis* 26: 1335–1342, 2005.

114. Y Cai, GA Martens, SA Hinke, H Heimberg, D Pipeleers, and M Van de Casteele. Increased oxygen radical formation and mitochondrial dysfunction mediate beta cell apoptosis under conditions of AMP-activated protein kinase stimulation. *Free Radic Biol Med* 42: 64–78, 2007.

115. A Saito, T Hayashi, S Okuno, T Nishi, and PH Chan. Oxidative stress affects the integrin-linked kinase signaling pathway after transient focal cerebral ischemia. *Stroke* 35: 2560–2565, 2004.

116. JS Liou, C-Y Chen, JS Chen, and DV Faller. Oncogenic Ras mediates apoptosis in response to protein kinase c inhibition through the generation of reactive oxygen species. *J Biol Chem* 275: 39001–39011, 2000.

117. WJ Huang, CW Tung, C Ho, JT Yang, ML Chen, PJ Chang, PH Lee, CL Lin, and JY Wang. Ras activation modulates methylglyoxal-induced mesangial cell apoptosis through superoxide production. *Ren Fail* 29: 911–921, 2007.

118. S Wang, SS Leonard, J Ye, N Gao, L Wang, and X Shi. Role of reactive oxygen species and Cr(VI) in Ras-mediated signal transduction. *Mol Cell Biochem* 255: 119–127, 2004.

119. C-L Lin, J-Y Wang, J-Y Ko, K Surendran, Y-T Huang, Y-H Kuo, and F-S Wang. Superoxide destabilization of {beta}-catenin augments apoptosis of high glucose-stressed mesangial cells. *Endocrinology* 149: 2934–2942, 2008.

120. T Sato, T Machida, S Takahashi, S Iyama, Y Sato, K Kuribayashi, K Takada, T Oku, Y Kawano, T Okamoto, R Takimoto, T Matsunaga, T Takayama, M Takahashi, J Kato, and Y Niitsu. Fas-mediated apoptosome formation is dependent on reactive oxygen species derived from mitochondrial permeability transition in jurkat cells. *J Immunol* 173: 285–296, 2004.

121. L Wang, N Azad, L Kongkaneramit, F Chen, Y Lu, B-H Jiang, and Y Rojanasakul. The Fas death signaling pathway connecting reactive oxygen species generation and FLICE inhibitory protein down-regulation. *J Immunol* 180: 3072–3080, 2008.

122. H Lin, C-C Hou, C-F Cheng, T-H Chiu, Y-H Hsu, Y-M Sue, T-H Chen, H-H Hou, Y-C Chao, T-H Cheng, and C-H Chen. Peroxisomal proliferator-activated receptor-alpha (PPAR-{alpha}) protects renal tubular cells from adriamycin-induced apoptosis. *Mol Pharmacol* 72: 1238–1245, 2007.

123. RA Kirkland, GM Saavedra, and JL Franklin. Rapid activation of antioxidant defenses by nerve growth factor suppresses reactive oxygen species during neuronal apoptosis: Evidence for a role in cytochrome *c* redistribution. *J Neurosci* 27: 11315–11326, 2007.

124. I Petrache, TR Medler, AT Richter, K Kamocki, U Chukwueke, L Zhen, Y Gu, J Adamowicz, KS Schweitzer, WC Hubbard, EV Berdyshev, G Lungarella, and RM Tuder. Superoxide dismutase protects against apoptosis and alveolar enlargement induced by ceramide. *Am J Physiol Lung Cell Mol Physiol* April 2008; 10.1152/ajplung.00448.2007.

125. G Hua, Q Zhang, and Z Fan. Heat shock protein 75 (TRAP1) antagonizes reactive oxygen species generation and protects cells from granzyme M-mediated apoptosis. *J Biol Chem* 282: 20553–20560, 2007.

126. D Martinvalet, DM Dykxhoorn, R Ferrini, and J Lieberman. Granzyme A cleaves a mitochondrial complex I protein to initiate caspase-independent cell death. *Cell* 133: 681–692, 2008.

127. E Bonfoco, D Krainc, M Ankarcrona, P Nicotera, and SA Lipton. Apoptosis and necrosis: Two distinct events induced, respectively, by mild and intense insults with *N*-methyl-D-aspartate or nitric oxide/superoxide in cortical cell cultures. *Proc Natl Acad Sci USA* 92: 7162–7166, 1995.

128. S Hortelano, B Dallaporta, N Zamzami, T Hirsch, SA Susin, I Marzo, L Bosca, and G Kroemer. Nitric oxide induces apoptosis via triggering mitochondrial permeability transition. *FEBS Lett* 410: 373–377, 1997.

129. C Ward, TH Wong, J Murray, I Rahman, C Haslett, ER Chilvers, and AG Rossi. Induction of human neutrophil apoptosis by nitric oxide donors: Evidence for a caspase-dependent, cyclic-GMP-independent, mechanism. *Biochem Pharmacol* 59: 305–314, 2004.

130. S Hortelano, AM Alvarez, and L Boscá. Nitric oxide induces tyrosine nitration and release of cytochrome *c* preceding an increase of mitochondrial transmembrane potential in macrophages. *FASEB J* 13: 2311–2317, 1999.

131. A Pautz, R Franzen, S Dorsch, B Boddinghaus, VA Briner, J Pfeilschifter, and A Huwiler. Cross-talk between nitric oxide and superoxide determines ceramide formation and apoptosis in glomerular cells. *Kidney Int* 61: 790–796, 2002.

132. RR Nazarewicz, WJ Zenebe, A Parihar, SK Larson, E Alidema, J Choi, and P Ghafourifar. Tamoxifen induces oxidative stress and mitochondrial apoptosis via stimulating mitochondrial nitric oxide synthase. *Cancer Res* 67: 1282–1290, 2007.

133. DA Langer, A Das, D Semela, N Kang-Decker, H Hendrickson, SF Bronk, ZS Katusic, GJ Gores, and VH Shah. Nitric oxide promotes caspase-independent hepatic stellate cell apoptosis through the generation of reactive oxygen species. *Hepatology* 47(6): 1983–1993, 2008.

134. S Heigold, C Sers, W Bechtel, B Ivanovas, R Schafer, and G Bauer. Nitric oxide mediates apoptosis induction selectively in transformed fibroblasts compared to nontransformed fibroblasts. *Carcinogenesis* 23: 929–941, 2002.

135. M Steinmann, N Moosmann, M Schimmel, C Gerhardus, and G Bauer. Differential role of extra- and intracellular superoxide anions for nitric oxide-mediated apoptosis induction. *In Vivo* 18: 293–309, 2004.

136. P Ghafourifar, U Schenk, SD Klein, and C Richter. Mitochondrial nitric-oxide synthase stimulation causes cytochrome c release from isolated mitochondria. Evidence for intramitochondrial peroxynitrite formation. *J Biol Chem* 274: 31185–31188, 1999.

137. S Levrand, C Vannay-Bouchiche, B Pesse, P Pacher, F Feihl, B Waeber, and L Liaudet. Peroxynitrite is a major trigger of cardiomyocyte apoptosis in vitro and in vivo. *Free Radic Biol Med* 41: 886–895, 2006.

138. P Ascenzi, M Marino, and E Menegatti. CO(2) impairs peroxynitrite-mediated inhibition of human caspase-3. *Biochem Biophys Res Commun* 349: 367–371, 2006.

139. JB Mannick, XQ Miao, and JS Stamler. Nitric oxide inhibits fas-induced apoptosis. *J Biol Chem* 272: 24125–24128, 1997.

140. L Rossig, B Fichtlscherer, K Breitschopf, J Haendeler, AM Zeihert, A Mulsch, and S Dimmeler. Nitric oxide inhibits caspase-3 by S-nitrosation in vivo. *J Biol Chem* 274, 6823–6826, 1999.

141. Y-M Kim, ME de Vera, SC Watkins, and TR Billiar. Nitric oxide protects cultured rat hepatocytes from tumor necrosis factor-α-induced apoptosis by inducing heat shock protein 70 expression. *J Biol Chem* 274: 1402–1411, 1999.

142. S-J Lee, K-M Kim, S Namkoong, C-K Kim, Y-C Kang, H Lee, K-S Ha, J-A Han, H-T Chung, Y-G Kwon, and Y-M Kim. Nitric oxide inhibition of homocysteine-induced human endothelial cell apoptosis by down-regulation of p53-dependent Noxa expression through the formation of S-nitrosohomocysteine. *J Biol Chem* 280: 5781–5788, 2005.

143. K Yuyama, H Yamamoto, I Nishizaki, T Kato, I Sora, and T Yamamoto. Caspase-independent cell death by low concentrations of nitric oxide in PC12 cells: Involvement of cytochrome c oxidase inhibition and the production of reactive oxygen species in mitochondria. *J Neurosci Res* 73: 351–363, 2003.

144. A Bobba, A Atlante, L Moro, P Calissano, and E Marra. Nitric oxide has dual opposite roles during early and late phases of apoptosis in cerebellar granule neurons. *Apoptosis* 12: 1597–1610, 2007.

145. A Cheng, SL Chan, O Milhavet, S Wang, and MP Mattson. P38 MAP kinase mediates nitric oxide-induced apoptosis of neural progenitor cells. *J Biol Chem* 276: 43320–43327, 2001.

146. S Pervin, R Singh, WA Freije, and G Chaudhuri. MKP-1-induced dephosphorylation of extracellular signal-regulated kinase is essential for triggering nitric oxide-induced apoptosis in human breast cancer cell lines: Implications in breast cancer. *Cancer Res.* 63: 8853–8860, 2003.

147. P Song, Y Wu, J Xu, Z Xie, Y Dong, M Zhang, and M-H Zou. Reactive nitrogen species induced by hyperglycemia suppresses Akt signaling and triggers apoptosis by upregulating phosphatase PTEN (phosphatase and tensin homologue deleted on chromosome 10) in an LKB1-dependent manner. *Circulation* 116: 1585–1595, 2007.

6 ROS and RNS Signaling in Senescence and Aging

The study of free radical-mediated signaling processes in aging and senescence is a relatively new field of current investigations. Attention was drawn worldwide to the role of free radicals in aging and senescence in 1956 by Dr. Harman's work "Aging: A Theory Based on Free Radical and Radiation Chemistry" [1]. At that time the structures of free radicals, which are formed in cells and tissues, were unknown, and all free radicals were considered toxic species capable of destroying biomolecules and stimulating aging of living organisms. This hypothesis led to important practical conclusions, pointing out for the first time the possibility of age regulation by antioxidants capable of suppressing free-radical formation.

Despite numerous subsequent works in which the role of free radicals and antioxidants in aging has been studied, there are still different views concerning many important questions. However, this is not surprising, taking into account numerous factors that regulate free radical formation in normal and aged cells and tissues, as well as different functions of free radicals in various enzymatic processes. At present, there are many reviews discussing the free-radical theory of aging. The latest ones are cited here [2–11]. (In this book the important studies performed with yeast, plants, worms, and fruit flies will not be discussed because they were already considered by many highly competent authors).

As already mentioned, at the time of Harman's proposal of the free-radical theory of aging, the structure of free radicals in biological systems was mainly unknown. Discoveries made from 1968 to 1980 about "physiological" free radicals superoxide ($O_2\cdot^-$) and nitric oxide ($\cdot NO$) completely changed this situation. It is now known that these radicals, being relatively unreactive and harmless in themselves, are the precursors of hydrogen peroxide (H_2O_2) and peroxynitrite ($ONOO^-$), compounds capable of forming highly reactive free radicals by the $O_2\cdot^-$-driven Fenton reaction (Reactions 6.1 and 6.2) and the decomposition of $ONOO^-$ (Reactions 6.3 and 6.4):

$$O_2\cdot^- + Fe^{3+} \Rightarrow O_2 + Fe^{2+} \tag{6.1}$$

$$Fe^{2+} + H_2O_2 \Rightarrow Fe^{3+} + HO\cdot + HO^- \tag{6.2}$$

$$O_2\cdot^- + (\cdot)NO \Rightarrow {}^-OONO \tag{6.3}$$

$$^-OONO + H^+ \Rightarrow HO\cdot + \cdot NO_2 \tag{6.4}$$

Thus, the overproduction of $O_2\cdot^-$ and NO can lead to the formation of reactive $HO\cdot$ and NO_2 radicals, the initiators of pathological disorders and aging. On the other hand, it became clear that $O_2 \cdot^-$ and NO are important signaling intermediates of many physiological and pathophysiological processes regulating aging and senescence. It is quite possible that an important factor of aging development is the disturbance of $O_2 \cdot^-$/NO balance in cells, which is responsible for mitochondria dysfunction, stimulation of apoptosis, and other pathological changes typical of senescence and aging. All these signaling effects of reactive oxygen (ROS) and nitrogen (RNS) species in aging and senescence will be discussed in this chapter.

6.1 FREE RADICALS IN AGED CELLS, TISSUES, AND WHOLE ORGANISMS

6.1.1 Enhancement of ROS Production in Mitochondria, Cells, and Tissue with Age

In 1972, Harman's [12] proposal that the mitochondrially formed $O_2 \cdot^-$ was an important determinant of aging development. A view of $O_2 \cdot^-$ as the important initiator of aging has been discussed in numerous works, and now more than 1900 works on this subject are cited by Medline.

Overproduction of $O_2 \cdot^-$ in aging cells and tissues, and animals and humans, has been widely studied. Sawada and Carlson [13] demonstrated a significant increase in $O_2 \cdot^-$ production in mitochondria from the brain and heart of aged rats. In parallel, an elevation in the level of lipid peroxidation was also found in the brain and liver homogenates. In a subsequent work, Sawada et al. [14] confirmed the enhancement of $O_2 \cdot^-$ formation and biochemical alterations in plasma membranes from the brain, heart, and liver in aging rats. $O_2 \cdot^-$ supposedly was involved in membrane breakdown in older rats. Schreiber et al. [15] found that there was a sharp increase in $O_2 \cdot^-$ formation in brain slices from old rats after 15 min of hypoxia and reoxygenation, which was almost nonexistent in the youngest animals. Chung et al. [16] suggested that an increase in $O_2 \cdot^-$ levels in the rat kidney with age was due to the enhanced conversion of xanthine dehydrogenase into xanthine oxidase (XO; a producer of $O_2 \cdot^-$).

Hamilton et al. [17] investigated the relationship between endothelial function, $O_2 \cdot^-$ production, and age in rats. It was found that $O_2 \cdot^-$ formation was enhanced in old rats compared to young animals, while NO bioavailability decreased in older rats. NADPH oxidase (Nox) was a probable source of $O_2 \cdot^-$ formation. Later on, these authors [18] confirmed that $O_2 \cdot^-$ production was enhanced in the brain tissues of normotensive old rats and, even to a greater extent, in the brain of spontaneously hypertensive stroke-prone old rats.

Moon et al. [19] showed that smooth muscle cells (SMC) from aged mice decreased proliferative capacity in response to α-thrombin stimulation and generated higher levels of ROS in comparison to cells from younger mice. These changes with age were explained by dysregulation of cell cycle-associated proteins such as cyclin D1 and p27Kip1 in SMC from aged mice. Chen et al. [20] demonstrated that Leydig cells (testicular cells responsible for synthesizing and secreting the essential steroid testosterone) from old rats produced significantly greater levels of mitochondrial $O_2 \cdot^-$ than those from younger rats.

Oudot et al. [21] investigated age-depended differences in myocardial ischemic recovery in rats and their possible relationship with cardiac and vascular oxidative stress. Myocardial function and adaptation to ischemia-reperfusion (I/R) during aging were related to a higher level of oxidative stress. These findings suggest that endothelial Nox is a major contributor to age-related cardiovascular damage. The renin–angiotensin system might be involved in the modulation of vascular $O_2{}^{\cdot-}$ production during the aging process.

Ali et al. [22] have studied the relationship between gender difference and ROS production in C57BL6 young and old mice, in which females are the shorter-lived gender. Both genders exhibited significant age-dependent increases in ROS, although females had a greater ROS increase with age measured by DHE oxidation but not by ESR spectroscopy in mitochondria. CuZnSOD and glutathione peroxidase (GPx) 1 protein levels were lower in old females. Antioxidant treatment with a superoxide dismutase (SOD) mimetic increased the life span to a greater degree in females. These data indicate that the differences in ROS formation might contribute to gender divergence in survival, although mitochondrial $O_2{}^{\cdot-}$ production may not be responsible for gender differences in the life span.

Newaz et al. [23] compared Nox and XO as sources of $O_2{}^{\cdot-}$ and vascular damage in aging. They found that there was increased XO-dependent free radical formation in old rats, while vascular Nox remained unchanged between both groups. De la Fuente et al. [24] investigated the changes with age of macrophage functions and ROS formation in mice. It was found that the adherence capacity and release of extracellular $O_2{}^{\cdot-}$ by macrophages increased with age. Thus, macrophages suffer from oxidative stress with aging, and the developed chronic oxidative stress can be linked to senescence.

Kozlov et al. [25] compared the ability of different rat tissues to produce ROS and RNS in old animals. In old rats, ROS and RNS formation increased significantly compared to young rats in the order of blood, skeletal muscle, lung, and heart, but was not changed in intestine, brain, liver, and kidney.

ROS and RNS were also enhanced in isolated heart mitochondria from old rats, while no ROS and RNS formation was detected in mitochondria from young rats. These findings identify heart, lung, and skeletal muscle as the tissues with increased ROS and RNS with age.

Jacobson et al. [26] studied $O_2{}^{\cdot-}$ production in young and old rats at normal and high intravascular pressure. In mesenteric arteries exposed to normal pressure, $O_2{}^{\cdot-}$ production was significantly higher in aged than in young vessels. High pressure enhanced $O_2{}^{\cdot-}$ production in vessels of both groups, but the enhancement was significantly greater in aged vessels. It was found that Nox was a major source of the enhanced $O_2{}^{\cdot-}$ production in high-pressure mesenteric arteries of both young and aged rats, but XO- and NOS-dependent $O_2{}^{\cdot-}$ production also contributed to $O_2{}^{\cdot-}$ enhancement in mesenteric arteries of aged rats.

In order to determine whether the production of ROS increases with age, Sazaki et al. [27] have studied $O_2{}^{\cdot-}$ generation in ex vivo brain slices from mammals and birds. It was found that the $O_2{}^{\cdot-}$-dependent chemiluminescence increased with age in the senescence-accelerated mouse (SAM) brain tissues. An increase in $O_2{}^{\cdot-}$ levels with age was also observed in C57/BL6 mice, Wistar rats, and pigeons. Importantly, the rate of age-related increase in $O_2{}^{\cdot-}$-dependent chemiluminescence was inversely related to the maximum life span of the animals.

6.1.2 Antioxidant Enzymes CuZnSOD and MnSOD in Aging and Senescence

ROS and RNS affect age and senescence by two major signaling pathways: through the enhancement ROS and RNS formation in the cells by outside and inside stimuli (diet components, environment contaminators, etc.), and a decrease in main antioxidant enzymes CuZnSOD and MnSOD. As early as 1976, Kellogg and Fridovich [28] showed that the activity of CuZnSOD in the brain and lung of long-living rats was higher than in short-living rats.

Ono and Okado [29] investigated the possible effects of SOD levels in the brain on the life span of 11 mammalian species. They found a positive correlation between SOD levels and life span of mammals. Didion et al. [30] demonstrated that loss of a single copy of the gene for CuZnSOD increased vascular $O_2{}^{\cdot-}$ levels and produced vascular dysfunction with aging. They found that vascular $O_2{}^{\cdot-}$ levels increased in aorta in wild-type CuZnSOD(+/+) old mice, and even more strongly in heterozygous CuZnSOD-deficient CuZnSOD(+/-) old mice with aging. In the last case, an increase in the $O_2{}^{\cdot-}$ level might be associated with the downregulation of CuZnSOD.

There is a common tendency of SOD decreasing with age. As early as 1976, Reiss and Gershon [31] showed that SOD activity decreased in the liver, heart, and brain of aged rats and mice.

Semsei et al. [32] demonstrated that the activities of CuZnSOD and catalase significantly diminished in rat liver between 6 and 29 months of age. Interestingly, that lifelong dietary restriction increased the expression of SOD and catalase in liver tissue from 18-month-old rats. Anisimov et al. [33] have shown that CuZnSOD activity in the brain of old rats decreased by 46.8% compared to young animals. Park et al. [34] found that mitochondrial CuZnSOD activity decreased in senescence-prone SAM-P/1 mice in all age groups.

Colombrita et al. [35] investigated the gene expression of manganese superoxide dismutase-2 (SOD-2) in the brains of young and aged rats. They found that there was significant downregulation of SOD2 mRNA expression in the hippocampus but there was upregulation in the cerebellum of aged rats.

Plymate et al. [36] have shown that MnSOD was a downstream mediator of the senescence-associated mac25/insulin-like growth-factor-binding-protein-related protein-1 (IGFBP-rP1) (a tumor growth suppressor) in human breast and prostate epithelial cells.

Weir and Robaire [37] examined the effect of age on antioxidant enzymatic activity and ROS production by epididymal spermatozoa from young (4-month-old) and old (21-month-old) Brown Norway rats. It was found that the activities of GPx and SOD decreased, while H_2O_2 and $O_2{}^{\cdot-}$ production increased, in aging spermatozoa.

Similarly, Sun et al. [38] found decreased activity of extracellular SOD in vessels of aged rats. At the same time, Van der Loo et al. [39] showed that the inhibition of CuZnSOD had no effect on $O_2{}^{\cdot-}$ production in aged rats although its expression decreased in an age-dependent manner. Furthermore, CuZnSOD lost its membrane association with increasing age and was relocated to the mitochondria, possibly to counterbalance age-associated oxidative stress. These findings can be important for understanding the redox mechanisms regulating age development.

Li et al. [40] investigated the effect of age on the regulation of MnSOD in vascular smooth muscle cells (VSMC) from old (24-months-old) versus young (6-months-old) rats grown in normal, high-glucose, or TNF-α conditions. It was found that MnSOD activity was reduced in VSMC from old rats supposedly due to the age-related Akt-induced phosphorylation and inactivation of FOXO3a (a member of the family of Forkhead transcription factors), which downregulated MnSOD transcription. Recently, Sarsour et al. [41] suggested that MnSOD activity can regulate a mitochondrial ROS-switch favoring an $O_2^{\cdot-}$-signaling regulating proliferation and a hydrogen peroxide-signaling supporting quiescence.

In spite of these quoted data, the changes in SOD activity with aging can differ in various cells and tissues. Although many works demonstrate the deterioration of SOD enzymes and decrease in their activities with age, in some cases, SOD activities were shown to enhance with age, probably in response to the augmentation of oxidative stress. In 1987, Scarpa et al. [42] showed that the concentrations of CuZnSOD and MnSOD increased with age in the brain of rats from 3 days before birth to 30 months of age.

Judge et al. [43] studied age-related changes in cardiac mitochondria, which exist in the myocardium as two distinct populations: subsarcolemmal mitochondria (SSM) and interfibrillar mitochondria (IFM). ROS production and antioxidant enzymes have been investigated in cardiac SSM and IFM isolated from young and old male Fischer-344 rats. There was a significant increase in oxidative stress levels measured by 4-hydroxy-2-nonenal-modified proteins, protein carbonyls, and malondialdehyde in IFM with age. In contrast, only protein carbonyls were elevated in SSM with age. Significant age-related increase in MnSOD activity was detected in both IFM and SSM. It was suggested that the enhancement of antioxidant enzyme activity occurred in response to increased mitochondrial production of $O_2^{\cdot-}$ and H_2O_2. Considering these contradictory data, one can propose that the difference in SOD response to aging processes might depend on the prooxidant/antioxidant balance in cells. SOD enzymes are inducible enzymes, and therefore, an increase in ROS formation with age can result in both suppression and induction of these enzymes.

6.1.3 SOD ACTIVITIES IN AGED HUMANS

The reduction of SOD activity with age has been demonstrated in humans. In 1995, Congy et al. [44] measured erythrocyte SOD and plasma and erythrocyte GPx in 52 elderly patients (a mean age 85 ± 6 years). It was found that SOD and erythrocyte and plasma GPx were lower in elderly patients than in young controls. Pansarasa et al. [45] investigated the relationship between muscle aging and oxidative damage in human muscle samples from individuals of different ages by measuring the total SOD and MnSOD activities, GPx, and catalase activities, and some other parameters of oxidative stress. The total SOD activity decreased significantly with age in the 66–75-year-old patients, although MnSOD activity increases significantly in the

76–85-year-old group. The activity of GPx and catalase did not change with age, demonstrating the exclusive role of $O_2\cdot^-$ and not H_2O_2 in age development. These data support the idea that ROS play an important role in the human muscle aging process. It is interesting that an increase in MnSOD activity has also been shown in the external intercostals of elderly healthy individuals [46].

Inal et al. [47] measured erythrocyte SOD, catalase, GPx, and plasma malondi-aldehyde (MDA) levels in 176 healthy subjects aged 0.2–1 to 69 years. SOD activity was significantly lower, while GPx and catalase activities were higher in the oldest subjects. Di Massimo et al. [48] studied modifications of extracellular superoxide dis-mutase (EC-SOD) activity and the impairment of plasma NO availability occurring with advancing age in healthy humans. They found that advancing age was related to decreased plasma values of NO and EC-SOD. Taken together, these findings demonstrate the mechanism of age-associated endothelial dysfunction, suggesting that the decreased EC-SOD activity may be the origin of the progressive reduc-tion of plasma NO availability with advancing age. In a subsequent work [49], these authors verified the involvement of the age-dependent modifications of EC-SOD in the impairment of plasma NO availability with advancing age. Investigation of 40 healthy men from 27 to 86 years old showed a significant age-related progressive decrease of plasma NO content and EC-SOD activity. Mariani et al. [50] reported that the activity of erythrocyte SOD in nonagenarians (persons 90 years old or between 90 and 100 years old) was higher than in the 80-year-old subjects.

Ksiazek et al. [51] found that human peritoneal mesothelial cells (HPMCs) from donors above 75 years had a lower proliferative capacity, an increased 8-OH-dG content, and a faster decline in SOD activity compared with cells from young indi-viduals. These results indicate that increased 8-OH-dG levels in HPMCs from aged individuals may indicate the in vivo presence of senescent cells with increased vulnerability to oxidative-stress-induced DNA damage.

6.1.4   ROS Signaling in Cellular Aging and Senescence

Chapter 4 demonstrated the role of free-radical signaling in numerous enzymatic processes catalyzed by protein kinases (PK), mitogen-activated protein kinases (MAPK), phosphatases, and other enzymes. It is important to look at the changes in these processes caused by cellular aging. One of the principal enzymatic pathways that is affected by aging is the "survival" Akt/ERK (extracellular signal-regulated kinase) cascade. Thus, Ikeyama et al. [52] showed that the lower survival of rat old hepatocytes was associated with the reduced activation of ERK and Akt kinases. This conclusion was supported by findings that the inhibition of ERK and Akt activi-ties in young cells markedly increased their sensitivity to H_2O_2; at the same time, caloric restriction increased the life span of rats. Smith and Hagen [53] demonstrated that loss of Akt activity caused a decrease in eNOS phosphorylation in aortas from old rats. Jin et al. [54] have also shown that PI3-kinase activity and Akt phosphoryla-tion (PI3-kinase/Akt pathway) were significantly reduced in the kidney of old rats after treatment with the $O_2\cdot^-$ producer menadione. In contrast, the activities of the other mitogen-activated protein kinases JNK1 and AMPK were higher in old rather than in young animals.

It is known that granulocyte-macrophage colony-stimulating factor (GM-CSF)-induced oxidative responses are defective in neutrophils of elderly humans. Tortorella et al. [55] investigated whether this phenomenon might be due to alterations in cytokine-dependent MAPK signaling. GM-CSF-stimulated neutrophils from elderly humans showed a significant reduction in the activation and phosphorylation of ERK1/2 kinase. No changes in GM-CSF-induced p38-MAPK phosphorylation were observed. Interestingly, there was no difference in $O_2\cdot^-$ production between elderly and young subjects. TNF-α-triggered $O_2\cdot^-$ production was not affected by age or ERK1/2 or p38-MAPK activation in TNF-α-stimulated neutrophils from elderly and young subjects. In accordance with the different potency of TNF-α in activating ERK1/2 and p38-MAPK, the TNF-α-induced oxidative responses were more sensitive to the inhibitory effects of SB203580 than to those of PD98059 in young as well as elderly subjects. Overall, a decline in ERK1/2 activation could potentially account for the GM-CSF-dependent impairment of the neutrophil respiratory burst that occurs with age.

As is seen from the foregoing data, the activity of Akt/protein kinase B (PKB) could be considered as a positive factor in the suppression of age development and senescence. However, the situation is not so simple because some findings contradict this conclusion. Thus, Miyauchi et al. [56] showed that Akt activity increased along with cellular senescence and that Akt inhibition increased the life span of primary cultured human endothelial cells. Constitutive Akt activation induced the senescence-like arrest of cell growth via a p53/p21-dependent pathway and inhibition of forkhead transcription factor FOXO3a. FOXO3a-stimulated p53 activity was mediated by ROS. These findings suggest the novel role of Akt in the regulation of cellular life span and indicate that Akt-induced senescence may be involved in vascular pathophysiology.

The negative effects of the Akt/Foxo1 pathway in the regulation of longevity were also shown in long-lived growth hormone (GH) receptor knockout mice. To study the favorable effects of reduced IGF-I/insulin signaling and caloric restriction (CR) on the extension of the life span and delay in age-related diseases, Al-Regaiey et al. [57] subjected normal and long-lived GH receptor knockout mice to CR for 20 months. Hepatic Akt phosphorylation was reduced by caloric restriction. The present results suggest a major role for the Akt/Foxo1 pathway in the regulation of longevity in rodents. As showed earlier, Li et al. [40] found that the age-related Akt-induced phosphorylation and inactivation of FOXO3a diminished MnSOD activity in VSMC from old rats.

Signaling processes mediated by $O_2\cdot^-$, which are catalyzed by the other MAPKs, can also be affected by aging. Rice et al. [58] studied the influence of aging on the parameters of oxidative stress in the aorta of adult (6-month), aged (30-month), and very aged (36-month) rats. ROS formation sharply increased in 36-month aortae. It was found that an increase in aortic $O_2\cdot^-$ with aging was significantly correlated with changes in the expression and/or regulation of AMPK-α kinase, MAPKs, and apoptotic (Bax, Bcl-2, Traf-2) and transcriptional (NFκB) activities.

Hsieh and Papaconstantinou [59] proposed that ROS produced by mitochondrial Complex I after rotenone treatment activated the p38-MAPK pathway in hepatocytes and that the activation of the p38 pathway depended on the ability of reduced

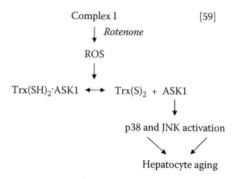

FIGURE 6.1 Mechanism of mitochondrial complex I ROS-mediated activation of the p38-MAPK pathway in hepatocytes depending on the reduction of thioredoxin and the inhibition of ASK 1 kinase [59].

thioredoxin to bind and inhibit ASK 1 kinase and to release from the complex after oxidation (Figure 6.1). This mechanism was supported by the ability of *N*-acetyl cysteine (NAC) to prevent dissociation of Trx-ASK1 and activation of the p38-MAPK pathway. It was also found that the ratio of ASK1/Trx-ASK1 increased in aged mouse livers. These findings suggest that enhanced ROS production may alter the ratio of ASK1 and Trx-ASK1 in aged liver.

Abidi et al. [60] investigated whether ROS overproduction exhibited its anti-steroidogenic action through the modulation of MAPK signaling pathways. They found that aging caused increased phosphorylation and activation of rat adrenal p38-MAPK but not ERK1/2 or JNK1/2 kinases. Taken together, these results indicate that p38-MAPK functions as a signaling effector in oxidative stress-induced inhibition of steroidogenesis during aging.

Mitochondrial ROS also stimulate physiological hypoxia, which extends the replicative life span of human cells in culture. Thus, Bell et al. [61] showed that the generation of mitochondrial ROS was necessary for hypoxic activation of the transcription factor hypoxia-inducible factor (HIF) in primary human lung fibroblasts.

Bachschmid et al. [62] investigated the role of protein kinase C (PKC)-mediated activation of the Nox system in age. In contrast to the involvement of this enzymatic cascade in hyperglycemia or hypertension, both PKC translocation, necessary for its activation, and expression of the cytosolic subunits of the Nox p47^{phox} and p67^{phox} remained unchanged with age. Therefore, it was suggested that oxidative stress-associated vascular aging mechanistically differed from endothelial dysfunction seen in the other cardiovascular risk factors, for which the PKC/Nox pathway was shown to be responsible.

Yan et al. [63] examined mice in which type 5 adenylyl cyclase (AC5) was knocked out (AC5 KO). These mice are resistant to cardiac stress and have an increased median life span of approximately 30%. They found significant activation of the Raf/MEK/ERK signaling pathway and the upregulation of SOD. Fibroblasts isolated from AC5 KO mice exhibited ERK-dependent resistance to oxidative stress. These findings suggest that adenylyl cyclase AC is a fundamentally important enzyme regulating life span and stress resistance.

Chaves et al. [64] have shown that ROS production by human granulocytes from people from 20 to 80 years depended on PKC activity.

6.1.5 RNS in Cellular Aging

Similar to ROS, RNS are important regulators of aging and senescence. The damaging activity of $ONOO^-$ in biological processes is a well-established fact, and therefore, its formation must promote aging development. The effects of NO on age are more complicated. In 1995, Mollace et al. [65] showed that NO synthase activity, as expressed by citrulline and nitrite formation in brain homogenates, decreased in 24-month-old rats compared to young rats. Similar results have been obtained in other works. Amrani et al. [66] concluded that, in rats, basal and stimulated release of NO by the coronary endothelium deteriorated with age. Gerhard et al. [67] have shown that endothelium-dependent vasodilation declined steadily with increasing age in healthy human subjects supposedly due to a decrease in NO bioavailability.

The effects of age on NO production might be cell- and tissue-dependent. Tschudi et al. [68] found that age differently altered endothelial NO release in rat aortas and pulmonary arteries; the initial rate of NO release and peak of NO concentrations significantly declined with the age only in aortas, while they increased in pulmonary arteries. van Der Loo et al. [69] demonstrated that the levels of NO decreased in old rats due to a threefold increase in endothelial $O_2\cdot^-$ production despite a sevenfold increase in the expression of NO-produced enzymes.

It is interesting that aging differently affected NO bioavailability in old normal and spontaneously hypertensive rats. Thus, Chou et al. [70] found that aging reduced the activity of endothelial nitric oxide synthase (eNOS) in old rats but not in old spontaneously hypertensive rats (SHR). Likewise, Ferrer et al. [71] have studied neuronal NO release induced by electrical field stimulation in isolated segments of endothelium-denuded mesenteric arteries from young and old SHR. It was found that $O_2\cdot^-$ production was greater in old than in young SHR rats, while $ONOO^-$ production induced by electrical field stimulation was only detected in old SHR. Neuronal NO and $O_2\cdot^-$ formation increased in mesenteric arteries from old SHR, decreasing NO bioavailability through $ONOO^-$ formation. The final effect was a decrease in neuronal NO after the electrical-field stimulated-vasomotor response in arteries from old SHR.

Payne et al. [72] studied the association of the late stages of aging with impaired NO-mediated vascular relaxation and enhanced vascular contraction in aging male SHR. It was found that the SHR nontreated or treated with the antioxidants tempol, vitamin E, and vitamin C showed the age-related inhibition of a vascular relaxation pathway involving not only NO production by endothelial cells but also NO bioavailability. The smooth-muscle response to NO was partially reversed during chronic treatment with the antioxidants tempol and vitamins E and C.

Csiszar et al. [73] have shown that flow-induced NO-mediated dilation of coronary arterioles was significantly diminished in aged rats due to increased $O_2\cdot^-$ production. It was suggested that the decreased expression of eNOS, the increased activity of Nox, and the increased expression of inducible NO synthase (iNOS) resulted in the enhancement of $ONOO^-$ formation. Blackwell et al. [74] also concluded that the aging-induced impairment of NO reactivity in mouse carotid arteries was due to increased $O_2\cdot^-$ formation. Woodman et al. [75] demonstrated that aging impaired vasodilator responses in soleus muscle feed arteries by attenuating NO- and prostacyclin-mediated, endothelium-dependent dilation in old rats.

Funovic et al. [76] proposed that the endothelium damaged in cardiovascular diseases can be protected during aging by treatment with beta-blockers via increasing NO, decreasing $ONOO^-$, and restoring the $NO/ONOO^-$ balance. In aortic endothelial cells from Wistar–Kyoto rats of different ages, eNOS underwent uncoupling with aging, manifested by NO decrease and a threefold $ONOO^-$ increase for old rats. The beta-blocker metoprolol reversed eNOS uncoupling and enhanced NO concentration; L-arginine and SOD were also beneficial.

Mayhan et al. [77] studied the effects of eNOS-dependent agonists (acetylcholine and ADP) and an independent agonist (nitroglycerin) on cerebral arterioles in adult and aged Fisher-344 rats before and during the application of tempol, apocynin, and diphenyleneiodonium chloride (DPI). They found that the acetylcholine- and ADP-induced dilations, but not the nitroglycerin-induced dilation of cerebral arterioles, were impaired in aged rats. Tempol, apocynin, and DPI diminished the impaired eNOS-dependent vasodilatation in aged rats but not in adults, and they did not affect responses to nitroglycerin. The proteins eNOS, p67^{phox}, and gp91^{phox} increased but SOD-1 protein decreased in aged rats. Basal and agonist-induced production of $O_2\cdot^-$ was elevated in aged rats. These findings suggest that aging impairs the eNOS-dependent reactivity of cerebral arterioles via an increase in $O_2\cdot^-$ produced by Nox.

A decrease in NO production with age was also mentioned in some previously discussed works. Thus, Di Massimo et al. [48,49] have shown that progressive reduction of plasma NO availability in healthy old men depended on a decrease in eSOD activity. Sun et al. [38] demonstrated that the increased production of $O_2\cdot^-$, the reduced activity of SOD, and the impaired shear-stress-induced activation of eNOS are the causes of decreased shear-stress-induced release of NO in vessels from aged rats. Smith and Hagen [53] reported that the reduction of endothelial-derived NO led to a loss of vasomotor function of the major conduit arteries due to age-related alterations in eNOS. They found that the levels of eNOS phosphorylation in aortas from aged rats were almost 50% lower than were those from young animals. Lower eNOS phosphorylation apparently depended on loss of constitutive Akt/protein kinase B activity.

6.1.6   Cyclooxygenase-Catalyzed Free Radical Overproduction in Age

In addition to ROS and RNA enzymatic producers, which have already been considered, cyclooxygenase-1 and cyclooxygenase-2 (COX-2) are enzymes also capable of producing free radicals during their catalytic cycles. Some works described an increase in the formation of cyclooxygenase products prostaglandins and prostanoids in old animals. Thus, Roberts and Reckelhoff [78] demonstrated a sharp increase in isoprostane levels in plasma and plasma lipids in aged rats. Mukai et al. [79] have shown that the inhibitors of COX-2 restored the enhanced endothelium-dependent relaxation in aortas from aged rats, suggesting the involvement of COX-2-derived vasoconstricting eicosanoids in aging development. Kim et al. [80] reported that the levels of prostaglandins E(2) (PGE(2)) and PGI(2) and thromboxane A(2) (TXA(2)) were elevated in old rats.

Ward et al. [81] have shown that the plasma free and total (free plus esterified) content of F_2-isoprostanes (F_2-isoPs) increased with age (185% and 66%, respectively), and that this increase was reduced by life-extending caloric restriction. In addition, they found that levels of esterified F_2-isoPs increased 68% with age in liver, and 76% with age in kidney. Caloric restriction modulated the age-related increase, reducing the esterified F_2-isoPs levels 27% in liver and 35% in kidney. Age-related enhancement of the esterified F_2-isoPs levels correlated well with DNA oxidation measured by 8-oxodeoxyguanosine production. It should be mentioned that an earlier work by Yamamoto et al. [82] is of special interest because it demonstrated a correlation between the life spans of humans and rats and the formation of non-enzymatic peroxidation products. The content of these products increased in the range humans < Sprague–Dawley rats < Nagase analbuminetic rats, and correlated well with the life spans of humans and rats.

6.1.7   GENE REGULATION OF FREE RADICAL FORMATION IN AGE

In 1999, Migliaccio et al. [83] concluded that no genes are known to increase individual life span. However, they demonstrated at the same time that targeted mutation of the mouse *p66shc* gene, a cytoplasmic signal transducer involved in the transmission of mitogenic signals, induced stress resistance and prolonged life span. Thus, *p66shc* could be a part of the signal transduction pathway that regulated stress apoptotic responses and life span in mammals. Later on, it was shown that there is a decrease in $O_2^{\cdot-}$ production and an increase in life span (about 30%) in old mice lacking *p66shc* (p66shc−/−) in comparison with control animals [84]. Another gene regulating aging development is the proapoptotic gene *gadd153/chop*, which is elevated in the liver of old rats [85].

At present, significant attention has been drawn to the discovery of the aging-suppressor gene *klotho*. Because high expression of *klotho* gene was detected in the brain, it was proposed that the gene is involved in the regulation of brain aging. Furthermore, it has been shown that *klotho* gene presents in the blood and that its serum level decreased with age in humans from 0 to 91 years. These findings suggest that *klotho* gene is a serum factor related to human aging [86].

Further studies demonstrated that klotho protein might be involved in the regulation of antioxidative defense. Saito et al. [87] showed that the free radical scavenger (T-0970), which effectively decreased plasma levels of 8-epi-prostaglandin, suppressed Angiotensin-II-induced downregulation of *klotho*. Nagai et al. [88] suggested that oxidative stress had a crucial role in the aging-associated cognition impairment in klotho mutant mice. Mitobe et al. [89] demonstrated that oxidant stress injury by H_2O_2 dose-dependently reduced *klotho* expression in mice. In 2005, a group of authors [90,91] showed that *klotho* is indeed an aging suppressor gene, which is able to extend life span when overexpressed in mice. It was also found that *klotho* protein increased resistance to oxidative stress at the cellular and whole-organism levels in mammals. *Klotho* protein activated the FoxO forkhead transcription factors, which were negatively regulated by insulin/IGF-1 signaling, thereby inducing expression of MnSOD. This, in turn, facilitated removal of ROS and increased oxidative stress resistance.

Ikushima et al. [92] demonstrated that *klotho* overexpression in mice decreased H_2O_2-induced apoptosis in endothelial cells. Caspase-3 and caspase-9 activities were lower in *klotho*-treated HUVEC than in control cells. *Klotho* protein interfered with H_2O_2-induced premature cellular senescence. Authors concluded that *klotho* acts as a humoral factor capable of reducing apoptosis and cellular senescence in vascular cells. Rakugi et al. [93] studied the effects of membrane-form *klotho* on MnSOD expression and NO production in HUVEC. It was found that *klotho* protein enhanced MnSOD expression partially via the activation of the cAMP signaling pathway. Furthermore, *klotho* increased NO production, which also contributed to the upregulation of MnSOD and inhibited Angiotensin-II-induced ROS production in HUVEC. These findings provide new insights into the mechanisms of *klotho* action and support the therapeutic potential of membrane-form *klotho* to regulate endothelial function.

6.1.8 FREE RADICAL-MEDIATED APOPTOSIS IN AGE

To the present time, ROS and RNS signaling in apoptotic processes is well documented. Kokoszka et al. [94] have shown that mitochondrial production of ROS, oxidative stress, functional decline, and the initiation of apoptosis appear to be central components of aging in normal mice and in mice with partial or complete deficiencies in the mitochondrial antioxidant enzyme MnSOD. Phaneuf and Leeuwenburgh [95] suggested that oxidative stress is an apoptosis-inducing signal that possibly increased in the aging rat heart. They found that the cytosolic cytochrome c content was significantly elevated in old animals. Furthermore, Bcl-2, an antiapoptotic protein, strongly decreased with age, whereas Bax, a proapoptotic protein, remained unchanged.

Liu et al. [96] investigated age-related differences in myocyte apoptosis and inflammatory response in a rat model of myocardial ischemia/reperfusion (MI/R). Aged (19 months) and young (4 months) male rats were subjected to MI/R. There was an enhanced myocyte apoptosis in the myocardium of aged rats compared to young rats following MI/R. Collectively, these findings suggested that MI/R was associated with the generation of $O_2^{\cdot-}$ in the heart of young rats, while aged rats exhibited an increase in the ratio of Bax mRNA to Bcl-2 mRNA and cardiomyocyte apoptosis.

It has been shown that advanced aging leads to the impaired endothelial NO synthesis and enhanced endothelial cell apoptosis. Correspondingly, Hoffmann et al. [97] investigated the sensitivity of aged endothelial cells toward apoptotic stimuli and NO generation. HUVECs were cultured until the 14th passage. It was found that, in aged cells, TNF-α-induced apoptosis and caspase-3-like activity were significantly enhanced compared to young cells (passage three). As NO is able to protect against endothelial cell death via S-nitrosylation of caspases, eNOS protein expression and the content of S-nitrosylated proteins were determined. Aged HUVECs were found to show significantly reduced eNOS expression and a decrease in the overall S–NO content, suggesting that eNOS downregulation may be involved in age-dependent increase of apoptosis. Thus, aging of endothelial cells depends on decreased NO synthesis and concomitantly increased apoptosis.

Numata et al. [98] investigated the relationship of NO-induced apoptosis to the acceleration of brain aging of senescence-accelerated mouse prone 10 (SAMP10). The expression of neuronal NOS increased in the cerebral cortex of the brain of SAMP10 in an age-dependent manner and significantly higher nNOS levels were observed in both young and old SAMP10 compared to age-matched controls. Moreover, a lower level of antiapoptotic protein Bcl-2 and a higher level of cyto-solic cytochrome c were in SAMP10 compared to the control. The present findings suggest that age-dependent increase in NO by upregulation of nNOS promotes the Bcl-2-linked apoptosis in the cerebral cortex of SAMP10 and that this may cause the acceleration of brain aging of SAMP10.

As follows from the discussed data, most studies agree that apoptotic pro-cesses increase with aging, being actually one major source of cell senescence and death. However, some contradictory results were obtained in the experiments with long-living dwarf mice. Thus, Kennedy and Rakoczy [99] showed that long-living Ames dwarf mouse hepatocytes readily underwent apoptosis. Tirosh et al. [100] also found that long-living alpha MUPA transgenic mice exhibited increased mito-chondrion-mediated apoptotic capacity.

6.2 MECHANISMS OF FREE RADICAL-MEDIATED DAMAGE IN AGING AND SENESCENCE

6.2.1 ROS Signaling and DNA Damage in Senescence and Aging

Different free radical-mediated mechanisms can be responsible for the development of aging and senescence in cells, tissues, and whole organisms. One of them is DNA damage by ROS and RNS. About 20 years ago, Ames and coworkers investigated the effect of oxygen stress on DNA damage as the source of cellular senescence. It is known that senescence, DNA damage, or differentiation induce an irreversible loss of cellular replication. To study the origin of growth termination in senescent human fibroblast cells, Chen et al. [101] determined the level of oxidative DNA damage in these cells. Senescent cells excised from DNA four times more 8-oxoguanine per day than the early-passage young cells. These data support the hypothesis that oxidative DNA damage contributes to replicative cessation in human diploid fibroblast cells.

Deshpande et al. [102] investigated the role and mechanism of the small GTPase rac1 in vascular endothelial cell senescence. They showed that the constitutive acti-vation of rac1, via the generation of ceramide, resulted in mitochondrial oxidative stress and premature endothelial cell senescence. However, endogenous rac1 activa-tion apparently did not induce mitochondrial oxidative stress and replicative senes-cence of endothelial cells.

In a recent review [103], the role of ROS in replicative and, in particular, prema-ture senescence has been considered. It was demonstrated in recent years that DNA damage is a common mediator for both replicative senescence, which is triggered by telomere shortening, and premature cellular senescence induced by oncogenic stress and oxidative stress. Extensive findings suggest that DNA damage accumulates with age, and that this might be due to an increase in ROS production and a decline in DNA repair capacity.

Martien and Abbadie [104] suggested that both telomere shortening and cumulative oxidative damage can contribute to senescence. It is also believed that, owing to DNA damage and irreversible cell-cycle arrest induced by shortened telomeres, senescence might be considered a tumor-suppressor mechanism that terminated the proliferation of genetically altered (tumor) cells. However, it is known that the incidence of the most frequent cancers in humans, carcinomas, increases with age, and inversely, when aging is delayed by caloric restriction, the cancer incidence decreases. Therefore, these authors proposed that senescence might play a protumoral role due to the importance of DNA oxidative damage. Such a mechanism could be particularly relevant for age-associated carcinomas because senescence in epithelial cells is driven more by oxidative stress than by telomere shortening.

Matthews et al. [105] demonstrated that human atherosclerosis is characterized by VSMC senescence and marked telomere shortening. Oxidative DNA damage was seen in vivo, and chronic oxidative stress accelerated telomere loss and VSMC senescence. These findings suggest that human VSMCs undergo a replicative senescence that is accelerated by ROS overproduction.

Zhang et al. [106] studied the mechanism of TNF-α inhibition of hematopoietic stem cell (HSC) hematopoiesis using the genomic instability syndrome Fanconi anemia mouse model. TNF-α stimulated premature senescence in bone marrow HSCs and progenitor cells as well as other tissues of Fanconi anemia mice. TNF-α-induced senescence correlated with increasing ROS levels and oxidative DNA damage. These results demonstrate an intimate link between ROS and DNA-damage-induced premature senescence in HSCs and progenitor cells, which may play an important role in aging and anemia.

6.2.2 MITOCHONDRIAL DNA DAMAGE AND NO/O$_2$ COMPETITION FOR CYTOCHROME C OXIDASE AS ORIGINS OF SENESCENCE AND AGING

At present, the most popular free radical theory of aging is based on the Harmon mitochondrial hypothesis, which emphasizes an importance of mitochondrial DNA (mtDNA) damage. mtDNA is believed to be especially vulnerable to oxidative damage because it is located near the inner mitochondrial membrane where oxidants are generated. Moreover, mtDNA lacks protective histones and has relatively little DNA repair activity. Harmon suggested that mtDNA mutations, which are accumulated progressively during life, are directly responsible for a deficiency in cellular oxidative phosphorylation activity, leading to enhanced oxygen radical production. In its turn, an increase in ROS production leads to an increased rate of mtDNA damage, causing a "vicious cycle" of exponentially increasing oxidative damage and mitochondrial dysfunction [1,107].

This hypothesis is widely discussed in literature (see, for example, Reference 108) and remains an important one, although it might overestimate the role of $O_2^{.-}$ production by mitochondria because its real levels under both physiological and pathophysiological conditions are still uncertain. In addition, the last data [109] seem to cast doubts at a direct relationship between mtDNA damage and aging. Moreover, Miro et al. [110] have shown earlier a progressive, significant increase of heart membrane lipid peroxidation with aging in hearts from human donors. Conversely, neither absolute nor relative enzyme activities of Complex I, II, III, and IV of mitochondrial respiratory chain decreased with age.

In 1994, several groups of authors showed that NO was able to inhibit mitochondrial cytochrome c oxidase (COX) reversibly [111–114]. Thus, Bolanos et al. [111] have shown that the induction of NOS resulted in the inhibition of COX activity in rat neonatal astrocytes. NADH-ubiquinone-1 reductase (Complex I) and succinate-cytochrome c reductase (Complex II–III) activities were not affected. These findings demonstrated that inhibition of the mitochondrial respiratory chain after induction of astrocytic NOS may represent a mechanism for NO-mediated neurotoxicity. Cleeter et al. [112] found that NO reversibly inhibited dioxygen consumption by rat skeletal muscle mitochondria due to inhibition of COX.

Brown and Cooper [113] also showed that NO reversibly inhibited oxygen consumption of brain synaptosomes. Inhibition was reversible, occurred at the level of COX, and was apparently competitive with dioxygen, with half-inhibition by 270 nM NO at dioxygen concentrations around 145 μM. Isolated COX was inhibited by similar levels of NO. These NO levels are within the measured physiological and pathological ranges for a number of tissues and conditions, suggesting that NO inhibition of COX and the competition with dioxygen may occur in vivo. Schweizer and Richter [114] reported that low NO concentrations can potently suppress respiratory activity of isolated liver and brain mitochondria at dioxygen concentrations typical for cells and tissues. These data demonstrate a direct action of NO on the mitochondrial respiratory chain and suggest that NO exerts some of its physiological and pathological effects by deenergizing mitochondria.

Taken together, these findings suggest that there is competition between dioxygen and NO that regulates dioxygen consumption and the formation of ROS. Subsequent studies showed how interplay between $O_2{\cdot}^-$ and NO stimulated aging development. Regulation of COX activity by NO could be an important factor of aging development. Adler et al. [115,116] demonstrated that NO-mediated control of cardiac dioxygen consumption by bradykinin or enalaprilat was markedly reduced in 23-month-old Fischer rats. They found that levels of eNOS protein increased in aging, whereas levels of CuZnSOD, MnSOD, EC-SOD, and p67phox were unchanged. In addition, there was an increase in $O_2{\cdot}^-$ formation in aging, which was inhibited by apocynin and supported by the upregulation of gp91^{phox}. An increase in $O_2{\cdot}^-$ may explain decreased bioactivity of NO in old rats. These findings suggest that increased $O_2{\cdot}^-$ production contributes to decreased NO bioavailability and possible cardiac dysfunction with age.

The deteriorating effects of $O_2{\cdot}^-$ on age demonstrate the importance of $O_2{\cdot}^-$ regulation of aging and senescence. One of the major factors affecting $O_2{\cdot}^-$ levels in aging is a decrease in the activities of SODs. Another important factor is the deregulation of the $O_2{\cdot}^-$/NO balance. Although some findings suggest that NO formation can be enhanced during aging, its principal effects probably depend on the interplay between NO and $O_2{\cdot}^-$. The formation of $ONOO^-$ in the reaction of NO with $O_2{\cdot}^-$ is an important pathway to the formation of ROS and RNS. However, it is not the only way for free radical-mediated damage in aging. These findings demonstrate that there is

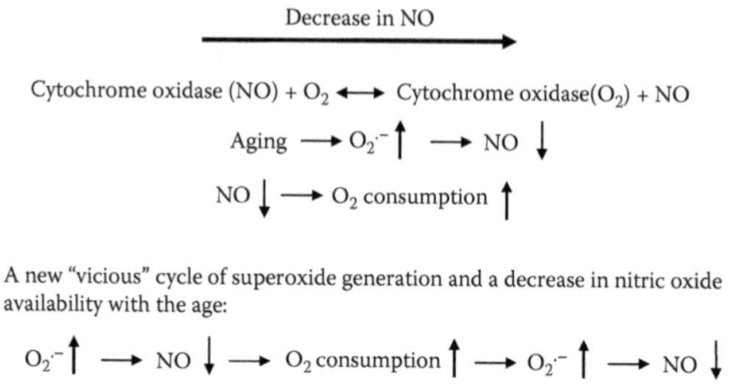

Decrease in NO

$\longrightarrow$

Cytochrome oxidase (NO) + O_2 $\longleftrightarrow$ Cytochrome oxidase(O_2) + NO

Aging $\longrightarrow O_2^{\cdot-}\uparrow \longrightarrow$ NO $\downarrow$

NO $\downarrow \longrightarrow O_2$ consumption $\uparrow$

A new "vicious" cycle of superoxide generation and a decrease in nitric oxide availability with the age:

$O_2^{\cdot-}\uparrow \longrightarrow$ NO $\downarrow \longrightarrow O_2$ consumption $\uparrow \longrightarrow O_2^{\cdot-}\uparrow \longrightarrow$ NO $\downarrow$

FIGURE 6.2 Reversible competition between dioxygen and NO for the interaction with cytochrome oxidase.

competition between dioxygen and NO, which regulates dioxygen consumption and the formation of ROS.

Recently, we proposed the mechanism of interaction between dioxygen and NO in mitochondria, resulting in the inhibition of COX, $O_2^{\cdot-}$ overproduction, mitochondrial damage, and cell death [117–121]. As $O_2^{\cdot-}$ overproduction in age decreases NO bioavailability [115,116], the equilibrium of reversible competition between dioxygen and NO for the interaction with cytochrome oxidase will be shifted to the right, increasing dioxygen consumption (Figure 6.2). An increase in dioxygen consumption will lead to the subsequent enhancement of electron leak from mitochondrial carriers and overproduction of $O_2^{\cdot-}$. Collectively, the reduction of NO formation can initiate a new "vicious" cycle of formation of ROS and RNS and a decrease in NO availability with age:

$$\text{NO} \downarrow \Rightarrow O_2 \text{ consumption} \uparrow \Rightarrow O_2^{\cdot-} \uparrow \Rightarrow \text{NO} \downarrow$$

6.2.3 WHAT ARE THE PRIMARY CAUSES OF AGING AND SENESCENCE?

All free radical theories of aging agree that the starting point of aging development is the overproduction of ROS and RNS. There are numerous origins of this phenomenon: pathologies associated with free-radical overproduction, environmental contamination, irradiation, etc. However, we know that all human beings are mortal and will certainly die even in the absence of any external damaging factors. Therefore, it would be important to know if it is possible to find the origins of free radical-mediated initiation of physiological aging in hypothetical conditions without external damage.

One might suppose that important causes of physiological aging can be diet components responsible for the initiation of free radical-mediated damage [119]. It is not a question of good or bad diet; any diet contains some damaging compounds. Therefore, it is not by chance that caloric restriction is a known therapeutic intervention capable of attenuating aging in mammals (see the following text). Unsaturated acids are probably the most abundant oxidizable food components. In the past, it was erroneously believed that unsaturated acids are just products of lipid peroxidation.

Now it is known that they have a dietary origin and are important components of the human diet. They are easily oxidized into prostaglandins and isoprostanes in living organisms. These products of enzymatic and nonenzymatic lipid peroxidation are highly toxic compounds, and their formation under physiological conditions may start aging. (A recent publication by Aiken et al. [122] points out the possibility of regulation of MnSOD, the primary defense enzyme against $O_2\cdot^-$, by nutrient sensing through MnSOD expression mediated by essential amino acid depletion.)

6.3 ANTIOXIDANT TREATMENT AGAINST AGING AND SENESCENCE: POSSIBILITY OF ENLARGEMENT OF LIFE SPAN

Free radical mechanisms of aging and senescence indicate the possibility of anti-oxidant treatment for suppression of aging development and even the enhancement of life span of humans and animals. It is true that experimental findings are not always encouraging, but there are definite results showing the principal possibility of affecting aging and life span by the use of antioxidants. The application of classic antioxidants and special treatment of aged organisms to increase an organism's antioxidant defense has been described [123]. Now we will consider the most important examples of antioxidant treatment of aging processes.

6.3.1 COMMENTS ON LONG-LIVING ANIMALS

Interesting data might be received from the study of long-living animals. It is usually accepted that, in animals, longevity (maximal life span) is inversely related to mass-specific basal metabolic rates. However, it has recently been pointed out by Rottenberg [124] that, in several mammals, exceptional longevity is associated with high basal metabolic rate and fast evolution of mtDNA-coded proteins. This is supposedly connected with the acceleration of basal respiration and the inhibition of ROS production. Rottenberg also suggested that the highest rate of cytochrome *b* evolution and the highest values of exceptional longevity in the songbird genus *Serinus* (e.g., canaries) depended on clustering of substitutions around the ubiquinone reduction binding site that may increase the rate of ubiquinone reduction.

The importance of free radical regulation with age was also demonstrated in the study of the longest-living mollusk, an ocean quahog, *Arctica islandica*, whose maximum life span is close to 400 years [125]. It was found that SOD activities maintained at high levels through the lifetime of this mollusk might explain its high longevity.

6.3.2 CALORIE RESTRICTION

Calorie restriction (CR) feeding is most frequently applied as a mean of reducing mammal aging. It has been proposed that CR feeding slows the rate of oxidative damage due to decrease in mitochondrial $O_2\cdot^-$ generation [126]. Hall et al. [127] have shown that CR reduced cellular injury and improved heat tolerance in old rats by lowering radical production and preserving cellular ability to adapt to stress through antioxidant enzyme induction. No age-associated increase in mitochondrial protein, lipid peroxidation, or $O_2\cdot^-$ generation was detected in calorically restricted old mice [128]. CR enhanced the transcripts of genes involved in reactive oxygen radical scavenging function, tissue development, and energy metabolism such as COX and superoxide dismutases SOD1 and SOD2 in old rats fed a calorie-restricted diet (60% of control diet) for 36 weeks [129]. Zou et al. [130] demonstrated that CR suppressed the elevated level of $O_2\cdot^-$-generating XO in the serum of old rats. Nisoli et al. [131] have shown that CR for 3 or 12 months induced eNOS expression in various tissues of male mice. Ferguson et al. [132] proposed that the genotype-specific extension of life span by CR may involve modulation of oxidative stress produced as a result of the interplay between metabolic rate and energy balance during aging.

6.3.3 ANTIOXIDANTS

Two classes of antioxidants can be used for the suppression of free radical over-production in cells and tissues: free radical scavengers capable of directly reacting with reactive free radicals through H-abstraction (for example, vitamins E and C and flavonoids), and the compounds capable of oxidizing free radicals by the one-electron transfer mechanism (SOD mimetics and ubiquinones) [118]. A great number of works have been published, in which the effects of vitamins E and C on aging and senescence have been studied; obviously, their consideration is beyond the limits of this chapter. We will look only at the most interesting examples.

6.3.3.1 Vitamins E and C

It has been proposed that vitamin E (α-tocopherol) exhibits favorable action on aging processes and that it is even able to expand the life span of experimental animals. As early as 1996, Poulin et al. [133] showed that a high vitamin E diet prevented aging-related decline in lymphocytes and brain of old mice. Reckelhoff et al. [134] found that vitamin E decreased renal lipid peroxidation and the accumulation of F2-isoprostanes in aged rats. However, Sumien et al. [135] demonstrated that supplementation with vitamin E failed to attenuate oxidative damage in aged mice. Important results have been received by Navarro et al. [136], who showed that high doses of vitamin E improved neurological performance, brain mitochondrial function, and survival in aged mice; the median life span increased by 40%, and maximal life span by 17% in aged male mice.

Recently, Park et al. [137] studied transcriptional alterations in the heart and brain of 30-month-old mice supplemented with α-tocopherol and γ-tocopherol since middle age (15 months). In the heart, both tocopherol supplementations were effective in inhibiting the expression of genes associated with cardiomyocyte hypertrophy. In the brain, α-tocopherol and γ-tocopherol prevented the induction of genes encoding ribosomal proteins and proteins involved in ATP biosynthesis in aged mice. These findings demonstrate that middle-age-onset dietary supplementation with α-tocopherol and γ-tocopherol can partially prevent age-associated transcriptional changes.

Despite numerous works on vitamin C, its effects on aging development are uncertain, and the administration of vitamin C sometimes even leads to a negative outcome. For example, lifelong vitamin C supplementation in combination with cold exposure did not affect oxidative damage or life span in mice, and even decreased the expression of antioxidant genes [138].

The treatment of healthy old subjects with antioxidant vitamin E and C still has not been very successful. For example, Retana-Ugalde et al. [139] investigated the effects of ascorbic acid (AA) and α-tocopherol on parameters of oxidative stress and DNA damage in the elderly. It was found that the administration of high doses of vitamins during 6 months did not lead to significant changes in the antioxidant status of healthy old subjects. It should be noted that these negative results might, of course, be the consequence of a very short time of treatment.

6.3.3.2 α-Lipoic Acid, Ubiquinones, Metallothioneins, and Carnitine

Some other antioxidants were also applied in aging studies with various degrees of success, for example, α-lipoic acid [140], metallothioneins [141], or coenzyme Q10 (ubiquinone 10) [142]. For example, Yang et al. [141] investigated the effects of antioxidant metallothionein on cardiomyocyte function, $O_2{}^{.-}$ generation, aconitase activity, cytochrome c release, and expression of oxidative stress-related proteins, such as the GTPase RhoA and Nox protein p47^{phox} in young and aged mice. The cardiac-specific metallothionein transgenic mice showed a longer life span. Aging increased $O_2{}^{.-}$ generation, active RhoA abundance, cytochrome c release, and p47^{phox} expression, and suppressed aconitase activity. These age-induced changes in oxidative stress were diminished by metallothioneins. Collectively, these findings demonstrate that metallothioneins may decrease aging-induced cardiac defects and oxidative stress, and that their effects may be responsible for prolonged life span in metallothionein transgenic mice.

Ochoa et al. [143] studied the effect of Q10 supplementation on young and old rats fed for 24 months on a PUFA-rich diet. For the supplemental group, there was an age-associated decrease in the peroxidizability index, an increase in catalase activity, and modulation of aging-related changes in mitochondrial electron transport chain components in the skeletal muscle. These findings suggest a mechanism to explain the effect of Q10 in extending the life span of animals fed a PUFA-rich diet. Izgut-Uysal et al. [144] have shown that L-carnitine administration significantly reduced enhanced $O_2{}^{.-}$ production by peritoneal macrophages from aged rats. It was suggested that L-carnitine administration might be useful in reversing some age-related changes.

Recently, Thangasamy et al. [145] found that the levels of lipid peroxides increased and the activities of antioxidant enzymes decreased in aged male Wistar rats. Administration of L-carnitine for 21 days significantly decreased the levels of lipid peroxides and improved the activities of SOD, catalase, glutathione peroxidase, and glutathione reductase.

Avci et al. [146] studied the effects of ingesting garlic on plasma and erythrocyte antioxidant parameters of elderly (mean age 70.69 ± 4.23) subjects. Ingestion of garlic led to significantly lowered plasma and erythrocyte MDA levels and increased the activities of some antioxidant enzymes. Thus, consumption of garlic can probably decrease oxidative stress.

6.3.3.3 (-)Deprenyl (Selegiline)

Special interest is drawn to (-)deprenyl (selegiline) (N-methyl-N-(1-methyl-2-phenyl-ethyl)-prop-2-yn-1-amine), a selective MAO-B inhibitor, which supposedly exhibits exclusively high antiaging effects in experimental animals. In 1988, Knoll [147] discovered that the (-)deprenyl treatment of rats resulted in an increase in their average life span up to 197.98 ± 2.36 weeks, that is, higher than the estimated maximum age of death in the rat (182 weeks). This effect of (-)deprenyl could be due to its antioxidant properties and an increase in SOD activity in the striatum. Kitani et al. [148] found that (-)deprenyl administration sharply increased the activities of CuZnSOD, MnSOD, and catalase in striatum and substantia nigra as well as life expectancy of old male rats. Archer and Harrison [149] also observed an increase in the life span of old mice after treatment with (-)deprenyl. At present, the antioxidant properties of (-)deprenyl are considered to be major factors of its antiaging activity in animals and possibly in humans [150].

6.3.3.4 Antioxidant Enzymes and Their Mimetics

Antioxidant enzymes (SOD and catalase), in principle, could be even more effective regulators of aging compared to the other pharmaceutical agents capable of expanding the life span. For example, just recently, Akila et al. [151] reported that erythrocyte catalase activity was highly decreased in normal elderly subjects compared to normal young subjects. There are two possibilities for the enhancement of activities of these enzymes: the overexpression of enzymes and transfection into experimental animals, or the application of low-molecular-enzyme mimetics.

In 1994, Orr and Sohal [152] demonstrated that the life span of *Drosophila melanogaster* might be increased by the overexpression of SOD and catalase. Recently, Schriner et al. [153] showed that median and maximum life spans of transgenic mice were maximally increased with overexpressed human catalase localized to the peroxisome, the nucleus, or mitochondria. Zhang et al. [154] demonstrated that chronic systemic administration of the SOD/catalase mimetic (EUK-189) prevented heat stress-induced liver injury by decreasing oxidative damage in aged rats. Quick et al. [155] administered a small-molecule synthetic SOD mimetic to wild-type middle-age mice. They found that chronic treatment with the SOD mimetic not only reduced age-associated oxidative stress and mitochondrial radical production but significantly extended the life span of mice.

Clausen et al. [156] studied the effectiveness of the SOD/catalase mimetics EUK-189 and EUK-207 on age-related decline in cognitive function and increase in oxidative stress in mice aged 17 months. Both mimetic treatments resulted in significantly decreased lipid peroxidation, nucleic acid oxidation, and ROS levels. In addition, the treatments also significantly improved age-related decline in performance in the fear-conditioning task. These findings confirm a critical role for oxidative stress in age-related decline in learning and memory and suggest the potential usefulness of mimetics treatment in reversing age-related declines in cognitive function.

In conclusion, we would like to ask: Are there any hopes for increased longevity in humans? This question is, of course, of utmost importance for everybody, and all scientific hypotheses and theories are supposed to answer this question and possibly give some hope. Free radical theory is no exception. I think that the major benefits of this theory compared to the others, which are undoubtedly also very important, lie in the comprehension of the detailed mechanisms of pathological events during aging and senescence mediated by free radicals. Furthermore, an understanding of the importance of the regulation of the balance between physiological radicals

$O_2^{\cdot-}$ and NO, which are the metabolites of normal physiological processes and always presented in living organisms, might encourage the appearance of new ideas and treatments.

REFERENCES

1. D Harman. Aging: A theory based on free radical and radiation chemistry. *J Gerontology* 11: 298–300, 1956.
2. SM McCann, C Mastronardi, A de Laurentiis, and V Rettori. The nitric oxide theory of aging revisited. *Ann NY Acad Sci* 1057: 64–84, 2005.
3. A Terman and UT Brunk. Oxidative stress, accumulation of biological "garbage," and aging. *Antioxid Redox Signal* 8: 197–204, 2006.
4. A Navarro and A Boveris. The mitochondrial energy transduction system and the aging process. *Am J Physiol Cel. Physiol* 292: C670–C686, 2007.
5. S Taddei, AVirdis, L Ghiadoni, DVersari, and A Salvetti. Endothelium, aging, and hypertension. *Curr Hypertens Rep* 8: 84–89, 2006.
6. A Sanz, R Pamplona, and V Barja. Is the mitochondrial free radical theory of aging intact? *Antioxid Redox Signal* 8: 582–599, 2006.
7. DP Jones. Extracellular redox state: Refining the definition of oxidative stress in aging. *Rejuvenation Res* 9: 169–181, 2006.
8. D Harman. Free radical theory of aging: An update: Increasing the functional life span. *Ann. N. Y. Acad. Sci* 1067: 10–21, 2006.
9. SI Rattan. Theories of biological aging: Genes, proteins, and free radicals. *Free Radic Res* 40: 1230–1238, 2006.
10. AD De Grey. Free radicals in aging: Causal complexity and its biomedical implications. *Free Radic Res* 40: 1244–1249, 2006.
11. KC Kregel and HJ Zhang. An integrated view of oxidative stress in aging: Basic mechanisms, functional effects, and pathological considerations. *Am J Physiol Regul Integr Comp Physiol* 292: R18–R36, 2007.
12. D Harman. The biologic clock: The mitochondria? *J Am Geriatr Soc* 20: 145–147, 1972.
13. M Sawada and JC Carlson. Changes in superoxide radical and lipid peroxide formation in the brain, heart and liver during the lifetime of the rat. *Mech Ageing Dev* 41: 125–137, 1987.
14. M Sawada, U Sester, and JC Carlson. Superoxide radical formation and associated biochemical alterations in the plasma membrane of brain, heart, and liver during the lifetime of the rat. *J Cell Biochem* 48: 296–304, 1992.
15. SJ Schreiber, D Megow, A Raupach, IV Victorov, and U Dirnagl. Age-related changes of oxygen free radical production in the rat brain slice after hypoxia: On-line measurement using enhanced chemiluminescence. *Brain Res* 703: 227–230, 1995.
16. HY Chung, SH Song, HJ Kim, Y Ikeno, and BP Yu. Modulation of renal xanthine oxidoreductase in aging: Gene expression and reactive oxygen species generation. *J Nutr Health Aging* 3: 19–23, 1999.
17. CA Hamilton, MJ Brosnan, M McIntyre, D Graham, and AF Dominiczak. Superoxide excess in hypertension and aging: A common cause of endothelial dysfunction. *Hypertension* 37: 529–534, 2001.
18. D Antier, HV Carswell, MJ Brosnan, CA Hamilton, IM Macrae, S Groves, E Jardine, JL Reid, and AE Dominiczak. Increased levels of superoxide in brains from old female rats. *Free Radic Res* 38: 177–183, 2004.

19. SK Moon, LJ Thompson, N Madamanchi, S Ballinger, J Papaconstantinou, C Horaist, MS Ruuge, and C Patterson. Aging, oxidative responses, and proliferative capacity in cultured mouse aortic smooth muscle cells. *Am J Physiol Heart Circ Physiol* 280: H2779–H2788, 2001.

20. H Chen, D Cangello, S Benson, J Folmer, H Zhu, MA Trush, and BR Zirkin. Age-related increase in mitochondrial superoxide generation in the testosterone-producing cells of Brown Norway rat testes: Relationship to reduced steroidogenic function? *Exp Gerontol* 36: 1361–1373, 2001.

21. A Oudot, C Martin, D Busseuil, C Vergely, L Demaison, and L Rochette. NADPH oxidases are in part responsible for increased cardiovascular superoxide production. *Free Radic Biol Med* 40: 2214–2222, 2006.

22. SS Ali, C Hiong, J Lucero, MM Behrens, LL Dugan, and KL Quick. Gender differences in free radical homeostasis during aging: Shorter-lived female C57BL6 mice have increased oxidative stress. *Aging Cell* 5: 565–574, 2006.

23. MA Newaz, Z Yousefipour, and A Oyekan. Oxidative stress-associated vascular aging is xanthine oxidase-dependent but not NAD(P)H oxidase-dependent. *J Cardiovasc Pharmacol* 48: 88–94, 2006.

24. M de la Fuente, A Hernanz, N Guayerbas, and P Alvares, C Alvarado. Changes with age in peritoneal macrophage functions. Implication of leukocytes in the oxidative stress of senescence. *Cell Mol Biol. (Noisy-le-grand)* 2004, 50, Online Pub:OL683–90.

25. AV Kozlov, L Szalay, F Umar, K Kropik, K Staniek, H Niedermuller, S Bahrami, and H Nohl. Skeletal muscles, heart, and lung are the main sources of oxygen radicals in old rats. *Biochim Biophys Acta* 1740: 382–389, 2005.

26. A Jacobson, C Yan, Q Gao, T Rincon-Skinner, A Rivera, J Edwards, A Huang, G Kaley, and D Sun. Aging enhances pressure-induced arterial superoxide formation. *Am J Physiol Heart Circ Physiol* 293: H1344–H1350, 2007.

27. T Sasaki, K Unno, S Tahara, A Shimada, Y Chiba, M Hoshino, and T Kaneko. Age-related increase of superoxide generation in the brains of mammals and birds. *Aging Cell* 2008 Apr 14 [Epub ahead of print].

28. EW Kellogg and I Fridovich. Superoxide dismutase in the rat and mouse as a function of age and longevity. *J Gerontology* 31: 405–408, 1976.

29. T Ono and S Okado. Unique increase of superoxide dismutase level in brains of long living mammals. *Exp Gerontol* 19: 349–354, 1984.

30. SP Didion, DA Kinzenbaw, LI Schrader, and FM Fagaci. Heterozygous CuZn superoxide dismutase deficiency produces a vascular phenotype with aging. *Hypertension* 48: 1072–1079, 2006.

31. U Reiss and D Gershon. Comparison of cytoplasmic superoxide dismutase in liver, heart and brain of aging rats and mice. *Biochem Biophys Res Commun* 73: 255–262, 1976.

32. I Semsei, G Rao, and A Richardson. Changes in the expression of superoxide dismutase and catalase as a function of age and dietary restriction. *Biochem Biophys Res Commun* 164: 620–625, 1989.

33. VN Anisimov, AV Arutiunian, TI Oparina, SO Burmistrov, VM Prokopenko, and VKh Khvinson. [Age-related changes in the activity of free-radical processes in rat tissues and blood serum] [Article in Russian]. *Ross Fiziol Zh Im IM Sechenova* 85: 502–507, 1999.

34. JW Park, CH Choi, MS Kim, and MH Chung. Oxidative status in senescence-accelerated mice. *J. Gerontol A Bio. Sci Med Sci* 51: B337–B345, 1996.

35. C Colombrita, V Calabrese, AM Stella, F Mattei, DL Alkon, and G Scapagnini. Regional rat brain distribution of heme oxygenase-1 and manganese superoxide dismutase mRNA: Relevance of redox homeostasis in the aging processes. *Exp Biol Med. (Maywood)* 228: 517–524, 2003.

36. SR Plymate, KH Haugk, CC Sprenger, PS Nelson, MK Tennant, Y Zhang, LW Oberley, W Zhong, R Drivdahl, and TD Oberley. Increased manganese superoxide dismutase (SOD-2) is part of the mechanism for prostate tumor suppression by Mac25/insulin-like growth factor binding-protein-related protein-1. *Oncogene* 22: 1024–1034, 2003.

37. CP Weir and B Robaire. Spermatozoa have decreased antioxidant enzymatic capacity and increased reactive oxygen species production during aging in the Brown Norway rat. *J Androl* 28: 229–240, 2007.

38. D Sun, A Huang, EH Yan, Z Wu, C Yan, PM Kaminski, TD Oury, MS Wolin, and G Kaley. Reduced release of nitric oxide to shear stress in mesenteric arteries of aged rats. *Am J Physiol Heart Circ Physiol* 286: H2249–H2256, 2004.

39. B van der Loo, M Bachschmid, JN Skepper, R Laabugger, S Schildknecht, R Hahn, E Mussig, D Gygi, and TF Luscher. Age-associated cellular relocation of Sod 1 as a self-defense is a futile mechanism to prevent vascular aging. *Biochem Biophys Res Commun* 344: 972–980, 2006.

40. M Li, JF Chiu, BT Mossman, and NK Fukagawa. Down regulation of MnSOD through phosphorylation of FOXO3a by AKT in explanted VSMC from old rats. *J Biol Chem* 281: 40429–40439, 2006.

41. EH Sarsour, S Venkataraman, AL Kalen, LW Oberley, and PC Goswami. Manganese superoxide dismutase activity regulates transitions between quiescent and proliferative growth. *Aging Cell* 7: 405–417, 2008.

42. M Scarpa, A Rigo, P Viglino, R Stevanato, F Bracco, and I Battistin. Age dependence of the level of the enzymes involved in the protection against active oxygen species in the rat brain. *Proc Soc Exp Med* 185: 129–133, 1987.

43. S Judge, YM Jang, A Smith, T Hagen, and C Leeuwenburgh. Age-associated increases in oxidative stress and antioxidant enzyme activities in cardiac interfibrillar mitochondria: Implications for the mitochondrial theory of aging. *FASEB J* 19: 419–421, 2005.

44. F Congy, D Bonnefont-Rousselot, S Dever, J Delattre, and J Emerit. [Study of oxidative stress in the elderly] [Article in French]. *Presse Med* 24: 1115–1118, 1995.

45. O Pansarasa, L Bertorelli, J Vecchiet, G Felzani, and F Marzatico. Age-dependent changes of antioxidant activities and markers of free radical damage in human skeletal muscle. *Free Radic Biol Med* 27: 617–622, 1999.

46. E Barreiro, C Coronell, B Lavina, A Ramirez-Sarmiento, M Orozco-Levi, and J Gea. Aging, sex differences, and oxidative stress in human respiratory and limb muscles. *Free Radic Biol Med* 41: 797–809, 2006.

47. ME Inal, G Kanbak, and E Sunal. Antioxidant enzyme activities and malondialdehyde levels related to aging. *Clin Chim Acta* 305: 75–80, 2001.

48. C Di Massimo, P Scarpelli, ND Lorenzo, G Caimi, FD Orio, and MG Ciancarelli. Impaired plasma nitric oxide availability and extracellular superoxide dismutase activity in healthy humans with advancing age. *Life Sci* 78: 1163–1167, 2006.

49. C Di Massimo, R Lo Presti, C Corbacelli, A Pompei, P Scarpelli, D De Amicis, G Caimi, and MG Tozzi Ciancarelli. Impairment of plasma nitric oxide availability in senescent healthy individuals: Apparent involvement of extracellular superoxide dismutase activity. *Clin Hemorheol Microcirc* 35: 231–237, 2006.

50. E Mariani, V Cornacchiola, MC Polidori, F Mangialasche, M Malavolta, R Cecchetti, P Bastiani, M Baglioni, F Mocchegiani, and P Mecocci. Antioxidant enzyme activities in healthy old subjects: Influence of age, gender and zinc status: Results from the Zincage Project. *Biogerontology* 7: 391–398, 2006.

51. K Ksiazek, K Piatek, and J Witowski. Impaired response to oxidative stress in senescent cells may lead to accumulation of DNA damage in mesothelial cells from aged donors. *Biochem Biophys Res Commun* 373(2): 335–339, 2008.

52. S Ikeyama, G Kokkonen, S Shack, X-T Wang, and NJ Holbrook. Loss in oxidative stress tolerance with aging linked to reduced extracellular signal-regulated kinase and Akt kinase activities. *FASEB J* 10.1096/fj.01-0409fje, 2001.

53. AR Smith and TM Hagen. Vascular endothelial dysfunction in aging: Loss of Akt-dependent endothelial nitric oxide synthase phosphorylation and partial restoration by (R)-alpha-lipoic acid. *Biochem Soc Trans* 31: 1447–1449, 2003.

54. Q Jin, BS Jhun, SH Lee, J Lee, JH Cho, HH Baik, and I Kang. Differential regulation of phosphatidyl 3-kinase/Akt, mitogen-activated protein kinase, and AMP-activated protein kinase pathways during menadione-induced oxidative stress in kidney of young and old rats. *Biochem Biophys Res Commun* 315: 555–561, 2004.

55. C Tortorella, I Stella, G Piazzolla, O Simone, V Caappiello, and S Antonaci. Role of defective ERK phosphorylation in the impaired GM-CSF-induced oxidative response of neutrophils in elderly humans. *Mech Ageing Dev* 125: 539–546, 2004.

56. H Miyauchi, T Minamino, K Tateno, T Kunieda, H Toko, and I Komura. Akt negatively regulates the in vitro lifespan of human endothelial cells via a p53/p21-dependent pathway. *EMBO J* 23: 212–220, 2004.

57. KA Al-Regaiey, MM Masternak, M Bonkowski, L Sun, and A Bartke. Long-lived growth hormone receptor knockout mice: Interaction of reduced insulin-like growth factor i/insulin signaling and caloric restriction. *Endocrinology* 146: 851–860, 2005.

58. KM Rice, DL Preston, EM Walker, and ER Blough. Aging influences multiple indices of oxidative stress in the aortic media of the Fischer 344/NNia x Brown Norway/BiNia rat. *Free Radic Res* 40: 185–197, 2006.

59. C-C Hsieh and J Papaconstantinou. Thioredoxin-ASK1 complex levels regulate ROS-mediated p38 MAPK pathway activity in livers of aged and long-lived Snell dwarf mice. *FASEB J* 20: 259–268, 2006.

60. P Abidi, S Leers-Sucheta, Y Cortez, J Han, and S Azhar. Evidence that age-related changes in p38 MAP kinase contribute to the decreased steroid production by the adrenocortical cells from old rats. *Aging Cell* 7: 168–178, 2008.

61. EL Bell, TA Klimova, J Eisenbart, PT Schumacker, and NS Chandel. Mitochondrial reactive oxygen species trigger hypoxia-inducible factor-dependent extension of the replicative life span during hypoxia. *Mol Cell Biol* 27: 5737–5745, 2006.

62. M Bachschmid, B van der Loo, K Schuler, R Labugger, S Thurau, M Eto, J Kilo, R Holz, TF Luscher, and V Ullrich. Oxidative stress-associated vascular aging is independent of the protein kinase C/NAD(P)H oxidase pathway. *Arch Gerontol Geriatr* 38: 181–190, 2004.

63. L Yan, DE Vatner, JP O'Connor, A Ivessa, H Ge, W Chen, S Hirotani, Y Ishikawa, J Sadoshima, and SF Vatner. Type 5 adenylyl cyclase disruption increases longevity and protects against stress. *Cell* 130: 247–258, 2007.

64. MM Chaves and ALP Rodrigues, A Pereira dos Reis, NC Gerzstein, and JA Nogueiro-Machado. Correlation between NADPH oxidase and protein kinase C in the ROS production by human granulocytes related to age. *Gerontology* 48: 354–359, 2002.

65. V Mollace, P Rodino, R Massoud, D Rotiroti, and G Nistico. Age-dependent changes of NO synthase activity in the rat brain. *Biochem Biophys Res Commun* 215: 822–827, 1995.

66. M Amrani, AT Goodman, CC Gray, and MH Yacoub. Ageing is associated with reduced basal and stimulated release of nitric oxide by the coronary endothelium. *Acta Physiol Scand* 157: 79–84, 1996.

67. M Gerhard, MA Roddy, SJ Creager, and MA Creager. Aging progressively impairs endothelium-dependent vasodilation in forearm resistance vessels of humans. *Hypertension* 27: 849–853, 1996.

68. MG Tschudi, M Barton, NA Bersinger, P Moreau, F Cosentino, G Noll, T Malinski, and TF Luscher. Effect of age on kinetics of nitric oxide release in rat aorta and pulmonary artery. *J Clin Invest* 98: 899–905, 1996.

69. B van der Loo, R Labugger, JN Skepper, M Bachschmid, J Kilo, JM Powell, M Palacios-Callender, JD Erusalimsky, T Quaschning, T Malinski, D Gygi, V Ullrich, and TF Luscher. Enhanced peroxynitrite formation is associated with vascular aging. *J Exp Med* 192: 1731–1744, 2000.

70. T-C Chou, M-H Yen, C-Y Li, and Y-A Ding. Alterations of nitric oxide synthase expression with aging and hypertension in rats. *Hypertension* 31: 643–648, 1998.

71. M Ferrer, M Sanchez, N Minoves, M Salaices, and G Balfagon. Aging increases neuronal nitric oxide release and superoxide anion generation in mesenteric arteries from spontaneously hypertensive rats. *J Vasc Res* 40: 509–519, 2003.

72. JA Payne, JF Reckelhoff, and RA Khalil. Role of oxidative stress in age-related reduction of NO-cGMP-mediated vascular relaxation in SHR. *Am J Physiol Regul Integr Comp Physiol* 285: R542–R551, 2003.

73. A Csiszar, Z Ungvari, JG Edwards P Kaminski, MS Wolin, A Koller, and G Kaley. Aging-induced phenotypic changes and oxidative stress impair coronary arteriolar function. *Circ Res* 90: 1159–1166, 2002.

74. KA Blackwell, JP Sorenson, DM Richardson, LA Smith, O Suda, K Nath, and ZS Katusic. Mechanisms of aging-induced impairment of endothelium-dependent relaxation: Role of tetrahydrobiopterin. *Am J Physiol Heart Circ Physiol* 287: H2448–H2453, 2004.

75. CR Woodman, EM Price, and MH Laughlin. Selected contribution: Aging impairs nitric oxide and prostacyclin mediation of endothelium-dependent dilation in soleus feed arteries. *J Appl Physiol* 95: 2164–2170, 2003.

76. P Funovic, M Korda, R Kubant, RE Barlag, RF Jacob, RP Mason, and T Malinski. Effect of beta-blockers on endothelial function during biological aging: A nanotechnological approach. *J Cardiovasc Pharmacol* 51: 208–215, 2008.

77. WG Mayhan, DM Arrick, GM Sharpe, and H Sun. Age-related alterations in reactivity of cerebral arterioles: Role of oxidative stress. *Microcirculation* 15: 225–236, 2008.

78. LJ Roberts II and JF Reckelhoff. Measurement of F(2)-isoprostanes unveils profound oxidative stress in aged rats. *Biochem Biophys Res Commun* 287: 254–256, 2001.

79. Y Mukai, H Shimokawa, M Higashi, K Morikawa, T Matoba, Y Hiroki, HMA Talukder, and A Takeshita. Inhibition of renin-angiotensin system ameliorates endothelial dysfunction associated with aging in rats. *Arterioscler Thromb Vasc Biol* 22: 1445–1450, 2002.

80. JW Kim, Y Zou, S Yoon, JH Lee, YK Kim, BP Yu, and HY Chung. Vascular aging: Molecular modulation of the prostanoid cascade by calorie restriction. *J Gerontol A, Biol Sci Med Sci* 59: B876–B885, 2004.

81. WF Ward, W Qi, H Van Remmen, WE Zackert, LJ Roberts II, and A Richardson. Effects of age and caloric restriction on lipid peroxidation: measurement of oxidative stress by f_2-isoprostane levels. *J Gerontol A Biol Sci Med Sci* 60: 847–851, 2005.

82. Y Yamamoto, K Wakabayashi, and M Nagano. Comparison of plasma levels of lipid hydroperoxides and antioxidants in hyperlipidemic Nagase analbuminemic rats, Sprague-Dawley rats, and humans. *Biochem Biophys Res Commun* 189: 518–523, 1992.

83. E Migliaccio, M Giorgio, S Mele, G Pelicci, P Reboldi, PP Pandolfi, L Lanfrancone, and PG Pelicci. The p66shc adaptor protein controls oxidative stress response and life span in mammals. *Nature* 402: (6759), 309–313, 1999.

84. P Francia, C della Gatti, M Bachschmid, I Martin-Padura, C Savoia, E Migliaccio, PG Pelicci, M Schiavoni, TF Luscher, M Volpe, and F Cosentino. Deletion of p66shc gene protects against age-related endothelial dysfunction. *Circulation* 110: 2889–2895, 2004.

85. S Ikeyama, X-T Wang, J Li, A Podlutsky, JL Martindale, G Kokkonen, R van Huizen, M Gorospe, and NJ Holbrook. Expression of the pro-apoptotic gene *gadd153/chop* is elevated in liver with aging and sensitizes cells to oxidant injury. *J Biol Chem* 278: 16726–16731, 2003.

86. NM Xiao, YM Zhang, Q Zheng, and J Gu. Klotho is a serum factor related to human aging. *Chim Med J* (Engl) 117: 542–747, 2004.

87. K Saito, N Ishizaka, H Mitani, M Ohno, and R Nagai. Iron chelation and a free radical scavenger suppress angiotensin II-induced downregulation of klotho, an anti-aging gene, in rat. *FEBS Lett* 551: 58–62, 2003.

88. T Nagai, K Yamada, HC Kim, YS Kim, Y Noda, A Imura, Y Nabeshima, and T Nabeshima. Cognition impairment in the genetic model of aging klotho gene mutant mice: A role of oxidative stress. *FASEB J* 17: 50–52, 2003.

89. M Mitobe, T Yoshida, H Sugiura, S Shirota, K Tsuchiya, and H Nihei. Oxidative stress decreases klotho expression in a mouse kidney cell line. *Nephron Exp Nephrol* 101: e67–e74, 2005.

90. H Kurosu, M Yamamoto, JD Clark, JV Pastor, A Nandi, P Gurnami, OP McGuinness, H Chikuda, M Yamaguchi, H Kawaguchi, I Shimomura, Y Takayama, J Herz, CR Kahn, KP Rosenblatt, and M Kuro-O. Suppression of aging in mice by the hormone Klotho. *Science* 309: (5742), 1829–1833, 2005.

91. M Yamamoto, JD Clark, JV Pastor, P Gurnani, A Nandi, H Kurosu, M Miyoshi, Y Ogawa, DH Castrillon, KP Rosenblatt, and M Kuro-O. Regulation of oxidative stress by the anti-aging hormone Klotho. *J Biol Chem* 280: 38029–38034, 2005.

92. M Ikushima, H Rakugi, K Ishikawa, Y Maekawa, K Yamamoto, J Ohta, Y Chihara, I Kida, and T Ogihara. Anti-apoptotic and anti-senescence effects of Klotho on vascular endothelial cells. *Biochem Biophys Res Commun* 339: 827–832, 2006.

93. H Rakugi, N Matsukawa, K Ishikawa, J Yang, M Imai, M Ikushima, Y Maekawa, I Kida, J Miyazaki, and T Ogihara. Anti-oxidative effect of Klotho on endothelial cells through cAMP activation. *Endocrine* 31: 82–87, 2007.

94. JE Kokoszka, P Coskun, LA Esposito, and DC Wallace. Increased mitochondrial oxidative stress in the Sod2 (+/−) mouse results in the age-related decline of mitochondrial function culminating in increased apoptosis. *Proc Natl Acad Sci USA* 98: 2278–2283, 2001.

95. S Phaneuf and C Leeuwenburgh. Cytochrome c release from mitochondria in the aging heart: A possible mechanism for apoptosis with age. *Am J Physiol Regul Integr Comp Physiol* 282: R423–R430, 2002.

96. P Liu, B Xu, TA Cavalieri, and CE Hock. Age-related difference in myocardial function and inflammation in a rat model of myocardial ischemia-reperfusion. *Cardiovasc Res* 56: 443–453, 2002.

97. J Hoffmann, J Haendeler, A Aicher, L Rossig, M Vasa, AM Zeiher, and S Dimmeler. Aging enhances the sensitivity of endothelial cells toward apoptotic stimuli. Important role of nitric oxide. *Circ Res* 89: 709–715, 2001.

98. T Numata, T Sato, K Maekawa, Y Takahashi, H Saitoh, T Hosokawa, H Fujita, and M Kurasaki. Bcl-2-linked apoptosis due to increase in NO synthase in brain of SAMP10. *Biochem Biophys Res Commun* 297: 517–522, 2002.

99. MA Kennedy, SG Rakoczy, and HM Brown-Borg. Long-living Ames dwarf mouse hepatocytes readily undergo apoptosis. *Exp Gerontol* 38: 997–1008, 2003.

100. O Tirosh, B Schwartz, I Zusman, G Kossoy, S Yahav, and R Miskin. Long-living alpha MUPA transgenic mice exhibit increased mitochondrion-mediated apoptotic capacity. *Ann NY Acad Sci USA* 1019: 439–442, 2004.

101. Q Chen, A Fischer, JD Reagan, L Yan, and BN Ames. Oxidative DNA damage and senescence of human diploid fibroblast cells. *Proc Natl Acad Sci USA* 92: 4337–4341, 1995.

102. SS Deshpande, B Qi, YC Park, and K Irani. Constitutive activation of rac1 results in mitochondrial oxidative stress and induces premature endothelial cell senescence. *Arterioscler Thromb Vasc Biol* 23: e1–e6, 2003.

103. J-H Chen, CN Hales, and SE Ozanne. DNA damage, cellular senescence and organismal ageing: Causal or correlative? *Nucleic Acids Res* 35: 7417–7428, 2007.

104. S Martien and C Abbadie. Acquisition of oxidative DNA damage during senescence: The first step toward carcinogenesis? *Ann NY Acad Sci* 1119: 51–63, 2007.

105. C Matthews, I Gorenne, S Scott, N Figg, P Kirkpatrick, A Ritchie, M Goddard, and M Bennett. Vascular smooth muscle cells undergo telomere-based senescence in human atherosclerosis: Effects of telomerase and oxidative stress. *Circ Res* 99: 156–164, 2006.

106. X Zhang, DP Sejas, Y Qiu, DA Williams, and Q Pang. Inflammatory ROS promote and cooperate with the Fanconi anemia mutation for hematopoietic senescence. *J Cell Sci* 120: 1572–1583, 2007.

107. MF Alexeyev, SP Ledoux, and GL Wilson. Mitochondrial DNA and aging. *Clin Sci* (London) 107: 355–364, 2004.

108. S Judge and C Leeuwenburgh. Cardiac mitochondrial bioenergetics, oxidative stress, and aging. *Am J Physiol Cell Physiol* 292: C1983–C1992, 2007.

109. LA Loeb, DC Wallace, and GM Martin. The mitochondrial theory of aging and its relationship to reactive oxygen species damage and somatic mtDNA mutations. *Proc Natl Acad Sci USA* 102: 18769–18770, 2005.

110. O Miro, J Casademont, E Casals, M Merea, A Urbano-Marquez, P Rustin, and F Cardellach. Aging is associated with increased lipid peroxidation in human hearts, but not with mitochondrial respiratory chain enzyme defects. *Cardiovasc Res* 18: 624–631, 2000.

111. JP Bolanos, S Peuchen, SJ Heales, JM Land, and JB Clark. Nitric oxide-mediated inhibition of the mitochondrial respiratory chain in cultured astrocytes. *J Neurochem* 63: 910–916, 1994.

112. MW Cleeter, JM Cooper, VM Darley-Usmar, S Moncada, and AH Schapira. Reversible inhibition of cytochrome c oxidase, the terminal enzyme of the mitochondrial respiratory chain, by nitric oxide. Implications for neurodegenerative diseases. *FEBS Lett* 345: 50–54, 1994.

113. GC Brown and CE Cooper. Nanomolar concentrations of nitric oxide reversibly inhibit synaptosomal respiration by competing with oxygen at cytochrome oxidase. *FEBS Lett* 356: 295–298, 1994.

114. M Schweizer and C Richter. Nitric oxide potently and reversibly deenergizes mitochondria at low oxygen tension. *Biochem Biophys Res Commun* 204: 169–175, 1994.

115. A Adler, E Messina, B Sherman, Z Wang, H Huang, A Linke, and TH Hintze. NAD(P) H oxidase-generated superoxide anion accounts for reduced control of myocardial O2 consumption by NO in old Fischer 344 rats. *Am J Physiol Heart Circ Physiol* 285: H1015–H1022, 2003.

116. S Adler, H Huang, MS Wolin, and PM Kaminski. Oxidant stress leads to impaired regulation of renal cortical oxygen consumption by nitric oxide in the aging kidney. *J Am Soc Nephrol* 15: 52–60, 2004.

117. IB Afanas'ev. Mechanism of superoxide-mediated damage. Relevance to mitochondrial aging. *Ann NY Acad Sci* 1019: 343–345, 2004.

118. IB Afanas'ev. Interplay between superoxide and nitric oxide in aging and diseases. *Biogerontology* 5: 267–270, 2004.

119. IB Afanas'ev. Free radical mechanisms of aging processes under physiological conditions. *Biogerontology* 6: 283–290, 2005.

120. IB Afanas'ev. Superoxide and nitric oxide in pathological conditions associated with iron overload. The effects of antioxidants and chelators. *Curr Med Chem* 12: 2731–2739, 2005.

121. IB Afanas'ev. Interplay between superoxide and nitric oxide in thalassemia and Fanconi's anemia. *Hemoglobin* 30: 113–118, 2006.

122. KJ Aiken, JS Bickford, ML Kilberg, and HS Nick. Metabolic regulation of manganese superoxide dismutase expression via essential amino acid deprivation. *J Biol Chem* 283: 10252–10263, 2008.

123. BH Ames, MK Shigenaga, and TM Hagen. Oxidant, antioxidants, and the degenerative diseases of aging. *Proc Natl Acad Sci USA* 90: 7915–7922, 1993.

124. H Rottenberg. Exceptional longevity in songbirds is associated with high rates of evolution of cytochrome b, suggesting selection for reduced generation of free radicals. *J Exp Biol* 210: 2170–2180, 2007.

125. D Abele, J Strahl, T Brey, and EE Philipp. Imperceptible senescence: Ageing in the ocean quahog Arctica islandica. *Free Radic Res* 42: 478–480, 2008.

126. BJ Merry. Oxidative stress and mitochondrial function with aging—the effects of calorie restriction. *Aging Cell* 3: 7–12, 2004.

127. DM Hall, ID Oberley, PM Moseley, GR Buettner, LW Oberley, R Weindruch, and KC Kregel. Caloric restriction improves thermotolerance and reduces hyperthermia-induced cellular damage in old rats. *FASEB J* 14: 78–86, 2000.

128. A Lass, BH Sohal, R Weindruch, MJ Forster, and RS Sohal. Caloric restriction reverts age-associated accrual of oxidative damage to mouse skeletal muscle mitochondria. *Free Radic Biol Med* 25: 1089–1097, 1998.

129. R Sreekumar, J Unnikrishan, A Fu, J Nygren, KR Short, J Schimke, R Barazzoni, and KS Nair. Effects of caloric restriction on mitochondrial function and gene transcripts in rat muscle. *Am J Physiol Endocrinol Metab* 283: E38–E43, 2002.

130. Y Zou, KJ Jung, JW Kim, BP Yu, and HY Chung. Alteration of soluble adhesion molecules during aging and their modulation by calorie restriction. *FASEB J* 18: 320–322, 2004.

131. E Nisoli, C Tonello, A Cardile, V Cozzi, R Bracale, L Tedesco, S Falcone, A Valerio, O Cantoni, E Clementi, S Moncada, and MO Carruba. Calorie restriction promotes mitochondrial biogenesis by inducing the expression of eNOS. *Science* 310 (5746): 314–317, 2005.

132. M Ferguson, I Rebrin, MJ Forster, and RS Sohal. Comparison of metabolic rate and oxidative stress between two different strains of mice with varying response to caloric restriction. *Exp Gerontol* 2008 May 4. [Epub ahead of print]

133. JE Poulin, C Cover, MR Gustafson, and MB Kay. Vitamin E prevents oxidative modification of brain and lymphocyte band 3 proteins during aging. *Proc Natl Acad Sci USA* 93: 5600–5603, 1996.

134. JF Reckelhoff, V Kanji, LC Racusen, AM Schmidt, SD Yan, J Marrow, LJ Roberts II, and AK Salahudeen. Vitamin E ameliorates enhanced renal lipid peroxidation and accumulation of F2-isoprostanes in aging kidneys. *Am J Physiol* 274: R767–R774, 1998.

135. N Sumien, MJ Forster, and RS Sohal. Supplementation with vitamin E fails to attenuate oxidative damage in aged mice. *Exp Gerontol* 38: 699–704, 2003.

136. A Navarro, C Gomez, MJ Sanchez-Pino, H Gonzalez, MJ Bandez, AD Boveris, and A Boveris. Vitamin E at high doses improves survival, neurological performance, and brain mitochondrial function in aging male mice. *Am J Physiol Regul Integr Comp Physiol* 289: R1392–R1399, 2005.

137. S-K Park, GP Page, K Kim, DB Allison, M Meydani, R Weindruch, and TA Prolla. {alpha}- and {gamma}-tocopherol prevent age-related transcriptional alterations in the heart and brain of mice. *J Nutr* 138: 1010–1018, 2008.

138. C Selman, JS McLaren, C Meyer, JS Duncan, P Redman, AR Collins, GG Duthie, and JR Speakman. Life-long vitamin C supplementation in combination with cold exposure does not affect oxidative damage or lifespan in mice, but decreases expression of antioxidant protection genes. *Mech. Ageing Dev* 127: 897–904, 2006.

139. R Retana-Ugalde, E Casanueva, M Altamirano-Lozano, C Gonzalez-Torres, and VM Mendoza-Nunez. High dosage of ascorbic acid and alpha-tocopherol is not useful for diminishing oxidative stress and DNA damage in healthy elderly adults. *Ann Nutr Metab* 52: 167–173, 2008.
140. JH Suh, R Moreau, SH Heath, and TM Hagen. Dietary supplementation with (R)-alpha-lipoic acid reverses the age-related accumulation of iron and depletion of antioxidants in the rat cerebral cortex. *Redox Rep* 10: 52–60, 2005.
141. X Yang, TA Doser, CX Fang, JM Nunn, R Janardhanan, M Zhu, N Sreejayan, MT Quinn, and J Ren. Metallothionein prolongs survival and antagonizes senescence-associated cardiomyocyte diastolic dysfunction: Role of oxidative stress. *FASEB J* 20: 1024–1026, 2006.
142. N Ishii, N Senoo-Matsuda, K Miyake, K Yasuda, T Ishii, PS Hartman, and S Furukawa. Coenzyme Q10 can prolong C. elegans lifespan by lowering oxidative stress. *Mech Ageing Dev* 125: 41–46, 2004.
143. JJ Ochoa, JL Quiles, M López-Frías, JR Huertas, and J Mataix. Effect of lifelong coenzyme Q_{10} supplementation on age-related oxidative stress and mitochondrial function in liver and skeletal muscle of rats fed on a polyunsaturated fatty acid (PUFA)-rich diet. *J Gerontol Ser A: Biol Sci Med Sci* 62: 1211–1218, 2007.
144. VN Izgut-Uysal, A Agac, I Karadogan, and N Derin. Peritoneal macrophages function modulation by L-carnitine in aging rats. *Aging Clin Exp Res* 16: 337–341, 2004.
145. T Thangasamy, P Jeyakumar, S Sittadjody, AG Joyee, and P Chinnakannu. L-Carnitine mediates protection against DNA damage in lymphocytes of aged rats. *Biogerontology* 2008 Jul 16. [Epub ahead of print]
146. A Avci, T Atli, IB Erguder, M Varli, E Devrim, S Aras, and I Durak. Effects of garlic consumption on plasma and erythrocyte antioxidant parameters in elderly subjects. *Gerontology* 54: 173–176, 2008.
147. J Knoll. The striatal dopamine dependency of life span in male rats. Longevity study with (-) deprenyl. *Mech Ageing Dev* 46: 237–262, 1988.
148. K Kitani, K Miyasaka, S Kanai, MC Carrillo, and GO Ivy. Upregulation of antioxidant enzyme activities by deprenyl. Implications for life span extension. *Ann NY Acad Sci* 786: 391–409, 1996.
149. JR Archer and DE Harrison. L-deprenyl treatment in aged mice slightly increases life spans, and greatly reduces fecundity by aged males. *J Gerontol A, Biol Sci Med Sci* 51: B448–B453, 1996.
150. K Kitani, C Minami, K Isobe, K Maehara, S Kanai, GP Ivy, and MC Carrillo. Why (-)deprenyl prolongs survivals of experimental animals: Increase of anti-oxidant enzymes in brain and other body tissues as well as mobilization of various humoral factors may lead to systemic anti-aging effects. *Mech Ageing Dev* 123: 1087–1100, 2002.
151. VP Akila, H Harishchandra, V D'souza, and B D'souza. Age related changes in lipid peroxidation and antioxidants in elderly people. *Indian J Clin Biochem* 22: 131–134, 2007.
152. WC Orr and RS Sohal. Extension of life-span by overexpression of superoxide dismutase and catalase in *Drosophila melanogaster Science* 263: 1128–1130, 1994.
153. SE Schriner, NJ Linford, GM Martin, P Treuting, CE Ogburn, M Emond, PE Coskun, W Ladiges, N Wolf, H Van Remmen, DC Wallace, and PS Rabinovitch. Extension of murine life span by overexpression of catalase targeted to mitochondria. *Science* 308: (5730), 1909–1911, 2005.
154. HJ Zhang, SR Doctrow, L Xu, LW Oberley, B Beecher, J Morrison, TD Oberley, and KC Kregel. Redox modulation of the liver with chronic antioxidant enzyme mimetic treatment prevents age-related oxidative damage associated with environmental stress. *FASEB J* 18: 1547–1549, 2004.

155. KL Quick, SS Ali, R Arch, C Xiong, D Wozniak, and LL Dugan. A carboxyfullerene SOD mimetic improves cognition and extends the lifespan of mice. *Neurobiol Aging* 29: 117–128, 2008.

156. A Clausen, S Doctrow, and M Baudry. Prevention of cognitive deficits and brain oxidative stress with superoxide dismutase/catalase mimetics in aged mice. *Neurobiol Aging* Jun 2008.

7 Mechanisms of ROS and RNS Signaling in Enzymatic Catalysis

Reactive oxygen species (ROS) and nitrogen species (RNS) signaling in enzymatic biological processes, including superoxide ($O_2^{\cdot-}$)- and nitric oxide (NO)-producing enzymes (xanthine oxidase [XO], NADPH oxidases [Noxs], and NO synthases [NOSs]) as well as the enzymes catalyzing heterolytic reactions of hydrolysis or etherification without the generation of ROS or RNS (protein kinases, MAP kinases [MAPKs], phospholipases, phosphatases, and others) have been considered in previous chapters. A most intriguing question is the role of ROS and RNS signaling in the last enzymatic processes because the classic mechanisms of heterolytic reactions do not depend on free-radical formation. Nonetheless, numerous experimental data (Chapters 2–6) suggest that free radicals are important mediators of these processes and not just the side products. In Chapter 1, we discussed competition between damaging activity and signaling functions of ROS and RNS in biological processes; now we will consider the mechanisms of free radical signaling in enzymatic catalysis.

7.1 MAJOR REACTIVE SIGNALING MOLECULES: $O_2^{\cdot-}$, H_2O_2, $\cdot NO$, $ONOO^-$

Superoxide, hydrogen peroxide (H_2O_2), NO, and peroxynitrite ($ONOO^-$) are major signaling species identified in numerous enzymatic processes. Two of them, $O_2^{\cdot-}$ and NO, are paramagnetic free radicals, whereas H_2O_2 and $ONOO^-$ are diamagnetic molecules. As a rule, it is difficult to study separately all these reactive species because two pairs ($O_2^{\cdot-}$ and H_2O_2) and (NO and $ONOO^-$) are always formed together and are easily converted to each other. Of course, all of these are well-known facts, but they should be mentioned for the understanding of signaling functions of a particular ROS or RNS.

In some way, it is much easier to understand the signaling mechanisms of free radicals $O_2^{\cdot-}$ and $\cdot NO$, whose chemical reactions are well studied. It is much harder to understand the signaling functions of H_2O_2 and $ONOO^-$, which probably must be transformed into reactive free radicals by decomposition or electron transfer processes. However, at present, there are reliable data that allow distinguishing the roles of individual species in enzymatic processes.

It is frequently suggested that a major role of $O_2^{\cdot-}$ is the conversion by dismutation to H_2O_2, which is a genuine species responsible for ROS signaling. However,

189

numerous findings now show a separate and frequently more efficient direct signaling activity of $O_2{}^{\cdot-}$ in many processes compared to H_2O_2. Some of these works have already been discussed in previous chapters.

Larsson and Cerutti [1] have shown that treatment of cells with low concentrations of $O_2{}^{\cdot-}$, but not H_2O_2, resulted in the phosphorylation of cytoplasmic extracts containing protein kinase C (PKC). Baas and Berk [2] compared the effects of H_2O_2 and $O_2{}^{\cdot-}$ on cultured rat aortic vascular smooth-muscle cell (VSMC) growth and signal transduction. They found that there was a concentration-dependent increase in MAPK activity by the $O_2{}^{\cdot-}$ producer naphthoquinolinedione (LY83583), whereas H_2O_2 produced no effect. It was also suggested that the increased expression of phosphatase-1 (MKP-1) cannot be a cause of the inability of H_2O_2 to activate MAPK. Li et al. [3] demonstrated that H_2O_2 and $O_2{}^{\cdot-}$ exhibited different effects on VSMC proliferation and apoptosis: $O_2{}^{\cdot-}$ induced proliferation, and H_2O_2 caused apoptosis. On the other hand, Moreno-Manzano et al. [4] have shown that mesangial cells stimulated by TNF-α produced $O_2{}^{\cdot-}$ and underwent apoptosis, while H_2O_2 and $ONOO^-$ were not the mediators of the TNF-α-initiated apoptotic pathway.

Important data confirming the separate signaling pathways for $O_2{}^{\cdot-}$ and H_2O_2 have been received by Devadas et al. [5]. These authors showed that T-cell receptor (TCR) signaling activated two distinct pathways of ROS generation. H_2O_2 generation in T cells preceded $O_2{}^{\cdot-}$ generation, suggesting that it was not derived from dismutation of $O_2{}^{\cdot-}$. Furthermore, $O_2{}^{\cdot-}$ inhibitors did not affect H_2O_2 formation. Therefore, TCR-stimulated generation of $O_2{}^{\cdot-}$ and H_2O_2 was proceeded by distinct pathways: H_2O_2 regulated ERK phosphorylation (proliferative pathway), while $O_2{}^{\cdot-}$ mediated TCR-stimulated activation of the proapoptotic Fas ligand (FasL) promoter and subsequent cell death.

Lyng et al. [6] found that $O_2{}^{\cdot-}$, but not NO or H_2O_2, mediated the activation of cytosolic ERK and nuclear c-Jun kinases (JNK) by high glucose and advanced glycation end products (AGEs). Cui and Douglas [7] have shown that arachidonic acid (AA) activated JNK by means of Nox and $O_2{}^{\cdot-}$ generation in kidney epithelial cells, while H_2O_2 inhibited activation by AA. $O_2{}^{\cdot-}$ produced by menadione, but not H_2O_2, stimulated the phosphorylation of ERK1/2 and, by this, attenuated cell death [8]. It was found that H_2O_2 was an effective inhibitor of calcineurin only at concentrations several orders of magnitude higher than those of $O_2{}^{\cdot-}$, which obviously pointed at different mechanisms of calcineurin inhibition by $O_2{}^{\cdot-}$ and H_2O_2 [9]. Wang et al. [10] identified p42/44 MAPK as a downstream target of endothelial NOS III (eNOS)-derived $O_2{}^{\cdot-}$ signaling pathway; this pathway was apparently distinct from the effects of either H_2O_2 or NO.

Important data have been received from the study of competition between $O_2{}^{\cdot-}$ and H_2O_2 in the inactivation of phosphatases. Both oxygen species inactivated these enzymes through the oxidation of thiol groups. However, kinetic studies showed that the reaction rates of $O_2{}^{\cdot-}$ with the thiol groups of phosphatases were significantly higher than those for H_2O_2. In 1999, Barrett et al. [11] measured the rate constants for the $O_2{}^{\cdot-}$- and H_2O_2-mediated inhibition of PTP-1B phosphatase as 334 ± 45 $M^{-1}s^{-1}$ and 42.8 ± 3.8 $M^{-1}s^{-1}$, respectively. Thus, the rate constant for $O_2{}^{\cdot-}$ reaction was about eight times higher than that of H_2O_2. Recently, Wang et al. [12] confirmed these results by demonstrating that the rate constants for the inactivation of PTPs by aromatic quinones (producers of $O_2^{.-}$) were 10 to 100 times higher than the rate constants for H_2O_2-mediated inhibition.

Hongpaisan et al. [13] considered $O_2^{.-}$ and H_2O_2 signaling in the upregulation of two kinases CaMKII and PKA by mitochondria in hippocampal cells. They found that enhancing superoxide dismutase (SOD) activity by overexpression or by application of an SOD mimetic that should accelerate the conversion of $O_2^{.-}$ into H_2O_2 was not enhanced by the rates of enzyme upregulation. It means that $O_2^{.-}$ was a principal ROS modulator of calcium-dependent signaling cascades.

Mendes et al. [14] have studied effects of $O_2^{.-}$ and H_2O_2 on the activation of NFκB by interleukin 1β (IL-1) and expression of inducible enzyme NOS II (iNOS) in bovine articular chondrocytes. In response to IL-1, chondrocytes produced both H_2O_2 and $O_2^{.-}$, but only $O_2^{.-}$ mediated IL-1-induced IκB-α degradation and the consequent NFκB activation and iNOS expression, whereas H_2O_2 does not seem to participate in those IL-1-induced responses.

The foregoing data demonstrate that $O_2^{.-}$ is able to mediate enzymatic processes independently of H_2O_2 or can generate H_2O_2 by dismutation. However, to these two ROS signaling pathways a third one should be added: the generation of $O_2^{.-}$ by H_2O_2. H_2O_2 can be oxidized directly by SOD to produce free radicals, including $O_2^{.-}$ [15,16]. However, it seems that a more important pathway from H_2O_2 to $O_2^{.-}$ is the H_2O_2 activation of Nox. Thus, Li et al. [17] demonstrated that exogenous exposure of nonphagocytic cells (smooth-muscle cells and fibroblasts) to H_2O_2 activated these cells to produce $O_2^{.-}$ by Nox. Later on, Coyle et al. [18] examined the effects of H_2O_2 and $O_2^{.-}$ on porcine aortic endothelial cells (PAEC). H_2O_2 markedly increased $O_2^{.-}$ levels and produced cytotoxicity in PAEC. Overexpression of human manganese superoxide dismutase (MnSOD) or treatment with NG-nitro-1-arginine methyl ester (L-NAME), an inhibitor of NOS, or apocynin, an inhibitor of Nox, reduced $O_2^{.-}$ formation, suggesting that both NOS and Nox contributed to H_2O_2-induced $O_2^{.-}$ generation in PAEC. These findings suggested that H_2O_2 increased intracellular $O_2^{.-}$ levels through both NOS and Nox.

Signaling mechanisms by RNS significantly differ from those by $O_2^{.-}$ and H_2O_2. NO reacts with $O_2^{.-}$ with a diffusion-controlled rate to form $ONOO^-$ (Reaction 7.1); therefore, it is very difficult to distinguish the most effective signaling molecules when both RNS are present.

$$NO + O_2^{.-} \Rightarrow {}^-OONO \qquad (7.1)$$

Nonetheless, some studies suggest that NO itself is capable of activating or inactivating protein kinases and phosphatases. The most important physiological function of NO is the activation of soluble guanylyl cyclase (sGC). It was suggested that the interaction of NO with glutathione plays an important role in this process [19]. Kuhn and Arthur [20] found that NO inactivated tryptophan hydroxylase by selective oxidation of critical cysteine residues while sparing catalytic iron sites within the enzyme.

NO activated ERK kinase in rat mesangial cells, but the activation actually depended on the inhibition of phosphatases by S-nitrosation of their cysteine residues [21]. Mizuno et al. [22] proposed that exogenous NO transiently activated p42/44 (ERK) kinase through the induction of p53 and then inhibited it through the inactivation of the Ras and Raf cascade in smooth-muscle cells. Both NO and $O_2\cdot^-$ were also required for the activation of JNK kinase by shear stress in endothelial cells. However, it was supposed that, in this case, a major mediator was $ONOO^-$ [23].

Although these data seemingly suggest the ability of NO to activate or inactivate enzymatic processes, the mechanism of NO signaling has yet to be elucidated. It is unclear what properties of NO (except probably the transformation into $ONOO^-$) might be responsible for NO signaling. Recently, it has been demonstrated that a previous value for the reduction potential of the NO/NO^- pair equal to -0.39 V is wrong; a corrected value was determined as -0.68 V [24]. This correction is exclusively important because it shows that NO cannot be reduced to the highly reactive NO^- anion in biological systems. Therefore, the participation of the NO^- anion in nucleophilic substitution reactions catalyzed by protein kinases and phosphatases is practically impossible.

Another potential signaling way for NO could be nitrosation of tyrosine residues. However, NO is not an active nitration agent either; therefore, it must be converted into the other more active nitrogen compounds during nitrosation reaction, for example, into S-nitrosothiols. It was already mentioned that NO could react with the thiol groups of enzymes. Wink et al. [25] proposed a two-step mechanism of formation of S-nitrosothiols by NO:

$$4NO + O_2 \Rightarrow 2\,N_2O_3 \qquad (7.2)$$

$$N_2O_3 + GSH \Rightarrow GSNO + HNO_2 \qquad (7.3)$$

Gow et al. [26] suggested that NO can directly react with thiols:

$$NO + RSH \Rightarrow (RSNOH) \qquad (7.4)$$

$$(RSNOH)\cdot + O_2 \Rightarrow RSNO + O_2\cdot^- \qquad (7.5)$$

Nonetheless, the majority of authors believe that the most probable NO signaling mechanism is the conversion to $ONOO^-$.

Van der Vliet et al. [27] showed that $ONOO^-$ readily reacts with the thiol anions to form S-nitrosothiols, which are decomposed to NO:

$$GS^- + ONOOH \Rightarrow GSNO + HOO^- \qquad (7.6)$$

$$GSNO \Rightarrow NO + GS\cdot \;(?) \qquad (7.7)$$

Reiter et al. [28] proposed that $ONOO^-$ was able to nitrate tyrosine at neutral pH. Correspondingly, Balafanova et al. [29] have shown that NO induced $ONOO^-$-mediated tyrosine nitration of PKCε protein kinase in rabbit cardiomyocytes in vitro. Nitrotyrosine residues were also detected in the rabbit myocardium preconditioned with NO donors under in vivo conditions [30]. Jope et al. [31] demonstrated that $ONOO^-$ enhanced the phosphorylation activity of five tyrosine kinases of the src family from both human erythrocytes (lyn, hck, and c-fgr) and bovine synaptosomes

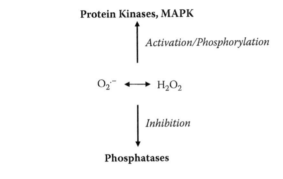

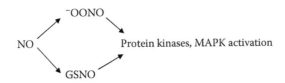

FIGURE 7.1   ROS and RNS signaling in the reaction of protein kinases, MAPKs, and phosphatases.

(lyn and fyn), and activated ERK1/2 and p38 protein kinases. Mallozzi et al. [32] have recently shown that ONOO⁻-induced tyrosine phosphorylation of serine/threonine protein phosphatase 1 alpha (PP1alpha) in human erythrocytes through activation of src family kinases. The effect of ONOO⁻ was completely reversed by dithiothreitol, suggesting that ONOO⁻ acted as an oxidizing agent via an SH-dependent or PP1-alpha-independent mechanism. Taken together, these data show that NO signaling is always an indirect process that occurs through the formation of other RNS such as ONOO⁻. Signaling pathways of ROS and RNS are presented in Figure 7.1.

7.2   ROS AND RNS SIGNALING BY REACTIONS WITH THE SULFHYDRYL GROUPS OF ENZYMES

Reactive sulfhydryl groups are able to react rapidly with ROS and RNS, resulting in the activation or inhibition of enzymes. We already considered the activation of PKC through the substitution of zinc cation and the oxidation of the S–H groups of a PKC regulatory domain by $O_2^{\cdot-}$ (Chapter 4). We suggested that only $O_2^{\cdot-}$, as an anion and a "supernucleophile," is able to remove Zn^{2+} cation with the subsequent oxidation of S–H groups of the regulatory domain. This exclusive role of $O_2^{\cdot-}$ suggests that the activation of PKC by the other prooxidants, for example, H_2O_2 or tamoxifen [33,34], should be mediated by $O_2^{\cdot-}$, whereas the prooxidants and "oxidized antioxidants" unable to produce $O_2^{\cdot-}$ inhibit PKC, reacting with the S–H groups of a catalytic domain (see Chapter 4, Figure 4.1). PKC activation by $O_2^{\cdot-}$ might also occur through the inhibition of phosphatases and the subsequent suppression of dephosphorylation of the enzyme. However, it has been shown [35] that PKC activation is apparently independent of phosphatase activity.

Activation of enzymes by the interaction of ROS with their sulfhydryl groups is important for many enzymes; the mechanism of these reactions was studied in detail for tyrosine phosphatases.

7.3 MECHANISM OF ROS SIGNALING DURING INACTIVATION OF PHOSPHATASES

The mechanism of inactivation of tyrosine phosphatases by $O_2\cdot^-$ and H_2O_2 has been thoroughly studied. Most phosphatases contain two major active centers, Fe–Zn centers and sulfhydryl groups, which can be attacked by ROS. As demonstrated earlier, at least one phosphatase—calcineurin—is inactivated by the reaction of $O_2\cdot^-$ with the Fe–Zn center. However, the other protein tyrosine phosphatases are oxidized and inactivated by the reaction of $O_2\cdot^-$ or H_2O_2 with cysteine sulfhydryl groups.

It has been shown that ROS can react with the cysteine residues of various proteins, for example, with two cysteines (Cys-Gly-Pro-Cys) of thioredoxin. Cysteine sulfhydryl groups (-SH) can be oxidized to sulfenic (-SOH), sulfinic (-SO$_2$H), and sulfonic (-SO$_3$H) residues. Two cysteines are easily oxidized into cyclic disulfide [36]. These transformations of sulfhydryl groups could be important in signaling mechanisms by ROS responsible for the inactivation of phosphatases. However, particular attention has been drawn to the role of the thiolate anion EnzS- in these processes.

In 1998, Denu and Tanner [37] proposed that, owing to the abnormally low pK_a value of 5.5 for "conserved" cysteine of phosphatases, it exists as the thiolate anion at physiological pH and can react with H_2O_2 by the mechanism of nucleophilic substitution:

$$EnzS^- + H_2O_2 \Rightarrow EnzSOH + HO^- \tag{7.8}$$

$$EnzSOH + RSH \Rightarrow EnzSSH + ROH \tag{7.9}$$

Such a mechanism explains a reversible inhibition of phosphatases by H_2O_2 because the EnzSSH form of the inactive enzyme can be oxidized back to the EnzS form with the other equivalent of thiol RSH. Meng et al. [38] confirmed that the reversible inactivation of phosphatases by H_2O_2 takes place under in vivo conditions.

Forman and coworkers discussed the mechanisms of inactivation of phosphatases by H_2O_2 in two reviews [39,40]. They proposed another mechanism of phosphatase inhibition by H_2O_2 for phosphatases of the Trx family containing two neighboring cysteines:

$$Trx(SH)(S^-) + H_2O_2 \Rightarrow Trx(SH)(SO^-) + H_2O \tag{7.10}$$

$$Trx(SH)(SO^-) \Rightarrow Trx(S_2) + HO^- \tag{7.11}$$

$$Trx(S_2) + NADPH \Rightarrow Trx(SH)(S^-) + NADP^+ \tag{7.12}$$

Reactions 7.8 and 7.9–7.12 explain well the mechanisms of phosphatase inhibition by H_2O_2; however, unfortunately, it is impossible now to estimate the real rates of these reactions. H_2O_2 is a rather inert molecule [40], and therefore the rates of nucleophilic substitution (Reactions 7.8 and 7.10) could not be very high. On the other hand, $O_2\cdot^-$, being "a super-nucleophile," must be a more efficient species in the inhibition of phosphatases. Therefore, it is not surprising that the rates of phosphatase inhibition by $O_2\cdot^-$ are about 10 times higher than those for H_2O_2 [11,12].

Barrett et al. [11] proposed the following mechanism for phosphatase inhibition by $O_2\cdot^-$, which is not very different from that for H_2O_2:

$$EnzSH + O_2\cdot^- \Rightarrow EnzSOH^- \quad\quad (7.13)$$

$$EnzSOH^- + GSH \Rightarrow EnzSSG \quad\quad (7.14)$$

$$EnzSSG \Rightarrow (thioltransferase) \Rightarrow EnzSH \quad\quad (7.15)$$

The efficiency of $O_2\cdot^-$ signaling can probably be enhanced by its chain formation in the reaction with thiols [41]:

$$O_2\cdot^- + RSH \Rightarrow [RSOOH]\cdot^- \Rightarrow RSO\cdot + HO^- \quad\quad (7.16)$$

$$RSO\cdot + RS^- \Rightarrow RSO^- + RS\cdot \quad\quad (7.17)$$

$$RS^- + RS\cdot \Rightarrow RSSR\cdot^- \quad\quad (7.18)$$

$$RSSR\cdot^- + O_2 \Rightarrow RSSR + O_2\cdot^- \quad\quad (7.19)$$

These findings demonstrate that $O_2\cdot^-$ is a more efficient signaling molecule compared to H_2O_2 in the reactions with thiols. On these grounds, we proposed [42] (see also Winterbourn and Metodiewa [41]) that signaling function of H_2O_2 might be explained by its transformation into $O_2\cdot^-$. (Conversion of H_2O_2 into $O_2\cdot^-$ by the oxidation of SOD or the activation of Nox has already been considered.)

7.4 MECHANISMS OF ROS SIGNALING IN REACTIONS CATALYZED BY PROTEIN KINASES

It follows from the foregoing data that ROS and RNS are able to activate protein kinases, MAPKs, and possibly other enzymes through reactions with the S–H groups of cysteine residues as has been shown in the case of ROS activation via the regulatory domain of PKC. However, if ROS and RNS activation is indeed starting by the attack on the thiol groups of enzymes, it is still completely unclear how this process can enhance the enzymatic activities of protein kinases and other enzymes catalyzing heterolytic reactions of hydrolysis and etherification.

As early as 1978, Niehaus [43] proposed a nucleophilic mechanism for deesterification of phospholipids by $O_2\cdot^-$, but later on, this proposal was forgotten. We considered the mechanisms of ROS signaling in enzymatic phosphorylation and dephosphorylation [44] during the catalysis by MAPK, phospholipase C, and other enzymes under physiological and pathophysiological conditions [45], the competition between $O_2\cdot^-$ and H_2O_2 as signaling molecules [42], and the signaling mechanisms of ROS and RNS in aging and diseases [46,47].

In accordance with the established mechanism, protein kinases catalyze phosphorylation reactions through the interaction of the phosphorylated enzyme with the threonine or serine residues of a protein [48]:

$$PK\text{-}O(phos) + HO(Ser)protein \Rightarrow PK\text{-}OH + (phos)O(Ser)protein \quad (7.20)$$

$$PK\text{-}O(phos) + HO(Thr)protein \Rightarrow PK\text{-}OH + (phos)O(Thr)protein \quad (7.21)$$

The catalytic cycle is completed by autophosphorylation or phosphorylation by another protein kinase:

$$PK\text{-}OH + (phos)O(X) \Rightarrow PK\text{-}O(phos) + HO(X) \quad (7.22)$$

Many studies (e.g., References [49,50]) suggest that ROS activate protein kinases through direct tirosine phosphorylation. We proposed the mechanism of ROS signaling in these reactions catalyzed by protein kinases. Reactions 7.20 and 7.21 are nucleophilic processes in which protons of serine or threonine residues are substituted by phosphate groups. As the pKa values for aliphatic hydroxyls of serine or threonine are very high, a strong nucleophile must participate in these reactions. It is quite possible that, in biological systems, there are no other intermediates with such strong nucleophilic properties as $O_2\cdot^-$, which is even a stronger nucleophile than hydroxyl anion HO^-. Therefore, we suggested [42,44,45] that a major role of $O_2\cdot^-$ in these phosphorylation processes is the deprotonation of serine and threonine residues that must sharply increase the rates of Reactions 7.20 and 7.21:

$$HO(Ser)X \text{ or } HO(Thr)X + O_2\cdot^- \Leftrightarrow {}^-O(Ser)X \text{ or } {}^-O(Thr)X + HOO\cdot \quad (7.23)$$

Reaction 7.23 is always followed by an electron transfer between $O_2\cdot^-$ and the perhydroxyl radical $HOO\cdot$ with a diffusion rate constant of $(8.86 \pm 0.43) \times 10^7 \text{ M}^{-1}\text{s}^{-1}$.

$$HOO. + O_2\cdot^- \Rightarrow HOO^- + O_2 \quad (7.24)$$

Because of that, Reaction 7.23 had to be kinetically completely shifted to the right. Accordingly, $O_2\cdot^-$ signaling will sharply accelerate Reactions 7.20 and 7.21 because the substitution reactions 7.25 and 7.26 with protein anions proceed with much greater rates than with neutral molecules:

$$PK\text{-}O(phos) + {}^-O(Ser)protein \Rightarrow PK\text{-}O^- + (phos)O(Ser)protein \quad (7.25)$$

$$PK\text{-}O(phos) + {}^-O(Thr)protein \Rightarrow PK\text{-}O^- + (phos)O(Thr)protein \quad (7.26)$$

$$PK\text{-}O^- + (phos)OX \Rightarrow PK\text{-}O(phos) + {}^-OX \quad (7.27)$$

Thus, $O_2\cdot^-$ signaling in reactions catalyzed by protein kinases can accelerate these reactions through deprotonation of the protein hydroxyl groups.

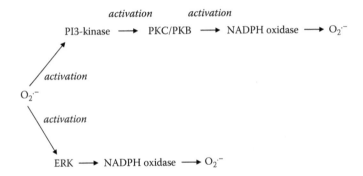

FIGURE 7.2   Chain mechanism of $O_2{}^{\cdot-}$-induced activation of enzymes.

7.5   CHAIN MECHANISM OF SUPEROXIDE SIGNALING

Numerous data already considered in Chapter 4 demonstrate that $O_2{}^{\cdot-}$ not only participates in the activation of protein kinases and other enzymes but is also formed during these enzymatic cycles. It has been found that the activation of protein kinases by $O_2{}^{\cdot-}$ resulted in the activation of Nox, phospholipase A2, and protein serine–threonine phosphatases. Under certain conditions, all these enzymes are able to stimulate $O_2{}^{\cdot-}$ production. For example, PKC and Czeta and ERK MAPK, stimulate oxygen radical production through direct activation and phosphorylation of the p47^{phox}, subunit of Nox [51–53].

The PI3-kinase/Akt pathway is another way to activate the neutrophilic Nox through the phosphorylation of serine304 and serine328 residues of subunit p47^{phox} and the induction of $O_2{}^{\cdot-}$ formation [54]. Moreover, phosphatidylinositol 3-kinase is also able to stimulate Nox-induced $O_2{}^{\cdot-}$ production in neutrophils [55,56].

Thus, there is a "chain" mechanism of $O_2{}^{\cdot-}$-induced activation of enzymes, in which $O_2{}^{\cdot-}$ is formed and reacts in the cyclic process (Figure 7.2).

7.6   NUCLEOPHILIC MECHANISM OF SUPEROXIDE SIGNALING IN GTPASE-CATALYZED REACTIONS

GTPases are a large family of hydrolase enzymes that can bind and hydrolyze guanosine triphosphate (GTP) to guanosine diphosphate (GDP). It was established that hydrolysis occurred by the SN2 mechanism of nucleophilic substitution. Therefore, it is of utmost importance that several authors demonstrated that $O_2{}^{\cdot-}$ mediated hydrolytic reactions catalyzed by GTPases. We suppose that this is an important example of $O_2{}^{\cdot-}$ signaling where $O_2{}^{\cdot-}$ reacts as "super-nucleophile."

Bhunia et al. [57] have shown that lactosylceramide (LacCer) stimulated endogenous $O_2{}^{\cdot-}$ production in human aortic smooth-muscle cells through the activation of Nox and that $O_2{}^{\cdot-}$ stimulated ERK (p44) activation through the p21raszGTP loading. It was also suggested that $O_2{}^{\cdot-}$ stimulated ERK (p44) activation through the p21raszGTP loading. This hypothesis is very important because it suggests that the hydrolysis of GTP to GDP can be mediated by $O_2{}^{\cdot-}$ (Chapter 4, Figure 4.7).

GTP binding and hydrolysis take place in the highly conserved G domain common to all GTPases. It was also demonstrated that the hydrolysis of the γ phosphate of GTP into GDP and Pi, inorganic phosphate, depended on the magnesium ion Mg2+.

REFERENCES

1. R Larsson and P Cerutti. Translocation and enhancement of phosphotransferase activity of protein kinase c following exposure in mouse epidermal cells to oxidants. *Cancer Res* 49: 5627–5632, 1989.
2. AS Baas and BC Berk, Differential activation of mitogen-activated protein kinases by H_2O_2 and $O_2\cdot^-$ in vascular smooth muscle cells. *Circ Res* 77: 29–36, 1995.
3. PF Li, R Dietz, and R von Harsdorf. Differential effect of hydrogen peroxide and superoxide anion on apoptosis and proliferation of vascular smooth muscle cells. *Circulation* 96: 3602–3609, 1997.
4. V Moreno-Manzano, Y Ishikawa, J Lucio-Cazana, and M Kitamura. Selective involvement of superoxide anion, but not downstream compounds hydrogen peroxide and peroxynitrite, in tumor necrosis factor-alpha-induced apoptosis of rat mesangial cells. *J Biol Chem* 275: 12684–12691, 2000.
5. S Devadas, L Zaritskaya, SG Rhee, L Oberlay, and MS Williams. Discrete generation of superoxide and hydrogen peroxide by T cell receptor stimulation: Selective regulation of mitogen-activated protein kinase activation and Fas ligand expression. *J Exp Med* 195: 59–70, 2002.
6. FM Lyng, P Maguire, B McClean, C Seymour, and C Mothersill. The involvement of calcium and MAP kinase signaling pathways in the production of radiation-induced bystander effects. *Radiation Res* 165: 400–409, 2006.
7. XL Cui and JG Douglas. Arachidonic acid activates c-jun N-terminal kinase through NADPH oxidase in rabbit proximal tubular epithelial cells. *Proc Natl Acad Sci USA* 94: 3771–3776, 1997.
8. L Conde de la Rosa, MH Schoemaker, TE Vrenken, M Buist-Homan, R Havinga, PL Jansen, and H Moshage. Superoxide anions and hydrogen peroxide induce hepatocyte death by different mechanisms: Involvement of JNK and ERK MAP kinases. *J Hepatol* 44: 918–929, 2006.
9. D Namgaladze and HW Hofer, V Ullrich. Redox control of calcineurin by targeting the binuclear Fe^{2+}-Zn^{2+} center at the enzyme active site. *J Biol Chem* 277: 5962–5969, 2002.
10. W Wang, S Wang, EV Nishanian, A Del Pilar Cintron, RA Wesley, and RL Danner. Signaling by eNOS through a superoxide-dependent p42/44 mitogen-activated protein kinase pathway. *Am J Physiol Cell Physiol* 281: C544–C554, 2001.
11. WC Barrett, JP DeGnore, YF Keng, ZY Zhang, MB Yim, and PB Chock. Roles of superoxide radical anion in signal transduction mediated by reversible regulation of protein-tyrosine phosphatase 1B. *J Biol Chem* 274: 34543–34546, 1999.
12. Q Wang, D Dube, RW Friesen, TG LeRiche, KP Bateman, L Trimble, J Sanghara, R Pollex, C Ramachandran, MJ Gresser, and S Huang. Catalytic inactivation of protein- tyrosine phosphatase CD4 and protein tyrosine phosphatase 1B by polyaromatic quinones. *Biochemistry* 43: 4294–4303, 2004.
13. J Hongpaisan, CA Winters, and SB Andrews. Strong calcium entry activates mitochondrial superoxide generation, upregulating kinase signaling in hippocampal neurons. *J Neurosci* 24: 10878–10887, 2004.
14. AF Mendes, MM Caramona, AP Carvalho, and MC Lopes. Differential roles of hydrogen peroxide and superoxide in mediating IL-1-induced NF-kappaB activation and iNOS expression in bovine articular chondrocytes. *J Cell Biochem* 88: 783–793, 2003.

15. DC Ramirez, SE Gomez Mejiba, and RP Mason. Mechanism of hydrogen peroxide-induced Cu,Zn-superoxide dismutase-centered radical formation as explored by immuno-spin trapping: The role of copper- and carbonate radical anion-mediated oxidations. *Free Radic Biol Med* 38: 201–214, 2005.
16. YM Kim, JM Lim, BC Kim, and S Han. Cu,Zn-superoxide dismutase is an intracellular catalyst for the H(2)O(2)-dependent oxidation of dichlorodihydrofluorescein. *Mol Cells* 21: 161–165, 2006.
17. WG Li, FJ Miller Jr, HJ Zhang, DR Spitz, LW Oberley, and NL Weintraub. H(2)O(2)-induced O(2) production by a non-phagocytic NAD(P)H oxidase causes oxidant injury. *J BiolChem* 276: 29251–29256, 2001.
18. CH Coyle, LJ Martinez, MC Coleman, DR Spitz, NL Weintraub, and KN Kader. Mechanisms of H(2)O(2)-induced oxidative stress in endothelial cells. *Free Radic Biol Med* 40: 2206–2213, 2006.
19. B Mayer, S Pfeiffer, A Schrammel, D Koesling, K Schmidt, and F Brunner. A new pathway of nitric oxide/cyclic GMP signaling involving S-nitrosoglutathione. *J Biol Chem* 273: 3264–3270, 1998.
20. DM Kuhn and R Arthur, Jr. Molecular mechanism of the inactivation of tryptophan hydroxylase by nitric oxide: Attack on critical sulfhydryls that spare the enzyme iron center. *J Neurosci* 17: 7245–7251, 1997.
21. D Callsen, J Pfeilschifter, and B Brune. Rapid and delayed p42/p44 mitogen-activated protein kinase activation by nitric oxide: The role of cyclic GMP and tyrosine phosphatase inhibition. *J Immunol* 161: 4852–4858, 1998.
22. S Mizuno, M Kadowaki, Y Demura, S Ameshima, I Miyamori, and T Ishizaki. p42/44 Mitogen-activated protein kinase regulated by p53 and nitric oxide in human pulmonary arterial smooth muscle cells. *Am J Respir Cell Mol Biol* 31: 184–192, 2004.
23. YM Go, RP Patel, MC Maland, H Park, JS Beckman, VM Darley-Usmar, and H Jo. Evidence for peroxynitrite as a signaling molecule in flow-dependent activation of c-Jun NH(2)-terminal kinase. *Am J Physiol* 277: H1647–H1653, 1999.
24. MD Bartberger, W Liu, E Ford, KM Miranda. C Switzer, JM Fukuto, PJ Farmer, DA Wink, and KN Houk. The reduction potential of nitric oxide (NO) and its importance to NO biochemistry. *Proc Natl Acad Sci USA* 99: 10958–10963, 2002.
25. DA Wink, JA Cook, SY Kim, Y Vodovotz, R Pacelli, MC Krishna, A Russo, JB Mitchell, D Jourd'heuil, AM Miles, and MB Grisham. Superoxide modulates the oxidation and nitrosation of thiols by nitric oxide-derived reactive intermediates. Chemical aspects involved in the balance between oxidative and nitrosative stress. *J Biol Chem* 272: 11147–11151, 1997.
26. AJ Gow, DG Buerk, and H Ischiropoulos. A novel reaction mechanism for the formation of S-nitrosothiol in vivo. *J Biol Chem* 272: 2841–2845, 1997.
27. A van der Vliet, PAC Hoen, PSY Wong, A Bast, and CE Cross. Formation of S-nitrosothiols via direct nucleophilic nitrosation of thiols by peroxynitrite with elimination of hydrogen peroxide. *J Biol Chem* 273: 30255–30262, 1998.
28. CD Reiter, R-J Teng, and JS Beckman. Superoxide reacts with nitric oxide to nitrate tyrosine at physiological pH via peroxynitrite. *J Biol Chem* 275: 32460–32466, 2000.
29. Z Balafanova, R Bolli, J Zhang, Y Zheng, JM Pass, A Bhatnagar, X-L Tang, O Wang, E Cardwell, and P Ping. Nitric oxide (NO) induces nitration of protein kinase Cε (PKCε), facilitating PKCε translocation via enhanced PKCε-RACK2 interactions. A novel mechanism of NO-triggered activation of PKCε. *J Biol Chem* 277: 15021–15027, 2002.
30. C Mallozzi, AM Di Stasi, and M Minetti. Activation of src tyrosine kinases by peroxynitrite. *FEBS Lett* 456: 201–206, 1999.
31. RS Jope, L Zhang, and L Song. Peroxynitrite modulates the activation of p38 and extracellular regulated kinases in PC12 cells. *Arch Biochem Biophys* 376: 365–370, 2000.

32. C Mallozzi, L De Franceschi, C Brugnara, and AM Di Stasi. Protein phosphatase 1alpha is tyrosine-phosphorylated and inactivated by peroxynitrite in erythrocytes through the src family kinase fgr. *Free Radic Biol Med* 38: 1625–1636, 2005.

33. U Gundimeda, Z-H Chen, and R Gopalakrishna. Tamoxifen modulates protein kinase c via oxidative stress in estrogen receptor-negative breast cancer cells. *J Biol Chem* 271: 13504–13508, 1996.

34. R Gopalakrishna and U Gundimeda. Antioxidant regulation of protein kinase c in cancer prevention. *J Nutr* 132: 3819S–3823S, 2002.

35. J Hongpaisan, CA Winters, and SB Andrews. Strong calcium entry activates mitochondrial superoxide generation, upregulating kinase signaling in hippocampal neurons. *J Neurosci* 24: 19878–19887, 2004.

36. VJ Thannickal and BL Fanburg. Reactive oxygen species in cell signaling. *Am J Physiol Lung Cell Mol Physiol* 279: L1005–L1028, 2000.

37. JM Denu and KG Tanner. Specific and reversible inactivation of protein tyrosine phosphatases by hydrogen peroxide: Evidence for a sulfenic acid intermediate and implications for redox regulation. *Biochemistry* 37: 5633–5642, 1998.

38. T-C Meng, T Fukada, and NK Tonks. Reversible oxidation and inactivation of protein tyrosine phosphatases in vivo. *Molecular Cell* 9: 387–399, 2002.

39. HJ Forman and M Torres. Reactive oxygen species and cell signaling: Respiratory burst in macrophage signaling. *Am J Respir Crit Care Med* 166: 54–58, 2002.

40. HJ Forman and JM Fukuto. Redox signaling: Thiol chemistry defines which reactive oxygen and nitrogen species can act as second messengers. *Am J Physiol Cell Physiol* 287: C246–C256, 2004.

41. CC Winterbourn and D Metodiewa. Reactivity of biologically important thiol compounds with superoxide and hydrogen peroxide. *Free Radic Biol Med* 27: 322–328, 1999.

42. IB Afanas'ev. Competition between superoxide and hydrogen peroxide signaling in heterolytic enzymatic processes. *Med Hypotheses* 66: 1125–1128, 2006.

43. WG Niehaus Jr. A proposed role of superoxide anion as a biological nucleophile in the deesterification of phospholipids. *Bioorg Chem* 7: 7–10, 1978.

44. IB Afanas'ev. Mechanism of superoxide-mediated damage. Relevance to mitochondrial aging. *Ann NY Acad Sci* 1019: 343–345, 2004.

45. IB Afanas'ev. On mechanism of superoxide signaling under physiological and pathophysiological conditions. *Med Hypotheses* 64: 127–129, 2005.

46. IB Afanas'ev. Interplay between superoxide and nitric oxide in aging and diseases, *Biogerontology* 5: 267–270, 2004.

47. IB Afanas'ev. Free radical mechanisms of aging processes under physiological conditions. *Biogerontology* 6: 283–290, 2005.

48. AC Newton. Regulation of protein kinase C. *Curr Opin Chem Biol* 9: 161–167, 1999.

49. H Konishi, M Tanaka, Y Takemura, H Matsuzaki, Y Ono, U Kikkawa, and Y Nishizuka. Activation of protein kinase C by tyrosine phosphorylation in response to H2O2. *Proc Natl Acad Sci USA* 94: 11233–11237, 1997.

50. AMN Kabir, JE Clark, M Tanno, X Cao, JS Hothersall, S Dashnyam, DA Gorog, M Bellahcene, MJ Shattock, and MS Marber. Cardioprotection initiated by reactive oxygen species is dependent on the activation of PKC{epsilon}. *Am J Physiol Heart Circ Physiol* 291: H1893–H1899, 2006.

51. LR Lopes, CR Hoyal, UR Knaus, and BM Babior. Activation of the leukocyte NADPH oxidase by protein kinase C in a partially recombinant cell-free system. *J Biol Chem* 274: 15533–15537, 1999.

52. PM Dang, A Fontayne, J Hakim, J El Benna, and A Perianin. Protein kinase C zeta phosphorylates a subset of selective sites of the NADPH oxidase component p47(phox) and participates in formyl peptide-mediated neutrophil respiratory burst. *J Immunol* 166: 1206–1213, 2001.

53. MA Laplante, R Wu, A El Midaoui, and J De Champain. NAD(P)H oxidase activation by angiotensin II is dependent on p42/44 ERK-MAPK pathway activation in rat's vascular smooth muscle cells. *J Hypertens* 21: 927–936, 2003.

54. CR Hoyal, A Gutierrez, BM Young, SD Catz, J-H Lin, PN Tsichlis, and BM Babior. Modulation of p47phos activity by site-specific phosphorylation: Akt-dependent activation of the NADPH oxidase. *Proc Natl Acad Sci USA* 100: 5130–5135, 2003.

55. NG Shenoy, GJ Gleich, and LL Thomas. Eosinophil major basic protein stimulates neutrophil superoxide production by a class I(A) phosphoinositide 3-kinase and protein kinase C-zeta-dependent pathway. *J Immunol* 171: 3734–3741, 2003.

56. M Oda, S Ikari, T Matsuno, Y Morimune, M Nagahama, and J Sakurai. Signal transduction mechanism involved in Clostridium perfringens alpha-toxin-induced superoxide anion generation in rabbit neutrophils. *Infect Immun* 74: 2876–2886, 2006.

57. AK Bhunia, H Han, A Snowden, and S Chatterjee. Redox-regulated signaling by lactosylceramide in the proliferation of human aortic smooth muscle cells. *J Biol Chem* 272: 15642–15649, 1997.

Index

A

AA. *See* arachidonic acid
Actinomycin D, 48, 137
Activation loop, 3, 04, 41, 114
Activators, 6, 29, 105, 109, 130
Adenine nucleotide translocase (ANT), 61
Adenocarcinoma, 19, 21, 46
Adenylyl cyclase AC, 106, 182
Adrenaline, 11, 42
Adriamycin, 145, 157
Adult respiratory distress syndrome, 6
Advanced glycation end-products, 44
Aging, 1, 2, 4, 9, 24, 32, 40, 48, 52, 70, 74, 153,
    159–188, 196, 200
AICAR. *See* 5-aminoimidazole-4-carboxamide
    riboside
AIDS dementia, 146
Akt/B protein kinase cascade, 12
Aldose reductase (AR), 111, 128
Alkalinization, 63
Alkalosis, 91, 118
Allopurinol, 7, 8, 9, 39
ALS. *See* familial amyotrophic lateral sclerosis
Alzheimer's disease, 25, 49, 146
5-Aminoimidazole-4-carboxamide riboside
    (AICAR), 144
5′-AMP-activated protein kinase (AMPK), 13,
    103, 104, 144, 164
AMPK. *See* 5′-AMP-activated protein kinase
Amyotrophic lateral sclerosis, 25, 66, 146
Ang-1. *See* angiopoietin-1
Ang II. *See* angiotensinII
Angiogenesis, 20, 21, 25, 47, 49, 86, 87, 112, 115,
    116, 155
Angiopoietin-1 (Ang-1), 1 20, 21, 23, 98
Angiotensin II (Ang II), 13, 16, 17, 19, 20, 22, 23,
    24, 35, 36, 67, 85, 86, 89, 92, 96, 102,
    105, 106, 124, 130
Animals, 17, 25, 26, 32, 134, 142, 160, 161, 162,
    168, 169, 170, 175, 176, 177, 178, 187
Anoxia, 89
ANT. *See* adenine nucleotide translocase
Antimycin A, 57, 83, 135, 152
Antioxidants, 3, 20, 29, 39, 51, 60, 66, 74, 77, 82,
    85, 87, 89, 90, 91, 93, 97, 100, 101,
    102, 105, 118, 122, 127, 132, 134, 136,
    138, 139, 140, 142, 143, 144, 148, 153,
    159, 167, 175, 176, 177, 183–187, 193

Aorta, 9, 17, 20, 45, 66, 85, 92, 98, 106, 115, 119,
    127, 162, 165, 183
Apaf-1. *See* apoptotic protease-activating factor 1
Apocynin, 18, 21, 29, 46, 93, 95, 102, 147, 166,
    173, 191
Apoptosis, 4, 21, 23, 29, 43, 48, 61, 64, 69, 73, 75,
    85, 87, 90, 93, 94, 96, 98, 103, 105,
    106, 114, 115, 116, 120, 126–158, 170,
    171, 184, 190, 198
Apoptosis signal-regulating kinase 1 (ASK1), 21,
    23, 64, 66, 76, 105, 106, 126, 139, 140,
    142, 143, 154, 166, 182
Apoptosome, 145, 157
Apoptotic protease-activating factor 1 (Apaf-1),
    145
AR. *See* aldose reductase
Arachidonic acid (AA), 7, 11, 15, 21, 39, 41, 42, 47,
    71, 85, 86, 114, 115, 122, 124, 190, 198
Arginase, 30, 37, 51
Artery, 7, 8, 19, 20, 23, 27, 37, 48, 64, 76, 89, 90,
    92, 94, 99, 104, 126, 183
Arthritis, 12, 38, 54, 94, 120
Ascorbic acid, 7, 133, 134, 152, 177, 187
ASK1. *See* apoptosis signal-regulating kinase 1
Astrocytes, 66, 77, 100, 106, 123, 173, 185
AT1 receptor, 47, 91
Atherogenesis, 11, 112
Atherosclerosis, 6, 9, 13, 36, 44, 112, 123, 124,
    128, 172, 185
Atherosclerotic lesions, 32
Atherosclerotic plaque, 23, 112
Atorvastatin, 23, 48
Autophosphorylation, 196

B

BaP. *See* benzo(a)pyrene
BCA. *See* 2′-benzoyl-oxycinnamaldehyde
Bcl-2 protein, 127, 129, 130, 132, 136, 148, 149,
    151, 165, 170, 171, 184
Benzo(a)pyrene (BaP), 142, 155
2′-Benzoyl-oxycinnamaldehyde (BCA), 132
Beta-blockers, 168, 183
Beta-catenin, 145
Bioavailability, 70, 79, 160, 167, 173, 174
Birds, 161, 180, 186
Blood pressure, 27, 91, 95
Bovine aortic endothelial cells (BAEC), 6, 112
Bradykinin, 89, 117, 173

203